Advances in Intelligent Systems and Computing

Volume 625

About this Series

The series "Advances in Intelligent Systems and Computing" contains publications on theory, applications, and design methods of Intelligent Systems and Intelligent Computing. Virtually all disciplines such as engineering, natural sciences, computer and information science, ICT, economics, business, e-commerce, environment, healthcare, life science are covered. The list of topics spans all the areas of modern intelligent systems and computing.

The publications within "Advances in Intelligent Systems and Computing" are primarily textbooks and proceedings of important conferences, symposia and congresses. They cover significant recent developments in the field, both of a foundational and applicable character. An important characteristic feature of the series is the short publication time and world-wide distribution. This permits a rapid and broad dissemination of research results.

More information about this series at http://www.springer.com/series/11156

Durgesh Kumar Mishra · Ahmad Taher Azar
Amit Joshi
Editors

Information and Communication Technology

Proceedings of ICICT 2016

Springer

Editors
Durgesh Kumar Mishra
Microsoft Innovation Centre
Sri Aurobindo Institute of Technology
Indore, Madhya Pradesh
India

Amit Joshi
Department of Information Technology
Sabar Institute of Technology for Girls
Ahmedabad, Gujarat
India

Ahmad Taher Azar
Faculty of Computers and Information
Benha University
Benha
Egypt

ISSN 2194-5357 ISSN 2194-5365 (electronic)
Advances in Intelligent Systems and Computing
ISBN 978-981-10-5507-2 ISBN 978-981-10-5508-9 (eBook)
https://doi.org/10.1007/978-981-10-5508-9

Library of Congress Control Number: 2017945682

Printed on acid-free paper

This Springer imprint is published by Springer Nature
The registered company is Springer Nature Singapore Pte Ltd.
The registered company address is: 152 Beach Road, #21-01/04 Gateway East, Singapore 189721, Singapore

Preface

This LNNS volume contains selected papers presented at the ICICT 2016: Second International Congress on Information and Communication Technology. The conference was held during December 12–13, 2016, Bangkok, Thailand and organized communally by G R Foundation, and Computer Society of India Division IV—Communication and Division V—Education and Research. It has targeted state of the art as well as emerging topics pertaining to ICT and effective strategies for its implementation for engineering and intelligent applications. The objective of this international conference was to provide opportunities for the researchers, academicians, industry persons, and students to interact and exchange ideas, experience, and expertise in the current trend and strategies for information and communication technologies. Besides this, participants were also enlightened about vast avenues, and current and emerging technological developments in the field of ICT in this era and its applications were thoroughly explored and discussed. The conference attracted a large number of high quality submissions and stimulated cutting-edge research discussions among many academic pioneering researchers, scientists, industrial engineers, and students from all around the world. Proposed new technologies shared their experiences and discussed future solutions for designing infrastructure for ICT. The deliberations enriched technocrats and academicians by presenting their innovative and constructive ideas and focused on innovative issues at international level by bringing together the experts from different countries. Research submissions in various advanced technology areas were received and after a rigorous peer-review process with the help of program committee members and external reviewers, 32 papers were accepted with an acceptance ratio of 0.21. The conference featured many distinguished personalities such as Dr. Somsak Choomchuay, Professor, Department of Electronics, King Mongkut's Institute of Technology Ladkrabang, Bangkok and Mr. Aninda Bose from Springer India. Separate invited talks were organized in industrial and academia tracks on both days. The conference also hosted few tutorials for the benefit of participants. We are indebted to G R Foundation and CSI Division IV and V for their immense support to make this conference possible in such a grand scale. A total of four sessions were organized as a part of *ICICT 2016* including two technical, one plenary, and one inaugural sessions. A total of 26 papers were presented in two technical sessions with high discussion insights. The total number of accepted submissions was 32 with a focal point on ICT and Intelligent Systems. Our sincere thanks to all sponsors, press, print, and electronic media for their excellent coverage of this conference.

Indore, India	Durgesh Kumar Mishra
Benha, Egypt	Ahmad Taher Azar
Ahmedabad, India	Amit Joshi
December 2016	

Organizing Committee

International Advisory Committee

Dr. Nobert Jere, Namibia University of Science and Technology, Namibia
Shalini Batra, Department of Computer Science and Engineering, Thapar University, Patiala, Punjab, India
Ernest Chulantha Kulasekere, Ph.D., University of Moratuwa, Sri Lanka
Shajulin Benedict, Director, HPCCLoud Research Laboratory, St. Xavier's Catholic College of Engineering, Chunkankadai District, Nagercoil, Tamil Nadu, India
James E. Fowler, Mississippi State University, Mississippi, USA
Dr. Majid Ebnali-Heidari, Shahrekord University, Shahrekord, Iran
affi_be@gtu.edu.i Rajendra Kumar Bharti, Assistant Professor, Kumaon Engineering College, Dwarahat, Uttarakhand, India
Prof. Murali Bhaskaran, Dhirajlal Gandhi College of Technology, Salem, Tamil Nadu, India
Pramod Parajuli, Nepal College of Information Technology, Nepal
Prof. Komal Bhatia, YMCA University, Faridabad, Haryana, India
Lili Liu, Automation College, Harbin Engineering University, Harbin, China
Brooke Fisher Liu, Department of Communication, University of Maryland, College Park, MD, USA
Prof. S.R. Biradar, Department of Information Science and Engineering, SDM College of Engineering and Technology, Dharwad, Karnataka, India
A.K. Chaturvedi, Department of Electrical Engineering, IIT Kanpur, India
Margaret Lloyd, Faculty of Education School of Curriculum, Queensland University of Technology, Queensland, Australia
Hoi-Kwong Lo, University of Toronto, Ontario, Canada
Pradeep Chouksey, Principal, TIT College, Bhopal, Madhya Pradesh, India
Shashidhar Ram Joshi, Ph.D., Institute of Engineering, Pulchowk Campus, Pulchowk, Nepal
Chhaya Dalela, Associate Professor, JSSATE, Noida, Uttar Pradesh, India
Jayanti Dansana, KIIT University, Bhubaneswar, Odisha, India
Desmond Lobo, Department of Computer Engineering, Faculty of Engineering at Kamphaeng Saen, Kasetsart University, Thailand
Sergio Lopes, Industrial Electronics Department, University of Minho, Braga, Portugal
Soura Dasgupta, Department of TCE, SRM University, Chennai, India
Dr. Apurva A. Desai, Veer Narmad South Gujarat University, Surat, India
V. Susheela Devi, Senior Scientific Officer, Department of Computer Science and Automation Indian Institute of Science, Bangalore, India
Subhadip Basu, Ph.D., Visiting Scientist, The University of Iowa, Iowa City, USA
Vijay Pal Dhaka, Jaipur National University, Jaipur, Rajasthan, India
Chih-Heng Ke, MIEEE, NKIT, Taiwan
Dr. Nobert Jere, Namibia University of Science and Technology, Namibia
Mr. Mignesh Parekh, Kamma Incoporation, Gujarat, India
Kok-Lim Low, National University of Singapore, Singapore

Technical Program Committee

Prof. Dan Boneh, Computer Science Department, Stanford University, CA, USA
Prof. Alexander christea, University of Warwick, London, UK
Dr. Aynur Unal, Stanford University, USA
Prof. Ahmad Al-Khasawneh, The Hashemite University, Jordan
Dr. Bharat Singh Deora, JRNRV University, India
Prof. Jean Michel Bruel, Departement Informatique IUT de Blagnac, Blagnac, France
Prof. Ngai-Man Cheung, Assistant Professor, University of Technology and Design, Singapore
Prof. Yun-Bae Kim, SungKyunKwan University, South Korea
Prof. Ting-Peng Liang, National Chengchi University, Taipei, Taiwan
Prof. Sami Mnasri, IRIT Laboratory Toulouse, France
Prof. Lorne Olfman, Claremont, CA, USA
Prof. Anand Paul, The School of Computer Science and Engineering, South Korea
Dr. Krishnamachar Prasad, Department of Electrical and Electronic Engineering, Auckland, New Zealand
Prof. Brent Waters, University of Texas, Austin, TX, USA
Er. Kalpana Jain, CTAE, Udaipur, India
Prof. (Dr.) Avdesh Sharma, Jodhpur, India
Er. Nilay Mathur, Director, NIIT Udaipur, India
Prof. Philip Yang, Price Water House Coopers, Beijing, China
Mr. Jeril Kuriakose, Manipal University, Jaipur, India
Prof. R.K. Bayal, Rajasthan Technical University, Kota, Rajasthan, India
Prof. Martin Everett, University of Manchester, England
Prof. Feng Jiang, Harbin Institute of Technology, China
Prof. Prasun Sinha, Ohio State University Columbus, Columbus, OH, USA
Dr. Savita Gandhi, Professor, Gujarat University, Ahmedabad, India
Prof. Xiaoyi Yu, National Laboratory of Pattern Recognition, Institute of Automation, Chinese Academy of Sciences, Beijing, China
Prof. Gengshen Zhong, Jinan, Shandong, China
Prof. A.R. Abdul Rajak, Department of Electronics and Communication Engineering, Birla Institute of Dr. Nitika Vats Doohan, Indore, India
Dr. Harshal Arolkar, Immd. Past Chairman, CSI Ahmedabad Chapter, India
Mr. Bhavesh Joshi, Advent College, Udaipur, India
Prof. K.C. Roy, Principal, Kautaliya, Jaipur, India
Dr. Mukesh Shrimali, Pacific University, Udaipur, India
Mrs. Meenakshi Tripathi, MNIT, Jaipur, India
Prof. S.N. Tazi, Government Engineering College, Ajmer, Rajasthan, India
Shuhong Gao, Mathematical Sciences, Clemson University, Clemson, South Carolina
Sanjam Garg, University of California, Los Angeles, CA
Garani Georgia, University of North London, UK

Dr. S.A.D Dias, Department of Electronics and Telecommunication Engineering, University of Moratuwa, Sri Lanka
David Diez, Leganés, Spain, Europe
Dr. Zhang Dinghai, Gansu Agricultural University, Lanzhou, China
Dr. P.D.D. Dominic, Department of Computer and Information Science, Universiti Teknologi Petronas, Tronoh, Perak, Malaysia
Aditya Khamparia, Lovely Professional University, Punjab, India
Dr. Nurbek P. Saparkhojayev, Kazakh National Research Technical University after K.I. Satpayev, Almaty, Kazakhstan

Organizing and General Committee Chairs

Dr. Malaya Nayak, UK
Ms. Bhagyesh Soneji, Chairperson, ASSOCHAM
Mr. Sachin Manocha, TCEB, Thailand
Mr. Bharat Patel, Chairman—Startup Mission for ASSOCHAM Western Region
Dr. Durgesh Kumar Mishra, CSI, India
Dr. Dharm Singh, NUST, Namibia
Dr. Suresh Chandra Satapathy, ANITS, India
Dr. Aynur Unal, USA
Mr. Mihir Chauhan, GR Foundation

Organizing Secretary

Amit Joshi, Member ACM, IEEE, CSI, IACSIT—Singapore, IDES, ACEEE, MSA and Community Member—NPA

Contents

About the Editors

Dr. Durgesh Kumar Mishra has received M.Tech. Degree in Computer Science from DAVV, Indore, in 1994 and Ph.D. degree in Computer Engineering in 2008. Presently, he is working as a Professor (CSE) and Director, Microsoft Innovation Centre at Shri Aurobindo Institute of Technology, Indore, MP, India. He is also a visiting faculty at IIT Indore, MP, India. He has 24 years of teaching and 10 years of research experience. He has completed his Ph.D. under the guidance of Late Dr. M. Chandwani on "Secure Multi-Party Computation for Preserving Privacy". He has published more than 90 papers in refereed international/national journals and conferences including IEEE, ACM conferences. He has organized many conferences such as WOCN, CONSEG, and CSIBIG in the capacity of conference General Chair and editor of conference proceeding. His publications are listed in DBLP, Citeseer-x, Elsevier, and Scopus. He is a Senior Member of IEEE and held many positions like Chairman, IEEE MP-Subsection (2011–2012), and Chairman, IEEE Computer Society Bombay Chapter (2009–2010). Dr. Mishra has also served the largest technical and profession association of India, the Computer Society of India (CSI) by holding positions as Chairman, CSI Indore Chapter, State Student Coordinator-Region III MP, Member-Student Research Board, Core Member-CSI IT Excellence Award Committee. He has been recently elected as Chairman CSI Division IV Communication at National Level (2014–2016). Dr. Mishra has delivered his tutorials in IEEE international conferences in India as well as abroad. He is also the programme committee member, and reviewer of several international conferences. He visited and delivered his invited talk in Taiwan, Bangladesh, Singapore, Nepal, USA, UK, and France. He has authored a book on "Database Management Systems". He had been Chief Editor of Journal of Technology and Engineering Sciences. He has been also serving as member of Editorial Board of many national and international refereed journals. He has been a consultant to industries and government organizations like sales tax and labor department of government of Madhya Pradesh, India. Recently, he has been awarded with "Paper Presenter at International Level" by Computer Society of India. Currently, he is Chairman of Division IV Computer Society of India and Chairman of ACM Chapter covering Rajasthan and MP State.

Dr. Ahmad Taher Azar has received the M.Sc. degree (2006) in System Dynamics and Ph.D. degree (2009) in Adaptive Neuro-Fuzzy Systems from Faculty of Engineering, Cairo University (Egypt). He is currently serving as Assistant Professor in Faculty of Computers and Information, Benha University, Egypt. Dr. Azar is Associate Editor of IEEE Trans. Neural Networks and Learning Systems. Dr. Ahmad Azar has worked in the areas of Control Theory and Applications, Process Control, Chaos Control and Synchronization, Nonlinear control, and Computational Intelligence and has authored/coauthored over 170 research publications in peer-reviewed reputed journals, book chapters, and conference proceedings. He is an editor of many books in the field of Fuzzy Logic Systems, Modeling Techniques, Control Systems, Computational Intelligence, Chaos Modeling, and Machine Learning. Dr. Ahmad Azar is closely associated with several international journals as a reviewer. He serves as international programme committee member in many international and peer-reviewed conferences. Dr. Ahmad Azar is also a senior member in IEEE, Chair of IEEE Computational Intelligence Society (CIS) Egypt Chapter, and Vice-chair of IEEE Computational Intelligence Society Interdisciplinary Emergent Technologies Task Force. Also, he is the Vice-president (North) of System dynamics Africa Regional Chapter and an Academic Member of IEEE Systems, Man, and Cybernetics Society Technical Committee on Computational Collective Intelligence.

Mr. Amit Joshi is a young entrepreneur and researcher who has completed his graduation (B.Tech.) in Information Technology and M.Tech in Computer Science and Engineering and pursuing his research in the areas of Cloud Computing and Cryptography. He has an experience of around 6 years in academic and industry in prestigious organizations of Udaipur and Ahmedabad. Currently, he is working as an Assistant Professor in Department of Information Technology at Sabar Institute in Gujarat. He is an active member of ACM, CSI, AMIE, IACSIT-Singapore, IDES, ACEEE, NPA, and many other professional societies. He is an Honorary Secretary of CSI Udaipur Chapter and Secretary for ACM Udaipur Chapter. He has presented and published more than 30 papers in national and international journals and conferences of IEEE and ACM. He has also edited three books on diversified subjects including Advances in Open Source Mobile Technologies, ICT for Integrated Rural Development, and ICT for Competitive Strategies. He has organized more than 15 national and international conferences and workshops including International Conference ICTCS—2014 at Udaipur through ACM—ICPS. For his contribution toward the society, he has been given Appreciation Award by The Institution of Engineers (India), ULC, on the celebration of Engineers, September 15, 2014 and by SIG-WNs Computer Society of India on the Occasion of ACCE—2012 on February 11, 2012.

Automation of Railway Engine Pilot Security System Using Multimodal Biometric Identification

K. Sujatha, K. Senthil Kumar, Nallamilli P.G. Bhavani, V. Srividhya, T. Kalpalatha Reddy and K.S. Ramkumar

Abstract Railways are the most convenient mode of transport, but safety precaution is lagging. Train accidents due to an unknown person operating the engine will lead to the end of many lives and also loss of railway property. The golden solution to meet this problem here the proposed effective system is 'Automation of Railway Engine Pilot Security System using Multimodal Biometrics Identification' (AREPSS using MBI). Iris and fingerprint inputs are given by engine pilot from cabin to control room. In control room, identification takes place by fusing inputs, then passing the decision signal to automatically start the engine. It is the most commonly used unimodal biometric system, which can be seen in most of the places due to its popularity. Its reliability has decreased because it requires larger memory footprint and higher operational cost and it has slower processing speed. So, we are introducing Multimodal Biometric Identification System which uses iris and fingerprint for security purpose. The major advantage of this multimodal analysis is based on the template-matching phenomenon which utilizes less memory for storage as compared with footprint. User corroboration by multiple modality methods yields high output, high reliability, and high accuracy. So this technique enhances security in engine and thus saves lives and property.

Keywords Iris · Fingerprint · Unimodal · Multimodal · Wavelet transform · Neural network and biometric

K. Sujatha (✉) · K. Senthil Kumar · N.P.G. Bhavani · V. Srividhya · T. Kalpalatha Reddy · K.S. Ramkumar
Department of EEE/ECE, Center for Electronics Automation and Industrial Research (CEAIR), Dr. MGR Educational & Research Institute, Chennai, India
e-mail: drksujatha23@gmail.com

K. Sujatha · K. Senthil Kumar · N.P.G. Bhavani · V. Srividhya · K.S. Ramkumar
SKR Engineering College, Chennai, India

K. Sujatha · K. Senthil Kumar · N.P.G. Bhavani · V. Srividhya · K.S. Ramkumar
J.N.N Institute of Engineering, Chennai, India

D.K. Mishra et al. (eds.), *Information and Communication Technology*, Advances in Intelligent Systems and Computing 625,
https://doi.org/10.1007/978-981-10-5508-9_1

1 Introduction

Since Indian Railway is a cosmic system with more than 7,000 stations comprising of around 7000 engines, 34,000 coaches, 5,550 other railway vehicles, and 291,360 farm carts [1]. To control the network of whole railway is more complicated. So there is a need to increase the security in railways. Automation is one of the best approaches (solution) to increase the security in railways. Here, we introduce one of the automation approaches for pilot security in railway engine using effective multimodal biometric recognition [2].

The basic idea is that the iris and fingerprint biometric inputs are given by the driver from the engine cabin to the control room. In the control room, the verification process of fusing both the inputs takes place and verifies the detail about the driver and then passes the signal to the engine to operate [3]. Misidentification rate of iris is very less. So here fingerprint is combined with iris to increase the identification efficiency of pilot security scheme by means of fusion process. Here, password-based authentication is replaced by multimodal biometric authentication [4].

2 Existing System

Presently, pilot security in railway engine is not automated. According to the ongoing trend, the driver inserts the key which in turn produces the required voltage necessary to run the train [5]. The driver and the guard are the decision makers while the train is running, though they get the required signal from the control room. So when an unknown person operated the engine, it leads to the end of many lives and also the loss of railway property. So there is a need to increase the security system in railways [6].

2.1 Problems in Existing System

Problems in existing system are as follows: manual control, high risk of train passengers, loss of railway property, and less security. Since the pilot is the ultimate decision maker to operate the engine, all these problems are said to exist in the present system [7].

2.2 Existing Methodologies

Iris and fingerprint biometric authentication technique is more effective to improve the pilot security. The existing biometric identification techniques are unimodal biometric identification and multimodal biometric identification with two complete unimodal systems. The unimodal biometric system consists of its own unique feature extractor and classifier. Its reliability is decreased because it requires larger memory footprint and less accuracy and it has slower processing speed [8]. The pictorial representation of the unimodal biometric scheme is depicted in Fig. 1.

The block diagram of the existing multimodal biometric detection scheme is depicted (two complete unimodal schemes) in Fig. 2. Two or more unimodal schemes are combined to form the existing multimodal biometric systems [9]. Every unimodal method possesses its exclusive set of trait extractor and comparator, consequently blending their scores with a supplementary score for normalization process and a multifaceted blending approach. The customary multimodal method has a progress in the correctness and constancy of the structure over unimodal method. But this development increases the rate of system and

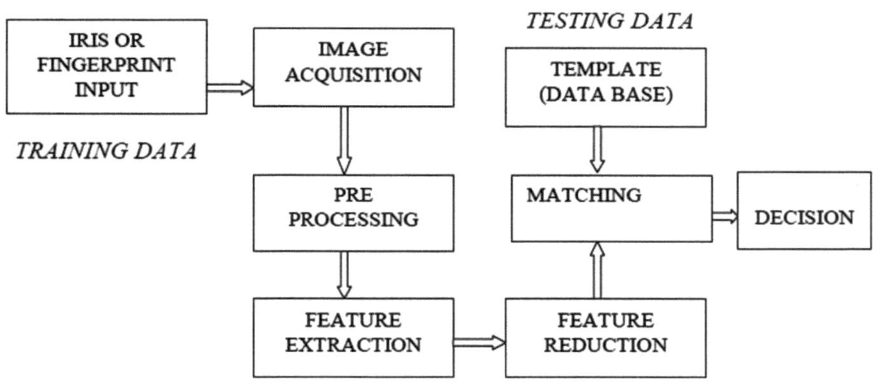

Fig. 1 Unimodal biometric system

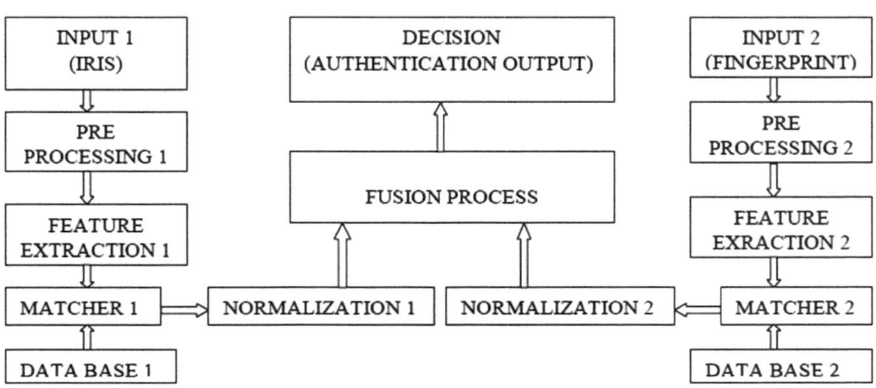

Fig. 2 Existing multimodal biometric system

requires larger memory for footprint storage [10]. In the majority cases, it needs either several algorithms or numerous sensors or both of them.

2.3 *Drawbacks in Existing Methodologies*

It requires larger memory footprint, high operational cost, multiple algorithms. Each unimodal system contains separate matcher (classifier), moderate fusion score level, and slower processing speed.

3 Proposed System

The main approach of the planned scheme involves developing a fully controlled and automated pilot security in railway engine using fusion-based multimodal biometric identification scheme devoid of the accessibility of two comprehensive unimodal structures [11, 12]. Here, iris and fingerprint are used as a part of biometric. Fusing these two input images reduces the chances of hacking. The schematic for the planned real-time system is illustrated in Fig. 3.

First, the input is given by the driver from the engine cabin to the control room; in the control room, the verification process of fusing both the inputs, i.e., iris and

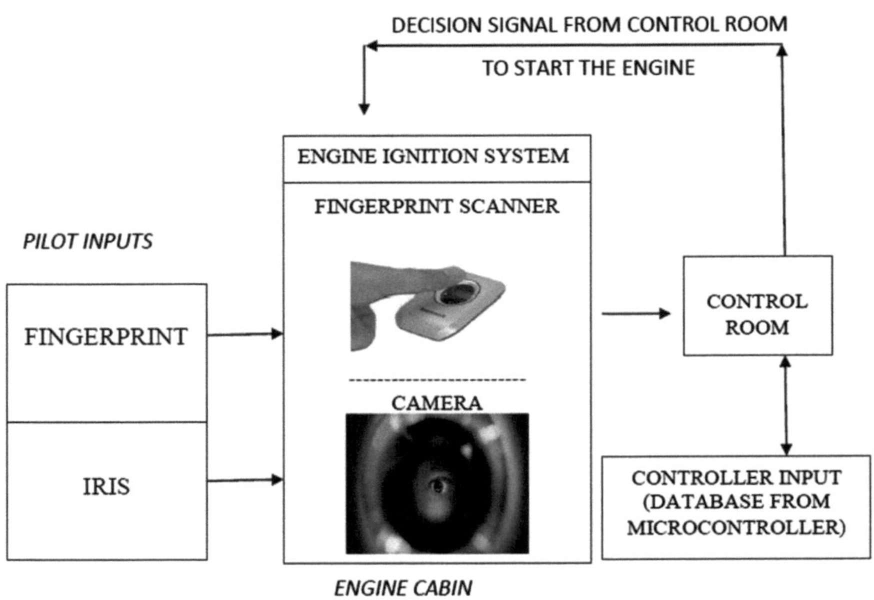

Fig. 3 Proposed automated pilot security system

fingerprint, takes place. For better safety, both the biometric inputs send to control room. Thus, the priority is given to the control room operator, who checks first, then verifies the detail about the driver, and then passes the signal to the engine to start. This process is repeated for each driver in break journey for long distance traveling trains, whereas for the short distance traveling like local trains, there are shifts for each driver or they may be used only for one-way journey.

3.1 Advantages of Proposed System

The merits include automated checkout and loss prevention. Safe journey for passengers eliminates the loss of railway property, provides high security, and is maintenance free.

3.2 Proposed Methodology

One of the key objectives for the growth of the planned methodology was to display that it is probable to devise an effectual multimodal biometric scheme devoid of two comprehensive unimodal systems. This proposed approach is used to overcome the problems occurred in existing methodologies and improve the pilot security in the existing system. The block diagram for the proposed methodology is shown in Fig. 4. Advancement in fingerprint- and iris-based multimodal biometric detection scheme with high score ranking for fusion makes use of a weighted Euclidean distance correlation or single Hamming distance correlation.

The first fingerprint biometric input is acquired and passed to the fingerprint trait extractor. The chosen base value is judged with the pattern in the information base using the distance correlation mentioned already. In the meantime, the subsequent input is obtained and moved frontward to the iris trait extractor. In the gap of computing the correlation value, the analysis of the initial biometric takes place and a correlation is obtained. Meanwhile, the second biometric input is processed and is prepared for evaluating the correlation coefficient value. The equivalent correlation technique is used to measure the reference value of iris biometric and compared with the templates to produce the corresponding output. The fusion is initiated only if the correlation values are presented. Proposed fusion-based multimodal biometric approach (Fig. 4) is for both modalities utilized same classifier is that both output scores will be same format. The overall proposed system will improve the security among pilot in railways.

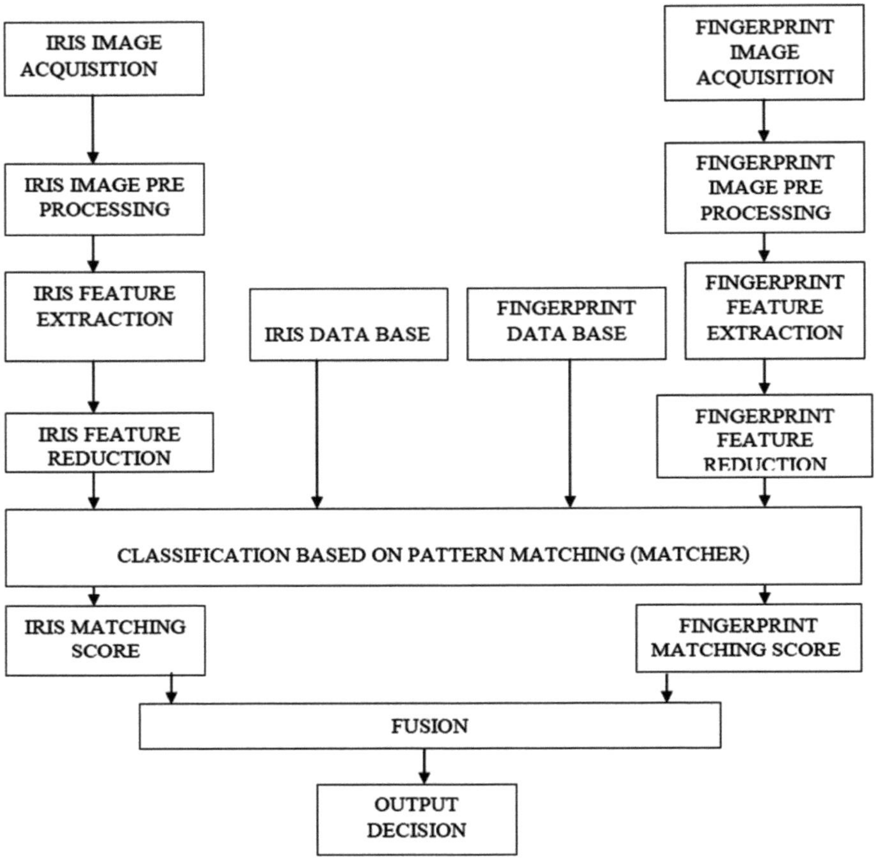

Fig. 4 Proposed multimodal biometric system

3.3 Advantages of Proposed Methodology

Eradicating the additional normalization functions, advances the handing speed, shrinks the storage space, simple design process and both modalities utilize a single classifier.

4 Recognition (Identification) Process

Planned multimodal biometric detection scheme with solo classifier for both methods is that both yielded scores will be in similar configuration, and it also simplifies the design process. The implementation of multimodal biometric system consists of the following processes: (1) iris recognition process, (2) fingerprint recognition process, and (3) fusion process. The implementation approach is involved in the procedure of obtaining an image of the region enclosing the text, preprocessing that image, dividing the individual lettering, relating the characters in a type appropriate for the processor to handle it, and recognizing it.

4.1 Iris Recognition Process

The iris detection is a very steadfast technique of individual recognition. The iris image is inimitable for all and does not get altered throughout the existence. The iris detection scheme is to robotically distinguish the individuality of anyone from a fresh image by matching up to the human iris samples denoted with uniqueness in the piled-up catalog. Iris detection method has quad stages. (1) First, a picture enclosing the user's eye is captured by CCD or infrared camera (image acquisition). (2) Then, the image is preprocessed to regularize the scale and lighting to confine to the iris in the image (image preprocessing). (3) The third step is the feature extraction indicating the extraction of iris patterns. It is done by using Daugman's approach. This approach includes iris marking, cropping, polar format, and wavelet encoding (converts polar to analog format). This process is shown in Fig. 5. (4) Finally, the choice is completed using classification (Classifier). Classifier refers to the ultimate decision for detection. This method not only is restricted to

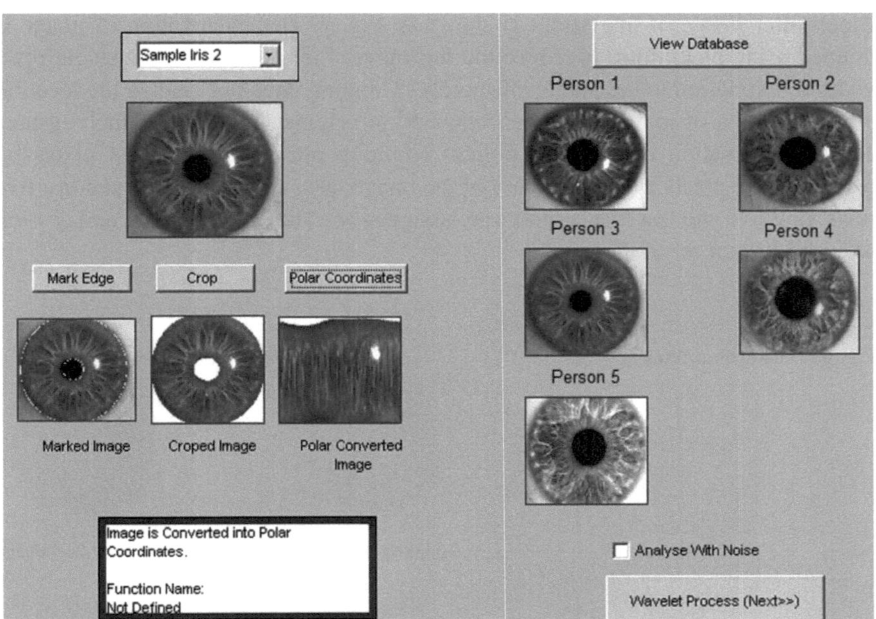

Fig. 5 Iris feature extraction process

comparison but also coordinates in obtaining the information from iris. In similar systems, it can be experienced by using the weighted Euclidean distance (WED) algorithm. WED can be used to judge against two templates, particularly if the template is of integer values. WED gives a gauge of similarity check over a collection of values between two templates.

4.2 Fingerprint Recognition Process

A fingerprint encompasses ridges and valleys. The ridges are the shady regions of the fingerprint, and the basins are the white regions flanked by the ridges.

4.2.1 Binarization

The fingerprint expressed as '0s' and '1s' is called as a binary image, as shown in Fig. 6. Binarization is used to convert grayscale image into binary image by setting up a threshold value. The pixel values higher than and lower than the threshold are set to 1 and 0, respectively.

4.2.2 Block Filter

Block filter image of fingerprint is shown in Fig. 7. The binarization of image is thinned using block filter to reduce the thickness of all ridge lines to a single pixel width to extract minutiae points effectively. Thinning does not change the location and orientation of minutiae points compared to original fingerprint which ensures accurate estimation of minutiae points. Thinning preserves outermost pixels by placing white pixels at the boundary of the image; as a result, first five and last five rows, first five and last five columns are assigned as value of 1. Dilation and erosion are used to thin the ridges.

Fig. 6 Binarization process

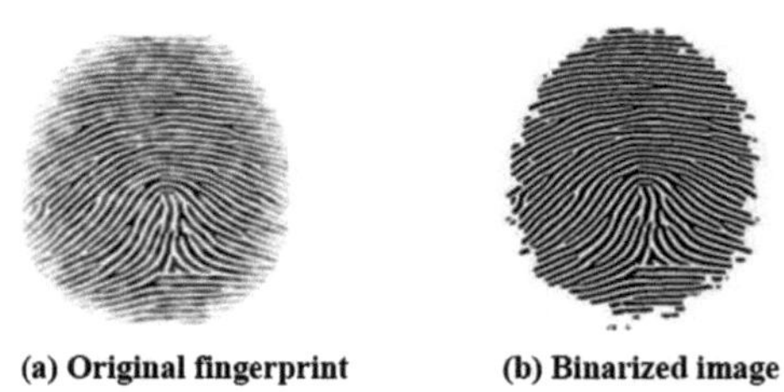

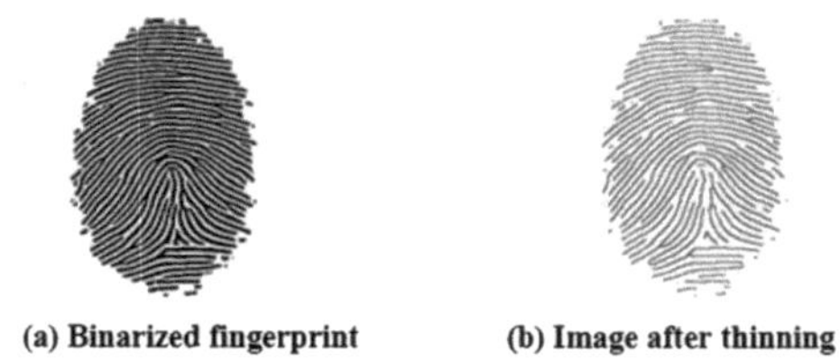

Fig. 7 Block filter process

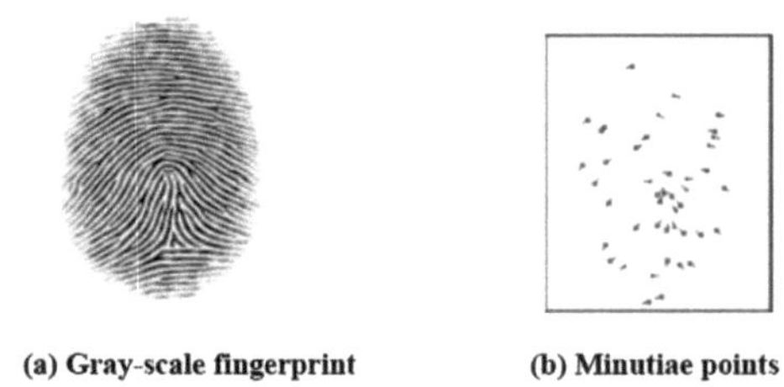

Fig. 8 Minutiae extraction process

4.2.3 Minutiae Extraction

Minutiae extraction of fingerprint is shown in Fig. 8. The minutiae location and minutiae angles are derived after minutiae extraction. The termination which lies at the outer boundaries is not considered as the minutiae points, and the crossing number is used to locate the minutiae points in the fingerprint image. The crossing number is defined as the half of the sum of difference between intensity values of the adjacent pixels. If the crossing number is 1, 2, and 3 or greater than 3, then the minutiae points are classified as termination, normal ridge, and bifurcation, respectively.

4.3 Fusion Process

The fusion of images is the procedure of joining two or more (iris and fingerprint) images into only image retaining important features from each. Recently, multi-resolution analysis is one of the acceptable methods to analyze the remotely sensed images. Several approaches to image fusion can be distinguished, depending on whether the images are fused in the spatial domain or they are transformed into another domain, and their transforms are fused. A new method based on discrete wavelet transform is proposed here. The DWT-based techniques became popular due to their multi-resolution properties. The wavelet transform—Heuristic additive rule-based method—is used given by $i_3 = ((i_1 + i_2)/2) + ((i_1 + i_2)/100)$. It can show a good position of a function (here this function is the image) in spatial and frequency space. It is used to display the efficient recognition rate of combined image. The iris–fingerprint fusion process has the following advantages such as high security, easy to implement, no training is required, universally accepted by all, and noise reduction.

5 Results and Discussion

Identification of engine pilot is based on the classifier output and fusion score level. In this section, we are going to discuss simulation output for classifier using MATLAB. The details regarding the identification database and false accept error rates are provided in MATLAB coding for all the three process. In this identification process, one fingerprint/iris image is selected as the verification image input from database. Then, input image is matched against the entire database images. If the given input matched with the database, then the user is identified as an authorized person to start the engine and the maximum matching score value will be displayed. If the given input not matched with the database, then the user is identified as unauthorized person and the minimum matching score value is displayed with error percentage. The simulation results of fingerprint and iris are shown in Figs. 9 and 10.

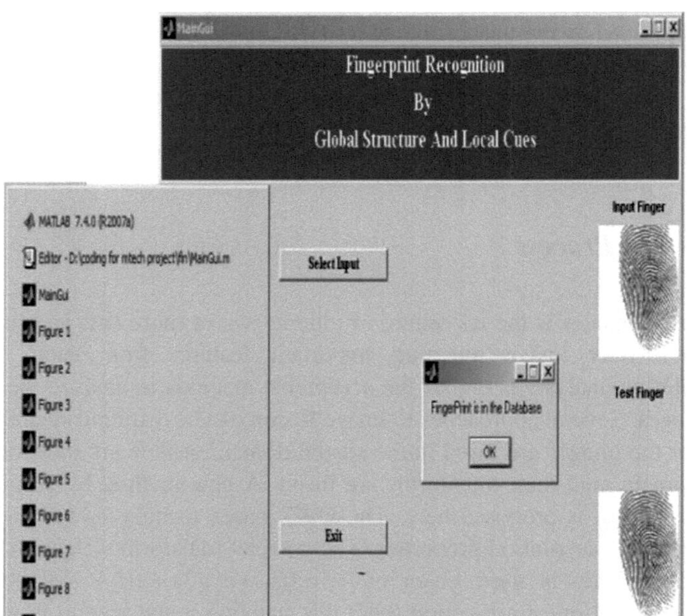

Fig. 9 Simulation matching result for fingerprint

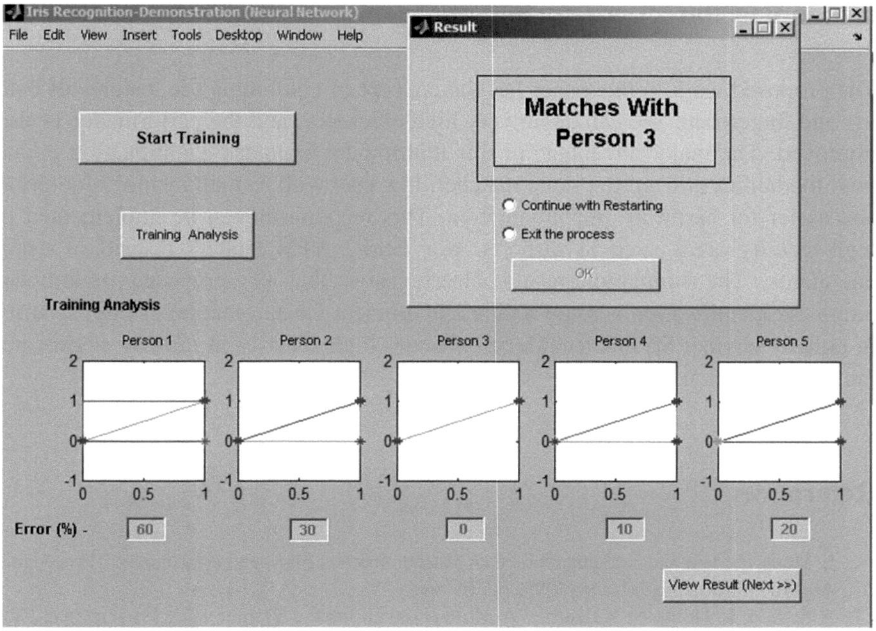

Fig. 10 Simulation matching result for iris

Fig. 11 Simulation result for fusion process

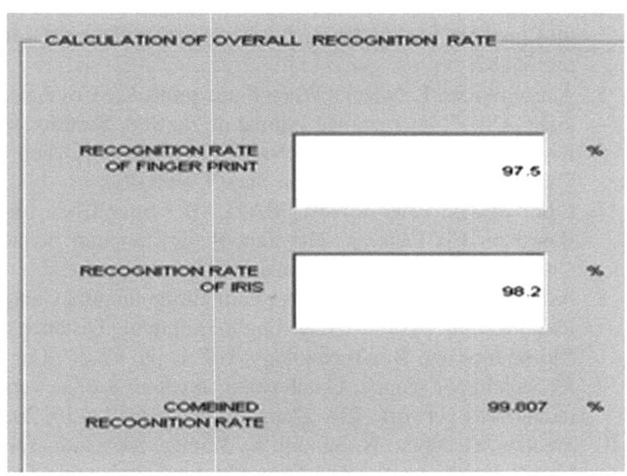

When the fusion process score value is compared with the individual score values of both input images, it will improve the efficiency of the average score value. The simulation result of fusion process is shown in Fig. 11.

The simulation result of the proposed fusion-based multimodal biometric identification clearly shows the improved recognition rate compared with existing identification method.

6 Conclusion

The proposed work in this paper has the concept of combining the features of both iris and fingerprint. We can attain very high efficiency, and the performance is also improved. The major advantage of this multimodal biometric identification is that both modalities utilized the same matcher, low cost with a small memory footprint, and easier for hardware implementation. This recognition can be implemented in high-security areas such as airports, war fields, ATM centers instead of using passwords. The simulation results clearly show that the proposed multimodal biometric identification is more secure and efficient for automation of pilot security in railway engine. So this technique enhances high security in railway engine and thus saves lives and property.

References

1. L. Hong, A. Jain and S. Pankanthi, "Can Multibiometrics Improve performance?" Proceeding AutoID'99, Summi, NJ, Oct 1999, pp. 59–64.
2. A.K. Jain, L. Hong, Y. Kulkarni, "A Multimodal Biometric System using Fingerprints, Face and Speech", 2nd International Conference Audio and Video-based Biometric Person Authentication. Washington, March 22–24, 1999, pp. 182–187.
3. A.k. Jain, S. Prabhakar and S. Chen, "Combing Multiple Matchers for a High Security Fingerprint Verification System", Vol. 20, No. 11–13, 1999, pp 1371–1379.
4. Fabio Roli, Josef Kittler, Giorgio Fumera, Daniele Muntoni, "An Experimental Comparison of Classifier Fusion Rules for Multimodal Personal Identity Verification System", 2002, pp. 76–82.
5. A. Lumini and L. Nanni, "When Fingerprints Are Combined with Iris – A case study: FVC2004 and CASIA", International Journal of Network Security, vol.4, Jan 2007, pp. 27–34.
6. Karthik Nandakumar, "MultiBiometric Systems: Fusion Strategies and Template Security", Ph.D. Thesis, 2008, Michigan State University,
7. Libor Masek, Peter Kovesi. MATLAB Source Code for a Biometric Identification System Based on Iris Patterns. The School of Computer Science and Software Engineering, the University of Western Australia. 2003.
8. Asim Baig, Ahmed Bouridane, faith kurugollu, and Gang Qu, "Fingerprint Iris Fusion based Identification System using single Hamming Distance Matcher," International journal of Bio-science and Bio-Technology, vol. 1, pp. 47–57, Dec-2009.
9. K. Sujatha, N. Pappa, Combustion monitoring of a water tube boiler using a discriminant radial basis network, ISA Transactions, 50, 101–110, 2011.
10. Sujatha, K. Pappa, N. Senthil, K. Kumar, Siddharth Nambi, U. and Raja Dinakaran, C. R. (2013) Intelligent Parallel Networks for Combustion Quality Monitoring in Power Station Boilers, Trans Tech Publications, Switzerland, Advanced Materials Research, Vol. 699, pp. 893–899.
11. Sujatha, K. Pappa, N. Senthil, K. Kumar, Siddharth Nambi, U. and Raja Dinakaran, C. R. (2013) Automation of Combustion Monitoring in Boilers using Discriminant Radial Basis Network, *Int. J. Artificial Intelligence and Soft Computing,* Vol. 3, No. 3.
12. Sujatha, K. (2015) "Smart Sensor for NO_x and SO_2 Emissions in Power Station Boilers ", Indian Journal of Science and Technology, Vol 8(27), pp: 1–7.

Extracting Hidden Patterns Within Road Accident Data Using Machine Learning Techniques

S. Vasavi

Abstract Road accidents may not be stopped altogether, but can be reduced. Driver emotions such as sad, happy, and anger can be one reason for accidents. At the same time, environment conditions such as weather, traffic on the road, load in the vehicle, type of road, health condition of driver, and speed can also be the reasons for accidents. Hidden patterns in accidents can be extracted so as to find the common features between accidents. This paper presents the results of the framework from the research study on road accident data of major national highways that pass through Krishna district for the year 2013 by applying machine learning techniques into analysis. These datasets collected from police stations are heterogeneous. Incomplete and erroneous values are corrected using data cleaning measures, and relevance attributes are identified using attribute selection measures. Clusters that are formed using K-medoids, and expectation maximization algorithms are then analyzed to discover hidden patterns using a priori algorithm. Results showed that the selected machine learning techniques are able to extract hidden patterns from the data. Density histograms are used for accident data visualization.

Keywords Machine learning techniques · Road accident data analysis · Preprocessing · Clustering · Association rule mining · Visualization

1 Introduction

Road safety means development and management of roads, provision of safer vehicles, and a comprehensive response to accidents [1]. Modern traffic management systems, such as real-time adjustment of traffic flow, model predictive control (MPC) technique in traffic light control, tolling strategy, etc., can be used in design and maintenance of roads, and also for producing safer vehicles. BRT system of Ahmadabad city has achieved its objective of providing a safe mode of transport

S. Vasavi (✉)
VR Siddhartha Engineering College, Kanuru, AP, India
e-mail: vasavi.movva@gmail.com

D.K. Mishra et al. (eds.), *Information and Communication Technology*,
Advances in Intelligent Systems and Computing 625,
https://doi.org/10.1007/978-981-10-5508-9_2

with more than 50% decrease in road traffic [2]. According to the National Crime Records Bureau [3], there were 39,344 road accidents, which resulted in the death of 14,966 persons. Another point of concern is that, while 8.9% of all accidents in the country occur in the state, the percentage of all deaths is higher at 10.8% t. Statistics also reveal that most accidental deaths involve people traveling in three-wheelers. More than 25% of accident deaths involving passengers of auto-rickshaws throughout the country are in Andhra Pradesh [3]. About 1,734 persons died in road accidents involving auto-rickshaws, and the state has the highest number of such deaths in the country [3]. According to the report given in [4], road accidents are the ninth leading cause of death in 2004 and expected to be fifth leading cause of death by 2030 worldwide. This paper proposes a framework that is based on the cluster analysis using K-medoids and expectation maximization (EM) and association rule mining using a priori algorithm. Association rule mining is further applied on these clusters to generate association rules. Performance is analyzed using precision and recall measures. The paper is organized as follows: Sect. 2 presents literature survey on various existing methods for accident data analysis. Methodology of proposed system is described in Sect. 3. Section 4 includes results obtained from our proposed system and analysis with respect to the performance measures. Conclusions and future work are given in Sect. 5.

2 Literature Survey

Results from the research study on applying large-scale data mining methods into analysis of traffic accidents on the Finnish roads are presented in [5]. The main intension is to show that the selected data mining methods are able to produce understandable patterns from the data, finding more fertilized information could be enhanced with more detailed datasets. The work of [6] emphasizes the importance of data mining classification algorithms in predicting the vehicle collision patterns occurred in training accident dataset. They followed a stepwise procedure which finally yields the required accident analysis results: data cleaning, data transformation, and relevance analysis. The feature selection algorithms have been explored to improve the classifier accuracy. The research work in [7] emphasizes the significance of data mining classification algorithms in predicting the factors which influence the road traffic accidents specific to injury severity. Further they applied feature selection methods to select the relevant road accident-related factors and Meta classifier Arc-X4 to improve the accuracy of the classifier. In order to improve road safety, the authors of [8] analyzed the Andalusia Complementary Road Network, by using advanced data mining techniques in order to discover hidden relationships between characteristic of the roads, ESM, and crashes. The research work in [9] is that accidents are not randomly scattered along the road network, and that drivers are not involved in accidents at random. Authors focused on the contribution of road-related factors to accident severity in Ethiopia. Work presented in [10] is about discovering interesting rules from a set of generated rules using both association rule algorithms. Work reported in [11] is to reduce the number of road accidents in main cities of Tamil Nadu. They used WEKA tool and H-DTANN techniques in order to predict the road accident injury levels.

3 Proposed System

The main objective of this research is to investigate the role of human-, vehicle-, and infrastructure-related factors in accident severity by applying machine learning techniques on road accident data. The overall architecture of the proposed system is shown in Fig. 1. The steps include data cleaning, data transformation, relevance analysis, clustering, association rules generation, and finally performance evaluation.

3.1 Database Creation

A total of 30 attributes that focus on various criteria, such as accident-specific attributes, driver-specific attributes, FIR details, circumstance-specific attributes, and other attributes given in the FIR report, form the input dataset.

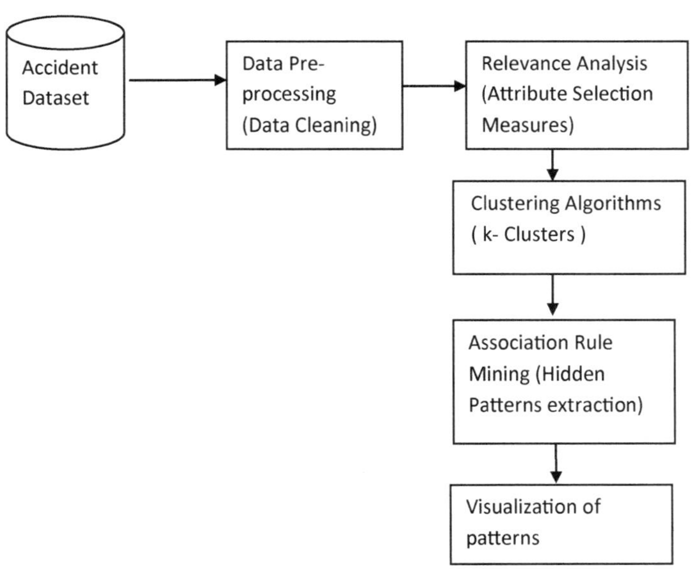

Fig. 1 Proposed system architecture

3.2 Data Preprocessing

Data preprocessing helps to remove noise, missing values, and inconsistencies. Missing values are replaced with NULL. Also each attribute data is discretized in order to make it appropriate for further analysis. Table 1 presents the data before and after transformation.

3.3 Attribute Selection Measures

Entropy measures information gain, and Gain ratio and Gini index are used to choose relevant attributes useful for performing analysis. Table 2 presents the

Table 1 Data transformation

Accident time		Accident place		Accident month		Deceased age	
Before	After	Before	After	Before	After	Before	After
8.30	Morning	Chittinagar	Chittinagar	January	January	50	Senior

Table 2 Top 20 attributes given by attribute selection measure for a Nunna dataset

Information gain	Gain ratio	Gini index	Attribute chosen
Place of accident	Any damage	Place of accident	Number injured
Any damage	Cost of damage	Any damage	Accident time
Cost of damage	Accused emotions	Cost of damage	Place of accident
Hospital reported	Place of accident	Hospital reported	Temperature
Month	Hospital reported	Month	Cost of damage
Accident type	Accident type	Accident type	Highway
Deceased emotions	No injured	Deceased emotions	Accident type
No injured	Ambulance used	No injured	Heavy traffic involved
Deceased age	Deceased emotions	Deceased age	Vehicles involved
Ambulance used	Month	Accident time	Deceased emotions
Accident time	Heavy traffic involved	Ambulance used	Accused emotions
Accused emotions	Highway	Accused emotions	Deceased age
Highway	Deceased age	Highway	Hospital reported
Vehicles involved	Vehicles involved	Weather	Month
Heavy traffic involved	Accident time	Vehicles involved	Ambulance used
Weather	Weather	Heavy traffic involved	Speed limit
Temperature	Temperature	Temperature	Road condition
Lightness	Lightness	Lightness	Lightness
Road condition	Road condition	Road condition	Weather

comparison of ranking of top 20 attributes given by each of the measure for sample dataset.

3.4 Clustering

K-medoids and expectation maximization algorithms are used for clustering, and the following clusters are formed.

Cluster 1 is the traffic cluster in which accidents happen because of low and high traffic. A total of 15% of the accidents occurred during high traffic, 76% of accidents occurred during low traffic, and 6% of accidents occurred during medium traffic.
Cluster 2 is the time of accident cluster in which accidents happen during morning, afternoon, evening, and night. A total of 32% of accidents occurred during morning time, 19.3% of accidents occurred in the afternoon, 18.5% of accidents occurred in the evening, and 29.2% of accidents occurred during nighttime.
Cluster 3 is the age of the drivers cluster in which 2.2% of accidents occurred to the age group children, 67.2% of accidents occurred to teenagers, 22.5% of accidents occurred to youth, 34.2% of accidents occurred to middle-aged people, 20.2% of accidents occurred to senior citizens, and 17% of accidents age value is missing.
Cluster 4 is the accident occurred month, in which 10.5% of accidents occurred in January, 7.7% of accidents occurred in February, 11.04% of accidents occurred in March, 9% of accidents occurred in April, 12.3% of accidents occurred in May, 11.3% of accidents occurred in June, 8.3% of accidents occurred in July, 10.24% of accidents occurred in August, 8.3% of accidents occurred in September, 7.4% of accidents occurred in October, 7.04% of accidents occurred in November, and 8.6% of accidents occurred in December.
Cluster 5 is the weather condition at the time of accident, in which 34.6% of accidents occurred when weather is cool, 33.5% of accidents occurred when weather is clear, and 31.9% of accidents occurred when weather is hot.
Cluster 6 is the lightening condition at the time of accident, in which 33% of accidents occurred when lightening is dark, 25.5% of accidents occurred in dim light, and 41.5% of accidents occurred in bright light.
Cluster 7 describes about type of accident, in which 69.5% of accidents occurred because of rash driving, 3.7% of accidents occurred because of single vehicle runoff, 0.33% of accidents occurred because of vehicle skidding, 0.33% of accidents occurred because of overlooking, 6.2% of accidents occurred because of overriding, 11.6% of accidents occurred because of hit by other vehicles, 8% of accidents occurred during lane change, and 0.34% of accidents are because of sudden turn back or animal hit, wrong direction.
Cluster 8 describes the speed limit of vehicles at the time of accident, in which 32% of accidents occurred at normal speed limit, 44.5% of accidents occurred at high speed limit, and 33.5% of accidents speed limit value is missing.

3.5 Discovery of Frequent Patterns and Association Rules

The next step is to extract hidden patterns and facts from road accident data using a priori algorithm. These hidden patterns may give analysis on various unknown risk factors behind fatal accidents and predict accident-prone areas. These rules are evaluated using support and confidence measures. Interesting measure, lift is used to rank the rules. From each of the cluster, top 20 rules are taken for analysis in this study. These rules are further visualized using density histograms.

3.6 Cluster Validation

F-measure is used for cluster analysis because it performs node-based analysis using Eqs. (1)–(3) [12].

$$\text{Precision} = \text{TP}/(\text{TP} + \text{FP}) \tag{1}$$

$$\text{Recall} = \text{TP}/(\text{TP} + \text{FN}) \tag{2}$$

$$\text{F-Measure} = (1+\alpha)/((1/\text{Precision}) + (\alpha/\text{Recall})) \quad \text{where } \alpha = 1. \tag{3}$$

3.7 Visualization

Graphical representation techniques will help in identifying the risk of the accident immediately by government officials. Density histograms for visualizing region-wise results are generated using MATLAB software as shown in Figs. 2 and 3 for sample dataset.

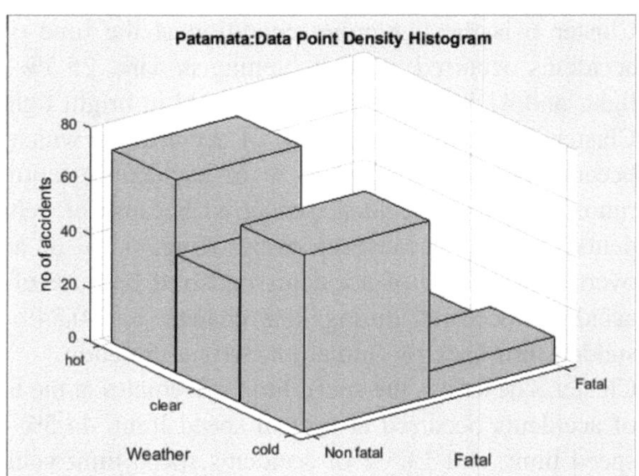

Fig. 2 Fatal versus weather

Fig. 3 Fatal versus time

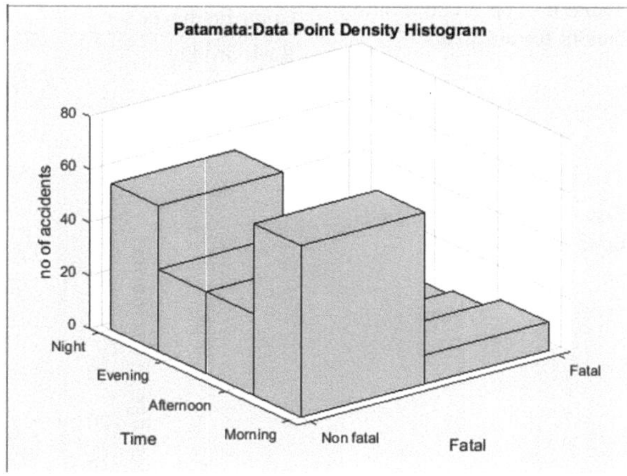

Table 3 Contributing factors for accident

Contributing factor	Percentage of accidents (%)
Human–vehicle	83.64
Human	16.3
Infrastructure	0.06

Similar graphs are generated for time versus day, fatal versus month, fatal versus traffic, and fatal versus age.

4 Results and Analysis

Road accidents and injuries occur because of human fault or vehicle fault or infrastructure fault or sometimes combinations of these factors. Each of these factors individually or in combination may cause accident. It was observed from the dataset that accidents mainly occurred because of combination of human fault and vehicle fault as shown in Table 3.

Human alone factors such as "helmet and seat belt not used" are not reported in the FIRs and as such are not known. Table 4 presents top 3 contributing factors for accidents, highest being rash driving of the people.

Analysis like type of vehicles (two-wheeler, car, bus, lorry, jeep, truck, etc.) is not given in the FIR report, and as such, analysis is not done. Figures 4 and 5 present percentage distribution of accidents on various criteria, speed limit, and injury severity.

Similar analysis is done on other criteria such as distribution of accidents by time of accidents and deceased age, distribution of accidents by month and weather during the accident, distribution of accidents by lightness and speed limit, distribution of accidents by accident type (human factors), distribution of accidents by day of accident and deceased age, distribution of accidents by deceased emotions,

Table 4 Top 3 contributing factors for accidents

Contributing factor	Percentage of accidents (%)
Rash driving	62.57
Object hit	26.67
Lane change	8.1

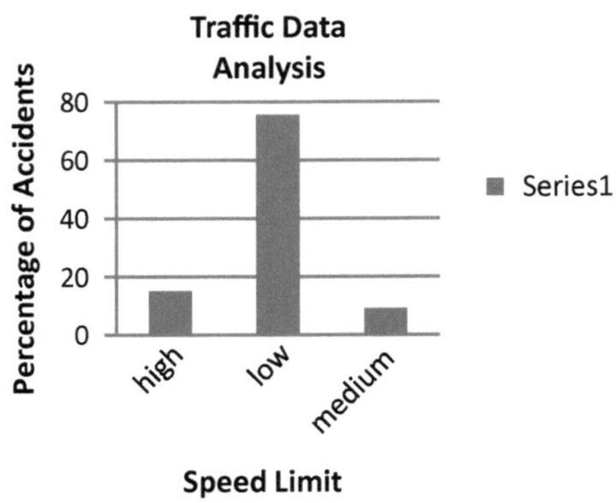

Fig. 4 Accidents by speed limit

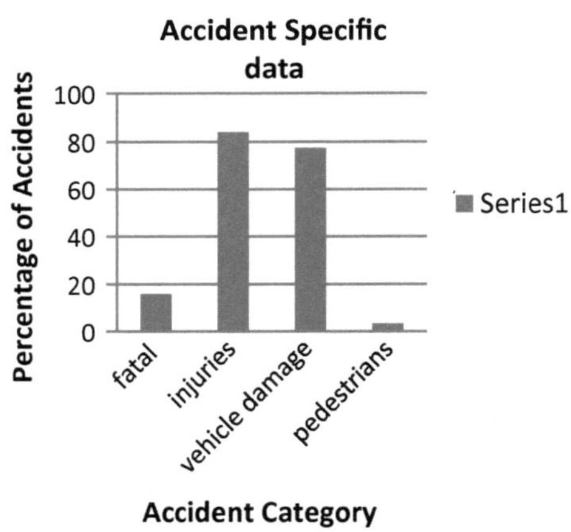

Fig. 5 Accidents by injury severity

distribution of accidents by hospital reported and ambulance used. Because of space limit, all graphs are not listed here.

K-medoids uses the cluster center to create clusters, whereas EM clustering uses the probabilities of the clusters to further calculate the optimized clusters. The results of two clusters formed from K-medoids and EM are given in Table 5.

The performance of the EM clustering is low compared to K-medoids clustering algorithm, because it uses probability measures to cluster the data. The number of iterations and runs taken to cluster the data using EM clustering is more when compared with the K-medoids clustering technique. Table 6 presents precision and recall values for both clustering algorithms.

Table 5 Comparison of clustering techniques based on emotion

K-medoids	C6 (age)	C	Y	M	S	NULL
	Obtained	1	28	35	43	11
	Expected	1	28	35	43	11
Expectation maximization	C6 (age)	C	Y	M	S	NULL
	Obtained	1	28	35	43	11
	Expected	1	28	35	43	11

Table 6 Performance measures for K-medoids and EM algorithm

Dataset	Precision		Recall		F-measure	
	EM	K-medoids	EM	K-medoids	EM	K-medoids
1504 tuples	0.5	0.8	0.4	0.6	0.45	0.69

From the data analysis, accident distribution is even in normal days, and it is observed to be higher in weekend. Accidents occurrence is high at cold nights compared to hot and clear conditions. Most accident-prone area is observed to be Kesarapalli village road and Venkateswara theater in Gannavaram. It is observed to be fatal accidents are high among the old-aged group and non-fatal in young-aged and middle-aged people. Accidents are high in the month of August and low in the month of June. Females involved in accidents are observed to be 20.16% of overall accidents to 73.45% of male.

5 Conclusions and Future Work

The aim of this paper is to generate association rules that will analyze how to discover hidden patterns that are the root causes for accidents among different combinations of attributes of a larger dataset. Density histograms for visualizing regionwise such as fatal versus weather, fatal versus time, time versus day, fatal versus month, fatal versus traffic, and fatal versus age are performed. Percentage distribution of accidents on various criteria, speed limit and injury severity, distribution of accidents by time of accidents and deceased age, distribution of accidents by month and weather during the accident, distribution of accidents by lightness and speed limit, distribution of accidents by accident type (human factors), distribution of accidents by day of accident and deceased age, distribution of accidents by deceased emotions, distribution of accidents by hospital reported and ambulance used is also made. Future work is to make analysis on road accidents' dataset by considering more features and clusters and also to use deep learning techniques so as to better cluster the records.

Acknowledgements I thank University Grants Commission (UGC), for funding this project. I also thank police authorities, Andhra Pradesh, for providing the required information. I am also thankful to the management of Siddhartha Academy for providing me resources and environment for successfully completing this project. Finally, I thank my students who helped me during the implementation of this project.

References

1. Road safety and traffic management : Report of the committee Planning Commission, Government of India in February 2007 (2007)
2. Rayle L, Pai M. Scenarios for future urbanization: carbon dioxide emissions from passenger travel in three Indian cities. Transportation Research Record: Journal of the Transportation Research Board, 2193:124–131 (2010)
3. http://www.deccanchronicle.com/130629/news-current-affairs/article/andhra-pradesh-ranked-3rd-road-accidents last accessed June 29th 2013
4. Road Accidents in India Issues & Dimensions, Ministry of Road Transport & Highways Government of India (2014)
5. SAMI AYRAMO, PASI PIRTALA, Mining road traffic accidents, Reports of the Department of Mathematical Information Technology Series C. Software and Computational Engineering No. C. 2/2009 (2009)
6. S. SHANTHI, DR.R. GEETHA RAMANI, Classification of Vehicle Collision Patterns in Road Accidents using Data Mining Algorithms, International Journal of Computer Applications (0975–8887) Volume 35– No.12, December 2011 (2011)
7. S. SHANTHI, R. GEETHA RAMANI, Feature Relevance Analysis and Classification of Road Traffic Accident Data through Data Mining Techniques, Proceedings of the World Congress on Engineering and Computer Science 2012 Vol I WCECS (2012)
8. Luis Martín, Leticia Baena, Laura Garach, Griselda López, Juan de Oña Using Data Mining Techniques to Road Safety Improvement in Spanish Roads, Volume 160, Pages 607–614, XI Congreso de Ingenieria del Transporte (CIT 2014)
9. Tibebe Beshah, Shawndra Hill, Mining Road Traffic Accident Data to Improve Safety: Role of Road- related Factors on Accident Severity in Ethiopia, AAAI Spring Symposium Series (2010)
10. Amira A. El Tayeb, Vikas Pareek, Abdelaziz Araar Applying Association Rules Mining Algorithms for Traffic Accidents in Dubai, International Journal of Soft Computing and Engineering (IJSCE) ISSN: 2231-2307, Volume-5 Issue-4, (2015)
11. K. Geetha, C. Vaishnavi Analysis on Traffic Accident Injury Level Using Classification, International Journal of Advanced Research in Computer Science and Software Engineering, Volume 5, Issue 2, (2015)
12. Jiawei Han and Micheline Kamber, *Data Mining Concepts and Techniques*, 2 ed, Elsevier publishers

Implementation of Smart Job First Dynamic Round Robin (SJFDRR) Scheduling Algorithm with Smart Time Quantum in Multi-core Processing System

Amit Kumar Gupta, Narendra Singh Yadav and Dinesh Goyal

Abstract Modern computer system is organized with multi-core processing system. The scheduling of processes in multiprocessing may turn into more complex task. In multi-core processing system, there are two or more cores embedded into a single chip. This architecture provides more efficiency in terms of throughput than single processor architecture. Previously, most of the work has been done in creating new scheduling algorithms for multi-core processing system, but small consideration has been given to merge user priority and system priority. In this paper, researcher has proposed Smart Job First Dynamic Round Robin Algorithm with smart Time Quantum (SJFDRR) in multi-core processing system in which a smart priority factor (SPF) is calculated for each process. The process which has lowest value of SPF is scheduled first. The time quantum is calculated dynamically for each processor. By this algorithm the average waiting time and average turnaround time and context switch is significantly decreases which lead to increase in performance of the system.

Keywords Operating systems · Multi-core processing system · Synchronous multi-core processor · Asynchronous multi-core processor · Scheduling algorithm · Time quantum · Round robin

A.K. Gupta (✉) · D. Goyal
Suresh Gyan Vihar University, Jaipur, India
e-mail: cseprof_amit@rediffmail.com

D. Goyal
e-mail: dinesh.goyal@mygyanvihar.com

N.S. Yadav
JECRC University, Jaipur, India
e-mail: narensinghyadav@yahoo.com

D.K. Mishra et al. (eds.), *Information and Communication Technology*,
Advances in Intelligent Systems and Computing 625,
https://doi.org/10.1007/978-981-10-5508-9_3

1 Introduction

In multi-core processing system, two or more processing units or cores are embedded on a single chip. There are two types of multi-core processors architecture exist in computer architecture. One is synchronous multi-core processors (SMPs) in which all the cores have similar architecture, and second is asynchronous multi-core processors (ASMPs) in which all the cores have different architecture. The multi-core processing unit uses the shared-memory architecture. Previously, the multi-core processors have been developed in the variety of multiprocessor system on chip (MPSoC), but they were restricted to use in a sector of applications such as networking. The simple accessibility of multi-core has required software developer to modify the way they imagine and write their applications. Unfortunately, the applications written so far are in order in nature [1]. The multi-core processor architecture is shown in Fig. 1.

Most of the multi-core processors have single shared cache storage which leads to mutual accesses in processing. The multi-core processing systems do not automatically provide performance improvements to applications. The applications are restructured to enhance their parallelism. Similarly, CPU schedulers are also restructured to enhance the performance of this new application parallelism [2]. Previously, most of the work has been done in creating new scheduling algorithms for multi-core processing system, but small consideration has been given to merge user priority and system priority.

The construction of the paper is as described below. Section 2 discusses various scheduling criteria. On the basis of these criteria, the performance of scheduling algorithm is evaluated. Section 3 discusses classical scheduling algorithms available in literature. Section 4 containing literature survey, in which various approaches of scheduling algorithm on multi-core processing system have been discussed. Section 5 discusses the proposed Smart Job First Dynamic Round Robin Algorithm with smart Time Quantum (SJFDRR) in multi-core processing system concept in detail. Section 6 demonstrated the proposed algorithm using one example and

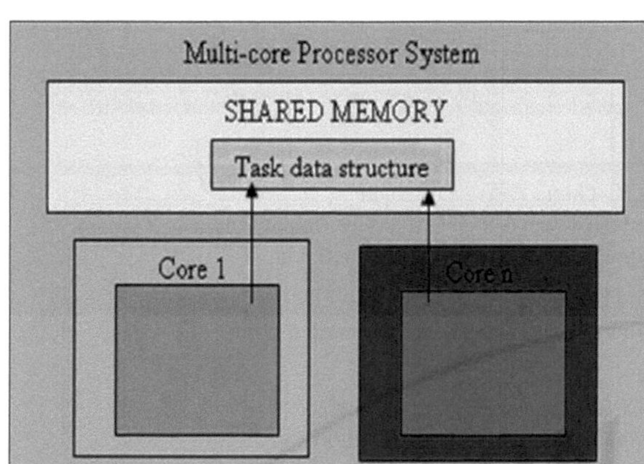

Fig. 1 Multi-core processing system [2]

discusses and compares the results of proposed algorithm with SJF-RR in multi-core system [3]. Section 7 illustrates the comparative results evaluated on simulator. Section 8 concludes the proposed algorithm with advantages and problem of proposed algorithm.

2 Scheduling Criteria

There are many CPU scheduling algorithms having different properties, and the selection of a particular algorithm may favour one class of processes over another. The algorithm is selected for a particular state; we must judge properties of a verity of algorithms. The criteria contain the following [4, 5]:

- **Context Switch**: A context switch occur when a process interrupt the normal execution sequence of another process. The CPU stores all relevant information of interrupted process in task control box (TCB). The context switch includes wastage of time, memory and scheduling overhead. So scheduling algorithm is designed in such way that it can minimize the number of context switches.
- **Throughput**: This term is defined as number of process completed per unit time. So scheduling algorithm is designed in such way that it can maximize the throughput.
- **CPU Utilization**: From the performance wise concern the CPU cannot be sit ideal. So scheduling algorithm is designed in such way that it cans maximum use of CPU as possible.
- **Turnaround Time**: It is the difference in the time of process when a process is ready to execute and when it complete its execution. So scheduling algorithm is designed in such way that it can minimize the turnaround time.
- **Waiting Time**: It is the sum of all waiting done by a process in ready queue for execution. So scheduling algorithm is designed in such way that it can minimize the waiting time.
- **Response Time**: Response time is the time it takes to start its execution not the time it takes to output the response.

3 Classical Scheduling Algorithm

There exist different scheduling algorithms, each of them has advantages and disadvantages and as follows:

- **First-Come-First-Served (FCFS)** FCFS is simple scheduling algorithm in which process are executed on the basis of their arrival time in ready queue. This scheduling algorithm is non-pre-emptive in nature. The disadvantages of this algorithm are long waiting time and response time for high-priority process.

- **Shortest-Job-First** (**SJF**) In this algorithm, the process which have minimum CPU burst time will schedule first. This algorithm can be implemented in two ways, one is pre-emptive and another one is non-pre-emptive. This is also known as shortest remaining time first (SRTF). This algorithm may lead to a problem that we cannot predict how long a job will be executed.
- **Priority Scheduling** In this algorithm, the process which has priority among the processes will schedule first. This algorithm may lead to a problem of starvation which is defined as if high-priority processes are regularly available in ready queue, then waiting time for low priority may become infinite.
- **Round Robin** (**RR**) algorithm which is the main concern of this research is one of the oldest, simplest and most widely used scheduling algorithms. This algorithm works on time-sharing phenomenon. A time slice is given to every process, and every process will be executed for particular defined time slice. New processes are added to the last of ready queue. The scheduler picks the process from the starting point of the ready queue and sets the timer to a defined time slice and also set an interrupt. If the process still not completed its complete execution within a time slice, it will be pre-empted after a time slice and added at the end of ready queue. The Round Robin scheduling gives the better response time, minimizes waiting time and turnaround time and maximizes throughput and CPU utilization [4, 5].

4 Literature Survey

In paper [1], Vaidya proposed a dynamic scheduling algorithm in which the scheduler is defined on all cores of a multi-core processing unit and accesses a shared task data structure (TDS) to select process which is in ready queue. This algorithm may lead to a problem that it senses additional wait times for cores due to it accessing shared task structure. The TDS is accessing using locks due to integrity of data. As the number of cores increases, then there will be more waiting time which may leads to problem of decreasing in performance of system. In paper [2], Jeet proposed scheduler algorithm which schedules the multiple tasks on multiple cores of a single-chip processor. One task is assigned to a single core, and the one is on another core processor within a single-chip processor for parallel processing. The proposed algorithm will allow each processor core to execute at least one task on single instance of time. This means if the number of processor cores is three, then three tasks will be execute simultaneously on all four cores. In paper [6], Li proposed that fair-share scheduling algorithm is inaccurate or inefficient and non-scalable for multiprocessors. He has defined a new scheduling algorithm DWRR which solve the problem of fair-share scheduling algorithm.

In [3], Mohsin et al. proposed scheduling in multi-core systems by minimizing average waiting time by merging (Round Robin with Shortest-Job-First Technique). In this paper, they have used two core processing systems. Each processor has its own queue. The queue can have maximum five processes at a time. The scheduler dispatched five processes in first processor queue (PQ1). Then, next five processes are dispatched to next processor queue (PQ2). The PQ1 and PQ2 assigned the priority to the processes in the manner that the process having shortest burst time has maximum priority and the process having highest burst has lowest priority. Then, the processes are dispatched by PQ1 to processor1 and by PQ2 to processor2. The processes are scheduled on processor using Round Robin algorithm by static time quantum. They have compared their algorithm with FCFS-RR scheduling in multi-core processing system and found that their algorithm has given better result in terms of decreasing average waiting time. The main drawback of this algorithm is that time quantum is taken static, and there is no use of user priority. To overcome these problems, we have defined a new algorithm SJFDRR with smart time quantum in multiprocessing system which is described in Sect. 5.

5 Proposed Work

In the proposed work, first described how the SJFDRR with smart time quantum algorithm work. In this work, there have been taken four assumptions which are described as:

(i) The processes are independent process.
(ii) The arrival time of all process is zero.
(iii) The synchronous multi-core processor (SMPs) architecture is used.
(iv) The two core processor architecture is used.

5.1 SJFDRR with Smart Time Quantum

In this algorithm, **first** calculate a smart priority factor (SPF) for every process. The process which has smallest 'SPF' value will be scheduled first. In this work, every process has two types of priority; one is user priority which is given by user itself (PRU), and second is the system priority which is defined by scheduling system in such a way that lowest burst time has highest system priority (PRS) (here, the researcher considers lower number represent higher system priority). The two important factors are also taken for calculating smart priority factor (SPF) which is user priority ratio (UPR) and system priority ratio (SPR). The user priority has more importance so the user priority ratio is given 55% weight and system priority ratio is given 45% weight. Suppose that all the processes has arrived at same time i.e. arrival time = 0. Then, smart priority factor 'SPF' is calculated as

$$\text{SPF} = \text{PRU} * \text{UPR} + \text{PRS} * \text{SPR} \tag{1}$$

So we calculate SPF for every process and decide which process will schedule first on the basis of SPF value.

Second, we will calculate smart time quantum (STQ) which is calculated as

(a) First, we will median for the set of processes in ready queue by given formula as given below

$$\text{Median}\,(M) = \begin{cases} Y_{\frac{n+1}{2}} & \text{if } n \text{ is odd} \\ \frac{1}{2}\left(Y_{\frac{n}{2}} + Y_{\frac{n}{2}+1}\right) & \text{if } n \text{ is even} \end{cases} \tag{2,3}$$

where

M median
Y number situated in the middle of a cluster of process numbers arranged in ascending order of burst time
n number of processes

(b) Then, the smart time quantum is calculated as follows:

$$\text{Smart Time Quantum (STQ)} = (Bt_{\max} + M)/2 \tag{4}$$

where $Bt_{\max}$ is maximum burst time among all the process in ready queue and M is median.

The smart time quantum is assigned to each process and is recalculated taking the remaining burst time in account after each cycle. This procedure goes on until the ready queue is empty.

5.2 *The Example on Uniprocessor Architecture*

Case: We Assume five processes arriving at time = 0, with increasing burst time (P1 = 9, P2 = 15, P3 = 27, P4 = 43, P5 = 82) as shown in Table 1. Table 2 shows the output SJFDRR. Figure 2 shows Gantt chart for the algorithm.

Table 1 Processes with burst time and priority

Processes	Arrival time	Burst time	User priority
P1	0	9	5
P2	0	15	2
P3	0	27	4
P4	0	43	1
P5	0	82	3

Table 2 Output using SJFDRR on uniprocessor system

Algorithm	Time quantum	Avg. TAT	Avg. WT	CS
SJFDRR	54.5, 27.5	82	46.8	5

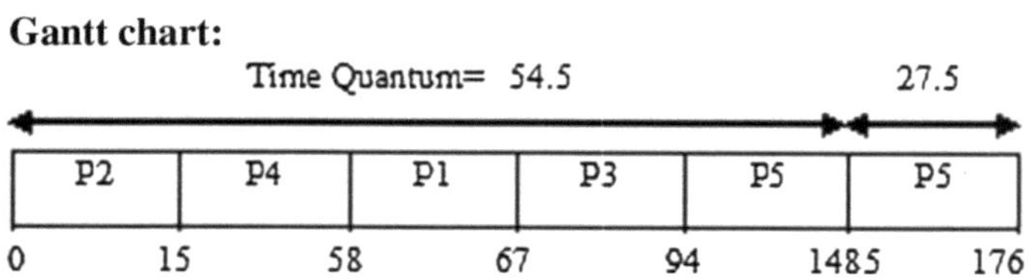

Fig. 2 Gantt chart using SJFDRR

5.3 SJFDRR in Multi-core Processing System

Now, we are defining the SJFDRR in multi-core processing system which is defined as

1. There are two cores and each core has two queues PQ1 and PQ2. The maximum processes assigned to each queue are 5.
2. The first five processes are selected from ready queue and dispatched to PQ1.
3. Then, next five processes are selected from ready queue and dispatched to PQ2.
4. If ready queue is not empty, then all remaining process will wait until PQ1 or PQ2 is empty.
5. Then, the processes are arranged in the ascending order of their burst time in PQ1 and PQ2.
6. In PQ1 and PQ2, first we will calculate the smart priority factor (SPF) by using Eq. (1).
7. Now, we will calculate the smart time quantum (STQ) for PQ1 and PQ2 by using Eq. (4).
8. Now, schedule the process on core1 and core2 by using SJFDRR algorithm described in Section B.

6 Implementation of SJFDRR with Example

Case: We assume ten processes arriving at time = 0, with increasing burst time (P1 = 4, P2 = 8, P3 = 6, P4 = 10, P5 = 2, P6 = 8, P7 = 12, P8 = 8, P9 = 6, P10 = 2) as shown in Table 3. Tables 4, 5 and 6 show the output comparison of SJFDRR in Uniprocessor, RR-SJF and SJFDRR in multi-core processor system.

Table 3 Processes with burst time and priority

Processes	Arrival time	Burst time	User priority
P1	0	4	1
P2	0	8	2
P3	0	6	1
P4	0	10	3
P5	0	2	4
P6	0	8	2
P7	0	12	1
P8	0	8	5
P9	0	6	4
P10	0	2	3

Table 4 Processes with the value of UPR, PRS and SPF in PQ1

Processes	UPR	PRS	SPF
P5	4	1	2.65
P1	1	2	1.45
P3	1	3	1.90
P2	2	4	2.90
P4	3	5	3.90

Table 5 Processes with the value of UPR, PRS and SPF in PQ2

Processes	UPR	PRS	SPF
P10	3	1	2.10
P9	4	2	3.10
P6	2	3	2.45
P8	5	4	4.55
P7	1	5	2.80

Table 6 Average waiting time comparison between RR-SJF and SJFDRR in multi-core processing system

Algorithm	Time quantum in PQ1	Time quantum in PQ2	Average waiting time in PQ1	Average waiting time in PQ2
RR-SJF [3]	4	4	10.4	14.8
SJFDRR in multi-core system	8, 2	10, 2	9.2	14.4
SJFDRR in uniprocessor system	9.5, 2, 0.5		25.65	

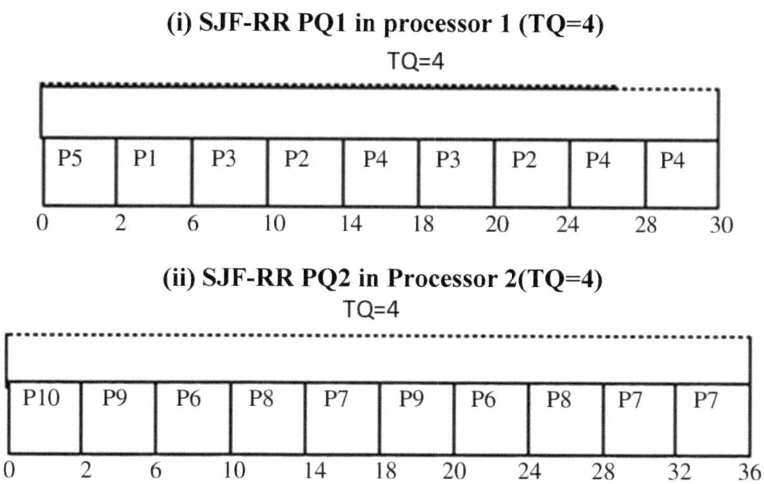

Fig. 3 Gantt chart showing waiting time for each processor PQ1 and PQ2 using RR-SJF in multiprocessors

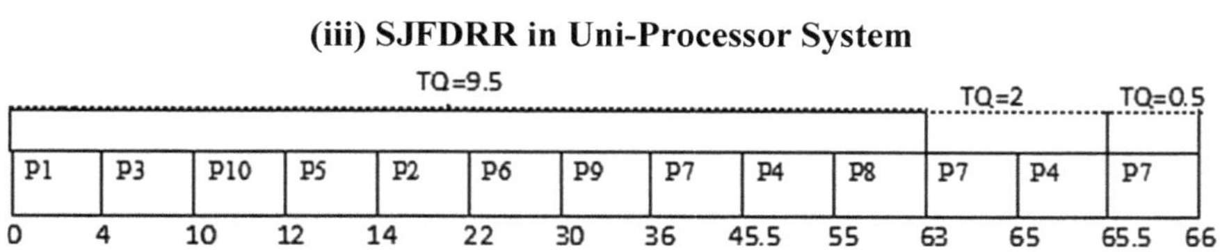

Fig. 4 Gantt chart showing waiting time using SJFDRR in uniprocessor system

Fig. 5 Gantt chart showing waiting time for each processor PQ1 and PQ2 using SJFDRR in multiprocessors

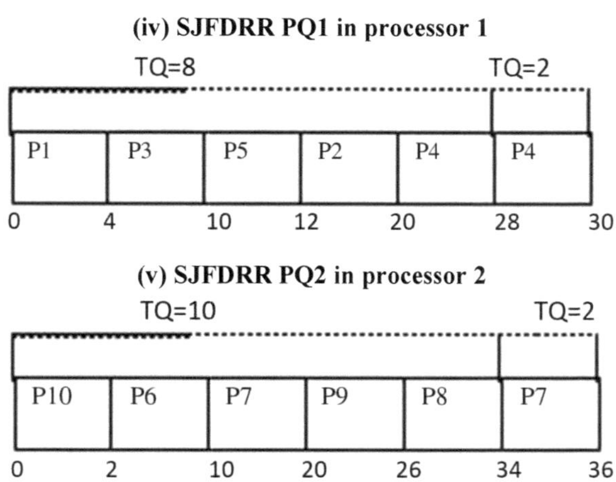

Figures 3, 4 and 5 show Gantt chart for the algorithm SJFDRR in Uniprocessor system, RR-SJF in multiprocessing and SJFDRR with smart time quantum in multiprocessing, respectively.

Now, here the researcher presented the evaluation of proposed algorithm.

1. First, the processes P1, P2, P3, P4 and P5 are allocated to queue PQ1, and the processes P6, P7, P8, P9 and P10 are allocated to queue PQ2 of core1 and core2, respectively, as per proposed algorithm.
2. Now, the processes are arranged in the ascending order of their burst time in the respective queues PQ1 and PQ2.
3. Now, the processes are assigned to the system priority (PRS) in PQ1 and PQ2 which is defined as the process which has lowest burst time will be assigned highest system priority.
4. Now, the smart priority factor (SPF) is calculated using Eq. (1) for each process in PQ1 and PQ2. Tables 4 and 5 show the process with system priority and SPF in PQ1 and PQ2, respectively.
5. As per the proposed algorithm, the process which has lowest value of factor SPF will be schedule first. So in PQ1, the process execution order in round 1 will be P1, P3, P5, P2 and P4. And in PQ2, the process execution order will be P10, P6, P7, P9 and P8.
6. Now, the researcher has calculated the smart time quantum (STQ) for first round in PQ1 and PQ2 using Eqs. (2), (3) and (4) which is calculated as follows:

   i. The STQ for PQ1 and PQ2 in first round.
   ii. First calculate median value which is $M = 6$ in PQ1 and $M = 8$ in PQ2 by using Eq. (1). The STQ for PQ1 is calculated as here the highest burst time in PQ1 is 10. So STQ for PQ1 $(10 + 6)/2 = 8$ by using Eq. (4). And here, the highest burst time in PQ2 is 12. So STQ for PQ2 $(12 + 8)/2 = 10$ by using Eq. (4).
   iii. So assign the calculated STQ for PQ1 and PQ2 to core1 and core2 in first round.
   iv. In first round, in core1, the processes P1, P2, P3 and P5 have finished their execution. The process P4 is not completed their execution. It remains 2 ms execution, so it will be executed in next round. Again, the smart time quantum is calculated, and the STQ will be 2 ms in next round in PQ1. Similarly in first round, in core2, the processes P6, P8, P9 and P10 have finished their execution. The process P7 is not completed their execution. It remains 2 ms execution, so it will be executed in next round. Again the smart time quantum is calculated, and the STQ will be 2 ms in next round in PQ 2 (Figs. 6, 7, and 8; Tables 7 and 8).

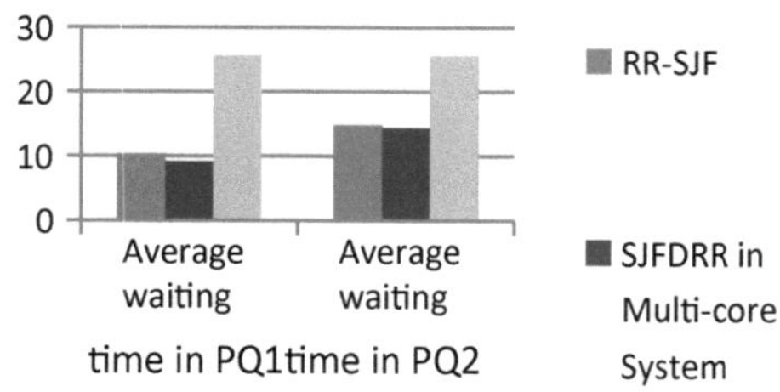

Fig. 6 Comparison graph of average waiting time among SJFDRR in uniprocessor, SJF-RR and SJFDRR in multi-core processing system

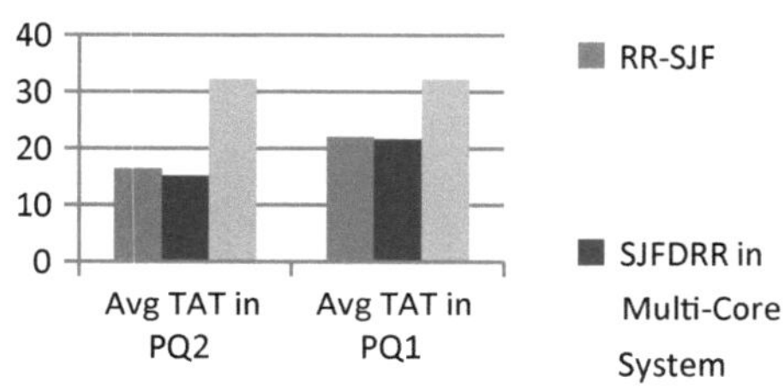

Fig. 7 Comparison graph of average turnaround time among SJFDRR in uniprocessor, SJF-RR and SJFDRR in multi-core processing system

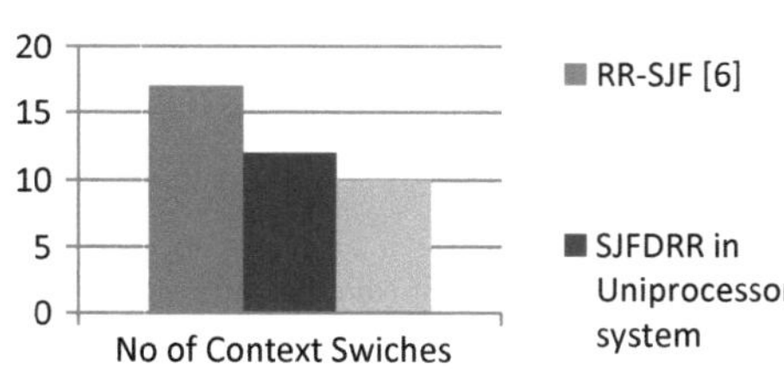

Fig. 8 Comparison graph of number of context switches among SJFDRR in uniprocessor, SJF-RR and SJFDRR in multi-core processing system

Table 7 Average turnaround time comparison between RR-SJF and SJFDRR in multi-core processing system

Algorithm	Time quantum in PQ1	Time quantum in PQ2	Avg TAT in PQ1	Avg TAT in PQ2
RR-SJF [3]	4	4	16.4	22
SJFDRR in multi-core system	8, 2	10, 2	15.2	21.6
SJFDRR in uniprocessor system	9.5, 2, 0.5		32.25	

Table 8 Number of context switches comparison between RR-SJF and SJFDRR in multi-core processing system

Algorithm	Number of context switches
RR-SJF [3]	17
SJFDRR in multi-core system	10
SJFDRR in Uniprocessor system	12

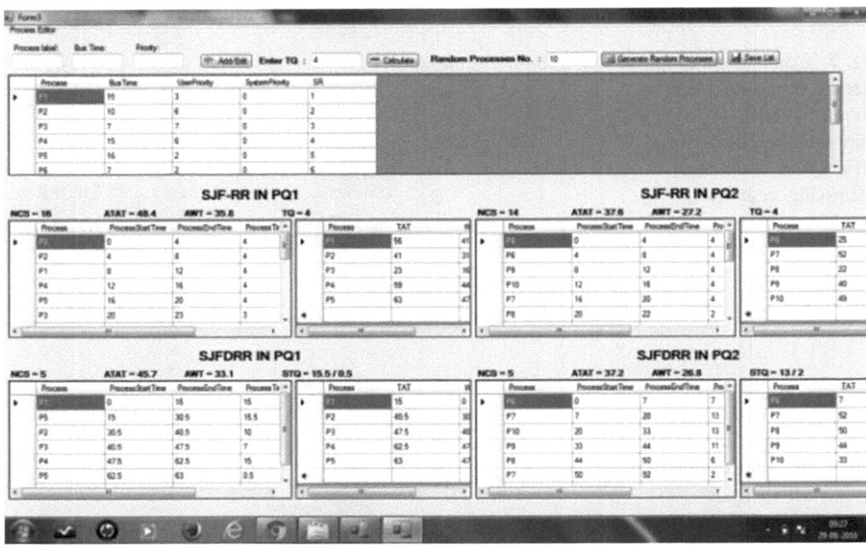

Fig. 9 Simulation result (number of process = 10)

7 Simulation Result

The researcher has designed a simulator in net framework which gives the comparison result of SJFDRR with smart time quantum and SJF-RR in multi-core processing system. Here in this paper, researcher has included some result on randomly generated process in Figs. 9, 10, 11, 12, 13, and 14. The number of process is taken: 10, 15, 20, 25, 40 and 50. Figure 15 shows the comparison graph. The simulation result gives that SJFDRR with smart time quantum is perform better in terms of decreasing the number of context switches, average waiting time and average turnaround time.

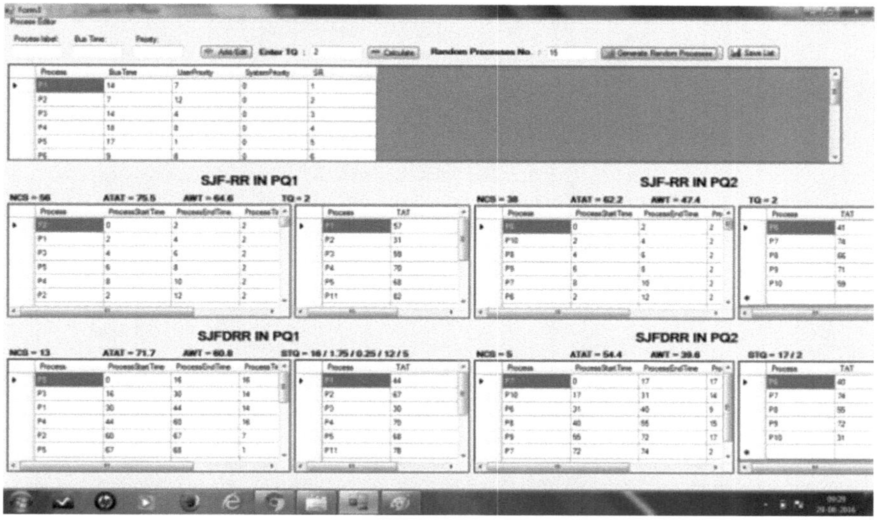

Fig. 10 Simulation result (number of process = 15)

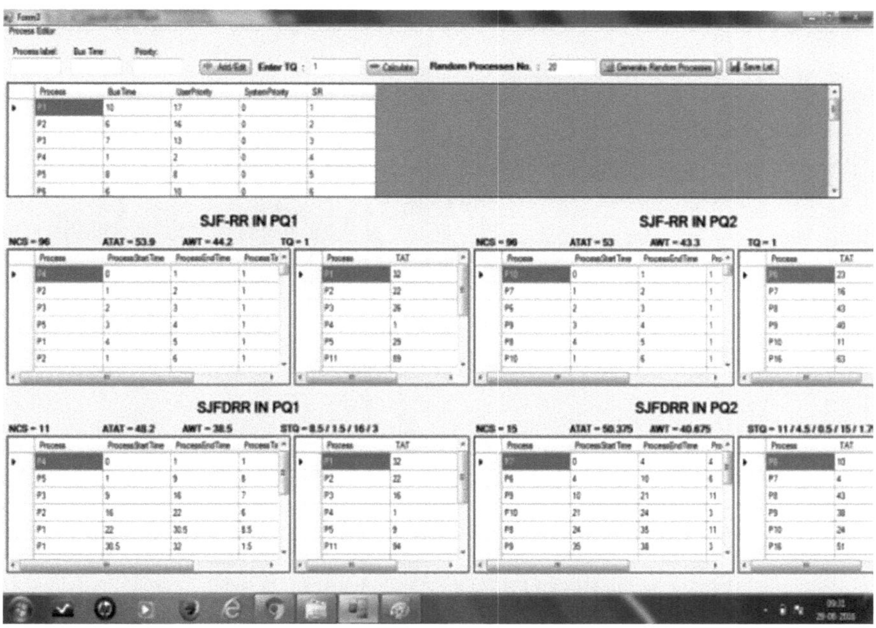

Fig. 11 Simulation result (number of process = 20)

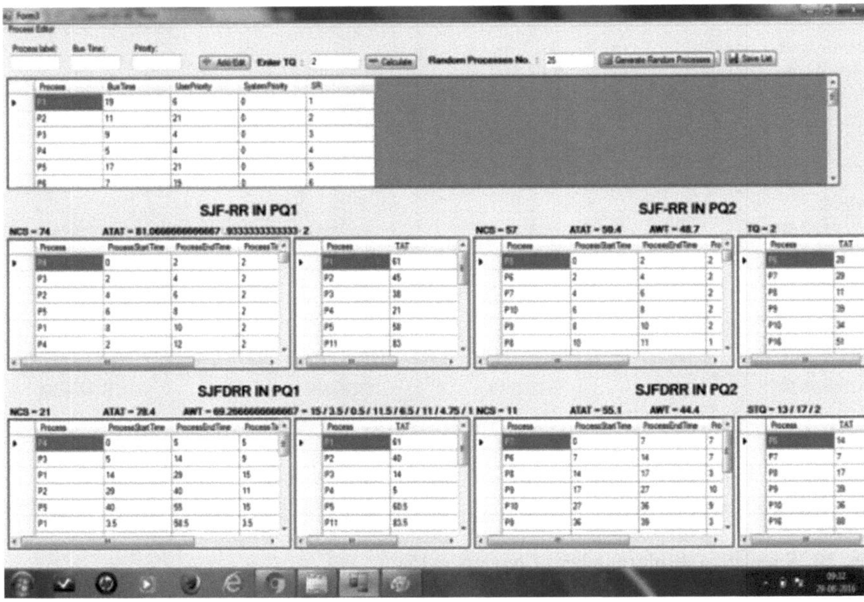

Fig. 12 Simulation result (number of process = 25)

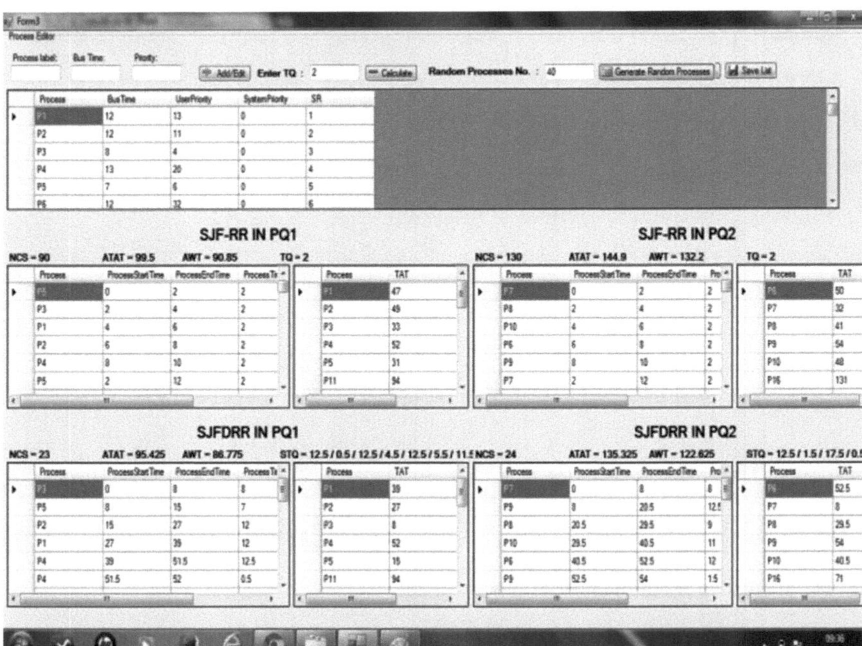

Fig. 13 Simulation result (number of process = 40)

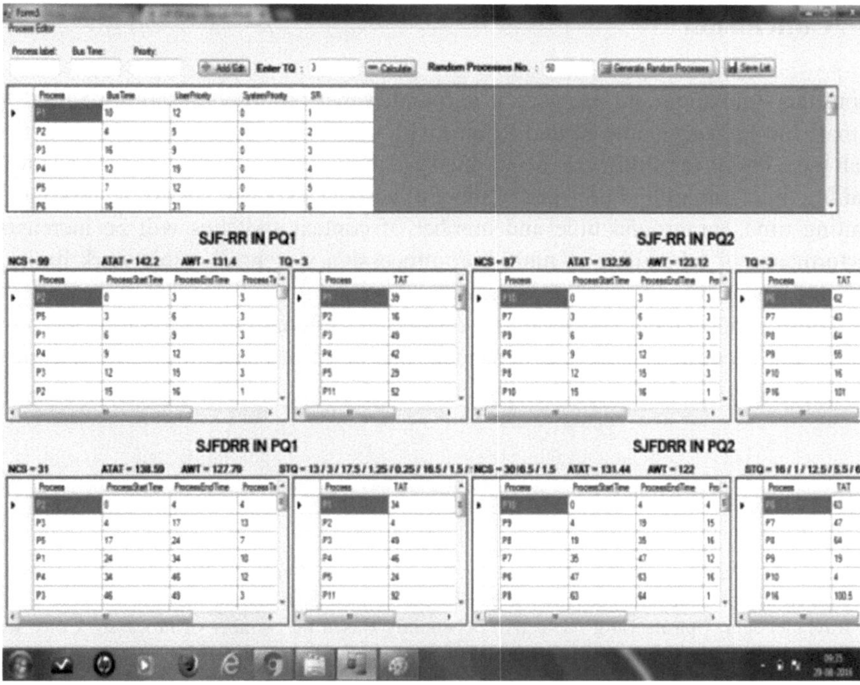

Fig. 14 Simulation result (number of process = 50)

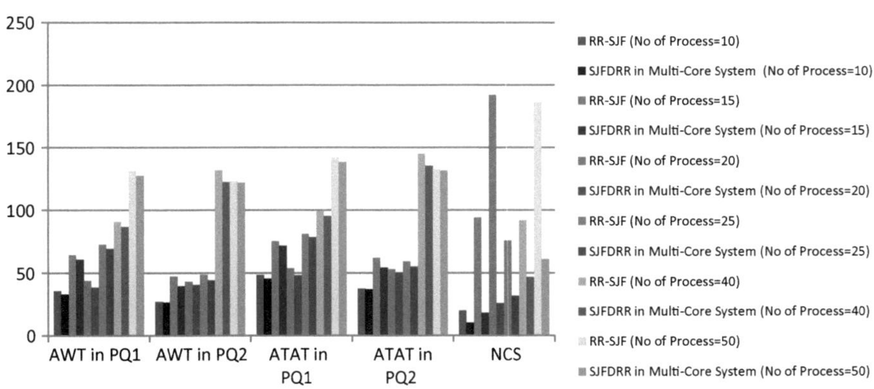

Fig. 15 Comparison graph (number of processes 10, 15, 20, 25, 40 and 50)

8 Conclusion

From the simulation and analysis, it is founded that proposed algorithm SJFDRR (Smart Job First Dynamic Round Robin) with smart time quantum performs better than the SJF-RR in multi-core processing system in terms of decreasing the average waiting time, turnaround time and number of context switches. So the decreasing in waiting time, turnaround time and number of context switches will be increases performance of scheduling in multi-core processing system. But this work limited only for when arrival time of all process are zero and for independent processes. So the future enhancement can be **first**, considering different arrival time of the processes and **second**, considering dependent processes in scheduling. So the future research work can be focused on these two problems.

References

1. Vinay G. Vaidya," Dynamic Scheduler for Multi-core Systems",2010 2nd International Conference on Software Technology and Engineering (ICSTE) 2010.
2. Ravinder Jeet, Upasna Garg "Selective Scheduling Based on Number of Processor Cores for Parallel Processing" International Journal of Science and Research (IJSR) Volume 4 Issue 1, January 2015.
3. Ahmad Mohsin, Muhammad Imran Rafique, Sumbul Aziz Khan, Qurratulain Munir, "Scheduling in Multi-core Systems: Minimizing Average Waiting Time by merging (Round-Robin with Shortest-Job-First Technique)", International Journal of Information Technology and Electrical Engineering (ITEE) Volume 2, Issue, 4 August 2013.
4. D.M. Dhamdhere operating Systems A Concept Based Approach,Second edition, Tata McGraw-Hill, 2006.
5. Silberchatz, Galvin and Gagne, 2003. Operating systems concepts.
6. Tong Li, Dan Baumberger, Scott Hahn, "Efficient and Scalable Multiprocessor Fair Scheduling Using Distributed Weighted Round-Robin", 2111 NE 25th Ave., Hillsboro, OR, USA.

Security Enhancement in MANETs Using Fuzzy-Based Trust Computation Against Black Hole Attacks

Ashish Kumar Jain, Vrinda Tokekar and Shailendra Shrivastava

Abstract Mobile ad hoc networks (MANETs) are much more susceptible to routing attacks as compared to wired networks and infrastructure-based wireless networks. MANETs lack infrastructure and trusted centralized authority, thereby vulnerable to black hole attacks. This paper presents novel approach based on weighted binary relational fuzzy trust model to mitigate black hole attacks in ad hoc on demand distance vector (AODV) protocol. It uses trust computing approach, which is by default a fuzzy approach. Direct trust computation method is applied to determine malicious nodes and thereby safe route in MANETs. The results show performance improvement of proposed protocol over AODV protocol.

Keywords Black hole attacks · Mobile ad hoc network · Trust computing · Fuzzy · Trust formulation

1 Introduction

The MANET is a peer-to-peer (p2p), a self-motivated, self-governing, multihop network [1]. MANETs have essential operation of distributed collaborations and information sharing [2]. These operations need that the competitive nodes should participate in trustworthy manner. MANETs are generally applied in unrestrained environment, thereby probability of nodes being compromised in MANETs

A.K. Jain (✉) · V. Tokekar
Institute of Engineering & Technology, Devi Ahilya University, Khandwa Road, Indore, MP, India
e-mail: ajain@ietdavv.edu.in

V. Tokekar
e-mail: vtokekar@ietdavv.edu.in

S. Shrivastava
Samrat Ashok Technological Institute, Vidisha, MP, India
e-mail: shailendrashrivastava@rediffmail.com

D.K. Mishra et al. (eds.), *Information and Communication Technology*, Advances in Intelligent Systems and Computing 625,
https://doi.org/10.1007/978-981-10-5508-9_4

increase substantially. So, aggressive nodes of MANETs need to be vigilant whenever they cooperate with each other.

In black hole attack, malicious node advertises that they have a path to the destination node by sending root reply (RREP) packet. Afterward, node simply drops the data packet. Furthermore, there is cooperative black hole attack, in which two nodes cooperate to attack and diminish the network performance [3].

Security and mobility are very vital topics for the design of routing protocol for MANETs. Security and mobility can be considered as reciprocal to each other [4]. Trust-based computation and trust management are challenging issues as nodes are mobile and autonomous in MANET which creates frequent topology changes [5].

This paper proposes fuzzy-based trust computation protocol (FTCP) to mitigate black hole attack in MANETs, thus enhancing security of AODV routing protocol. We discuss related work in Sect. 2. We propose fuzzy-based trust computation model to detect black hole attacks in Sect. 3. In Sect. 3, we also present several trust relation properties useful for trust-based computation in MANETs. Section 4 presents results and performance evaluation of simulated fuzzy-based trust computation system. Finally, conclusion is given in Sect. 5.

2 Related Work

Nguyen et al. [6] have described attacks in MANETs, and they have shown that the malicious node in black hole attack first occupies its position in the path and captures the data packets and then drops some or all the data packets it received.

Ayday et al. [7] have established a scheme for trust management for delay-tolerant networks. As a result, trust value of nodes reduces severely may be lower than the threshold value. Thereby normal nodes may be detected as malicious node, and it results in decay in the network performance. Hence, our approach uses maximum product scheme, in which lower trust values are automatically ignored.

Mohanty et al. [8] have proposed that devices taking part in MANETs offer huge attack space which is vulnerable. Secret malevolent functionality of MANET devices can be used to hack MANET software. Authors have developed trusted MANET module to overcome these attacks.

Chatterjee et al. [9] have proposed black hole node avoidance using triangular encryption in MANET. Authors were able to avoid black holes only not to detect them. Their protocol takes more runtime.

Liu et al. [10] have proposed B-AODV protocol which intends to find the route and local repair to rebuild the routing. Authors have shown that it improves routing repair capacity and network performance.

Guo et al. [11] have proposed trust management framework and able to detect abnormal trust behavior and also the parameter for abnormal node. Authors have applied Grey theory and fuzzy sets and calculate total trust value by using relational factors.

Sivagurunathan et al. [12] have proposed trust-based security model to withstand attacks in military ad hoc networks. They have calculated direct trust, indirect trust, situational trust, and stereo trust. On combining all these, they have computed overall trust of a node. They have taken three different trust levels to classify the nodes of network as untrusted, partially trusted, and trusted nodes.

3 Weighted Binary Relational Fuzzy Trust Model

Trust should be a necessary element of distributed systems which depends on relationship between different entities of the distributed systems [6]. Trust is a reliance of one entity to the other. It depends on the first entity that how much it believes in second. An entity may rely fully on the other entity. But in practical scenario, this is not possible. Therefore, trust can be modeled as a probabilistic value, which can be denoted as a fractional value between 0 and 1. Hence, trust computation approach is by default a fuzzy approach.

Trust can be modeled mathematically as a binary relation on $A \times A$, where A is a set of nodes in MANET. This binary relation is weighted as the weights of this relation are fractional trust values of one node to the other node. These weights represent the extent to which a node believes in other node.

Nodes of MANET are mobile in nature; therefore, computed trust should be based on spatial local information. Such kind of information can change rapidly [13]. Hence, trust should be dynamic. Dynamic variables should be treated as continuous variable. Hence, trust value is kept as continuous value ranges in [0, 1].

Trust has to be a reflexive relation as a node has to trust itself. Consider Cartesian product $A \times A$. T is a trust relation on this product. That means $T \subseteq A \times A$. Let us consider a node N. So, $T(N, N) = 1$ that means a node fully trusts itself. So, the relation has to be reflexive.

In trusted relation, consider two nodes N and M. Both of them may not trust each other with equal extent. So, the relation has to be asymmetric essentially.

In MANETs, trust value of strange node is calculated based on preceding trust values of neighboring nodes. As trust-based computation need inferences of neighboring nodes to compute trust of another node, the trust relation is considered as weighted transitive relation.

Trust value ranges between 0 and 1. A fuzzy discrimination table is defined to represent the behavior of node in terms of its trust value (Table 1).

Table 1 Fuzzy discrimination table for node behavior

Trust value	Fuzzy levels	Node behavior
0–0.2	Very small	Malicious node
0.2–0.4	Small	Selfish node
0.4–0.6	Medium	Normal node
0.6–0.8	Large	Cooperative node
0.8–1	Very large	Trustworthy and cooperative nodes

4 Trust-Based Computing

Nodes of MANET have to trust other nodes of same network. But, all the nodes are not equally trustworthy. Some nodes are selfish, some might be malicious, and others might be completely trustworthy. Hence, trusted computation should be used to detect the behavior of node.

Trust computation in static networks is straightforward as trust values vary only with the behavior of the node. After, some observations behaviors and trust values are predictable [2]. Trust computation in mobile networks is considerably difficult as compared to static networks, as compromised node may move after attack, and it will be very difficult to detect such malicious node [2]. Network topology significantly changes within time in a volatile manner. Hence, observations for neighboring node are difficult. Behavior of a node is predictable only after large number of observations. Furthermore, it is difficult to associate a mobile node with its location and gaining observations.

4.1 Trust Formulation

Trust is to be formulated by one entity for the other. This formulation might be guessing an opinion of other entity. Mathematically, trust is probability of trustworthiness of one entity about the other. In case of MANET, every node computes the trust value for the other node. MANET is an open network; node can enter and move out of the network at any time. When a new node joins the network, the trust with some default value is initialized. For the new node, the default value for the trust will depend upon the application where MANET is used. The trust will keep on changing over the time based on the feedback obtained from other nodes. Trust computation of node about neighboring node will depend upon certain parameters. These parameters may include packet delivery ratio of a node, percentage of energy exhaustion of node, percentage of buffer utilized by the node, and number of connection requests given by node.

Following parameters are identified to compute the trust.

1. **Packet forwarding ratio of a node**

$$P = \frac{\text{No of packets for warded by the node}}{\text{No of packets received at the node}} \tag{1}$$

Trustworthy node forwards all the packets received by it, whereas malicious node will drop few packets.

2. **Energy exhaustion**

The battery consumption of node depends on network activity. The more functional node consumes more battery power. Selfish and malicious nodes seem busy in network as they want to attract more nodes. So, these nodes will consume more power.

3. **Buffer Utilization**

The malicious nodes try to attack more often may be with high-speed data connection links. Because of this, their buffer may be full most of the time. Hence, this is an essential parameter for network trust-based computing.

4. **Number of connection request**

The malicious node communicates with multiple nodes simultaneously. So, it may try to establish connection with many nodes. Thus, the number of connection request generated by malicious node will be more than the legitimate nodes.

4.1.1 Direct Trust Computation

A node computes its trust value of its neighboring node directly by itself. Direct trust computation by one of the node A for the other node B may also be given as

$$T(A,B) = \sum_{i=1}^{n} (w_i x_i(A,B)) \tag{2}$$

where w_i is the ith parameter's weight required to compute trust by node A about node B. Such that $\sum_{i=1}^{n} w_i = 1$, and $x_i(A,B)$ is ith parameter observed by node A about node B. The four parameters $x_i(A,B)$ are calculated periodically among the nodes of MANET.

4.2 Malicious Node Detection Using Threshold

During routing in MANET, the numbers of hops are counted and denoted as hope_c. Let us consider round-trip time (RTT) denoted as δ_t. Hence, average round-trip time (ARTT) is estimated as:

$$\text{ARTT} = \frac{\delta_t}{\text{hope}_c} \tag{3}$$

There are N scenarios considered, and threshold value μ_{thr} is computed as follows:

$$\mu_{thr} = \frac{1}{N}\sum_{i=1}^{N} \text{ARTT} \tag{4}$$

After computing the threshold value μ_{thr}, if trust value $T(A, B) <$ threshold μ_{thr} then node B is regarded as malicious and thereby rejected from the path.

5 Performance Analysis

5.1 *Performance Metrics*

Three performance metrics are considered for analysis of FTCP protocol.

Packet delivery ratio (PDR) is defined as:

$$\text{PDR} = \frac{\sum_{i=1}^{\text{nodes}} \text{packets\_receive}(i)}{\sum_{i=1}^{\text{nodes}} \text{packets\_sent}(i)} \tag{5}$$

Average end-to-end delay (ETD): Packet takes certain time to be transmitted across the network from source to destination node. ETD also includes delay in route discovery and data packet transmission. Average ETD is defined as follows:

$$\text{Average ETD} = \frac{\sum_{p=1}^{\text{packets}} \{\text{receive\_time}(p) - \text{send\_time}(p)\}}{\text{Number of Connections}} \tag{6}$$

Throughput: Throughput is usually represented as bytes or bits per second. Throughput metrics is the result of total throughput over 'n' mobile nodes.

$$\text{Throughput} = \frac{\sum_{p=1}^{n} \text{packet\_transmitted}(p)}{\text{Total Time}} \tag{7}$$

5.2 *Simulation*

The simulation was carried out using NS2 as simulator with input parameters as given in Table 2. Network performance values are evaluated on each scenario as a function of number of malicious nodes. These network metrics are compared with Sivagurunathan model [12].

Table 2 Input parameters for simulation under AODV and FTCP

Simulation time	900 s
Area	1 km × 1 km
Total number of nodes	50
Mobility model	Random waypoint
Transmit range	250 m
Packets transmission rate	8/s
Packet size	512 bytes
Max. no of packet per connection	10,000
Traffic type	CBR
No. of malicious node	1–40 (2–80%)

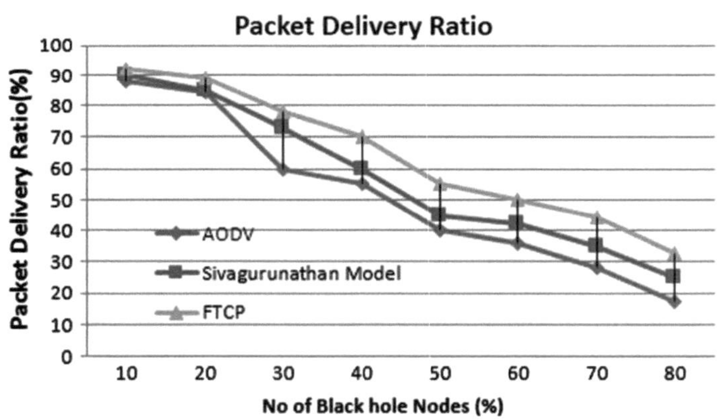

Fig. 1 Packet delivery ratio of FTCP as compared with AODV and Sivagurunathan model

5.3 *Results*

PDR, ETD, and throughput are computed for varying number of malicious nodes. Figure 1 shows PDR for FTCP as compared with AODV and Sivagurunathan model [12]. Figure 2 shows clear outperformance of FTCP for ETD over AODV and Shivagunathan model [12]. Figure 3 shows outperformance of FTCP for throughput over AODV.

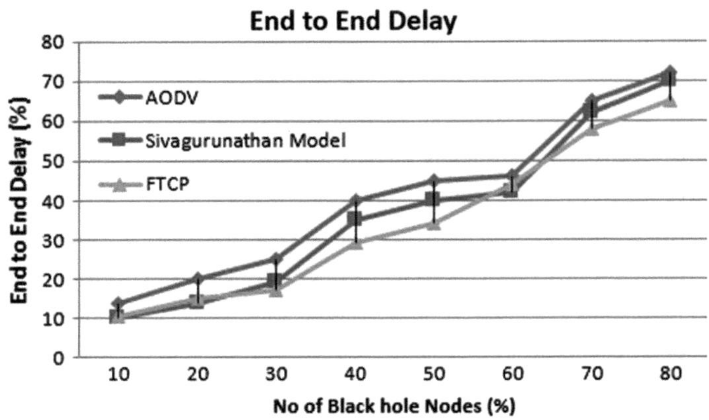

Fig. 2 End-to-end delay of FTCP as compared with AODV and Sivagurunathan model

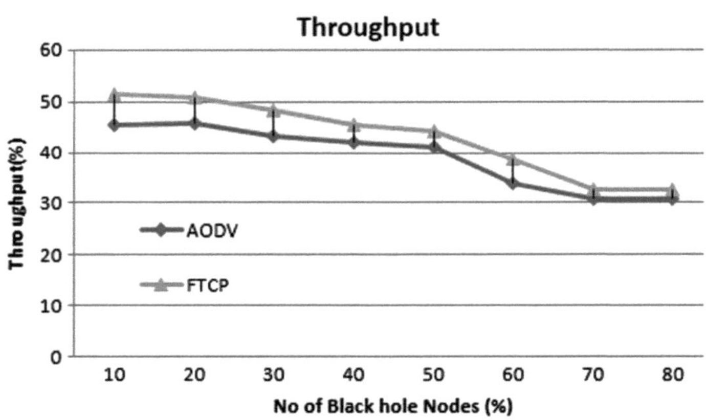

Fig. 3 Throughput of FTCP as compared with AODV

6 Conclusion

In the proposed work, we have given solution of black hole attack in an untrusted environment of MANET using trust computing approach. Our work uses fuzzy-based trust computation approach. Network performance of MANET varies because of varying number of malicious nodes in the terrain. The network performance is then compared with standard approach AODV and Sivagurunathan model [12].

The result shows that the proposed approach, FTCP have outperformed AODV and Sivagurunathan model [12] under all conditions in terms of network performance metrics.

References

1. Jain A. and Tokekar V.: Classification of denial of service attacks in mobile ad hoc networks, Proceedings of International Conference on Computational Intelligence and Communication Networks (CICN), pp 256–261, 2011
2. Govindan K., Mohapatra P.,:Trust Computations and Trust Dynamics in Mobile Adhoc Networks: A Survey, IEEE Communications Surveys & Tutorials (Volume: 14, Issue: 2, Second Quarter 2012), pp: 279–298,2012
3. Jain, A. K., & Tokekar, V. (2015).: Mitigating the effects of black hole attacks on AODV routing protocol in mobile ad hoc networks. In *Pervasive computing (ICPC), 2015 international conference on* (pp. 1–6)
4. Singhal, S. & Daniel, A. K., (2014).: Fuzzy logic based clusterhead selection protocol under Competence level, Goodness function and Mobility for mobile ad hoc network. In 2014 Conference on IT in business, industry and government (CSIBIG) (Vol. 8–9, pp. 1–6).
5. Qureshi B., Min G. and Kouvatsos D.:A distributed reputation and trust management scheme for mobile peer to peer networks, in Computer Communications pp 608–619
6. Nguyen H., Nguyen U.,: A study of different types of attacks in mobile ad hoc networks 25th IEEE Canadian conference on Electrical & Computer Engineering, 2012
7. Ayday E., Fekri F.: An Iterative algorithm for trust management and adversary detection for Delay-tolerant Networks, IEEE Transaction on Mobile Computing, Vol 11. No. 9, pp 1514 – 1531, 2012
8. Mohanty S., Thotakura V., Ramkumar M.: An efficient trusted computing base for MANET security, journal of information security, pp 202–206, 2014
9. Chatterjee N., Mandal J.: Detection of Blackhole behaviour using triangular encryption in NS2, 1st International conference on computational Intelligence (CIMTA-2013), pp 524–529, 2013
10. Liu S., Yang Y., Wang W.: Research of AODV routing protocol for ad hoc networks, AASRI conference on parallel and distributed computing and systems, 2013
11. Guo J., Marshal A., Zhou B: A New trust management framework for detecting malicious and selfish behaviour for mobile ad hoc networks, International joint conference of IEEE Trustcom -11, 2011
12. Sivagurunathan S, Prathapchandran K:Trust Based Security Model To Withstand Against Black Hole And Grey Hole Attacks In Military Based Mobile Ad Hoc Networks, International Journal of Mobile Network Communications & Telematics (IJMNCT) Vol. 6, No. 1, February 2016
13. Hall S. and McQuay W.: Fundamental features of a unified trust model for distributed systems, in Proceedings of National Aerospace and Electronics Conference, Dayton, Ohio, July, 20–22, 2011

An Energy-Efficient Dual Alternate Cluster Head-Based Routing Mechanism in Wireless Sensor Network

Nilayam Kumar Kamila and Sunil Dhal

Abstract Sensor nodes are deployed over specific critical area for gathering environmental data. These sensor nodes are constrained with low battery energy without any external power supply feasibility, minimal computational ability, and narrow communication bandwidth. However, in wireless sensor networks (WSN), the cluster (group)-oriented route mechanisms are applied to decrease extra power usages and hence extend the live period (lifetime) of the network. Here, in this research article, we suggest a routing method, namely energy-efficient dual alternate cluster head-based approach (DACH) where we consider two nodes in a single cluster to perform the cluster head role to balance the data gathering and network transmission load. Again, to access the effectiveness of the proposed approach, we show the analyzed result through simulation and mathematical studies.

Keywords Sensor network · Cluster head · Hierarchical routing

1 Introduction

The microsensors are nowadays inexpensive due to the technological advancement, and hence, it could be deployed over a large area where the periodic maintenance is almost impossible. These sensors are low power and low communication range, so it is very much necessary to deploy the sensors with moderate or high density so that they connect with each other (low connectivity range) to set up and maintain the network infrastructure among themselves [1]. The small and large number of nodes, with limited battery capacity [2], gather the target facts and forward the same

N.K. Kamila (✉)
Capital One, 802 Delaware Avenue, Wilmington, DE, USA
e-mail: nilayam_kamila@yahoo.com

S. Dhal
Sri Sri University, Godi Sahi, Cuttack, Odisha, India
e-mail: sunil.dhal@srisriuniversity.edu.in

D.K. Mishra et al. (eds.), *Information and Communication Technology*,
Advances in Intelligent Systems and Computing 625,
https://doi.org/10.1007/978-981-10-5508-9_5

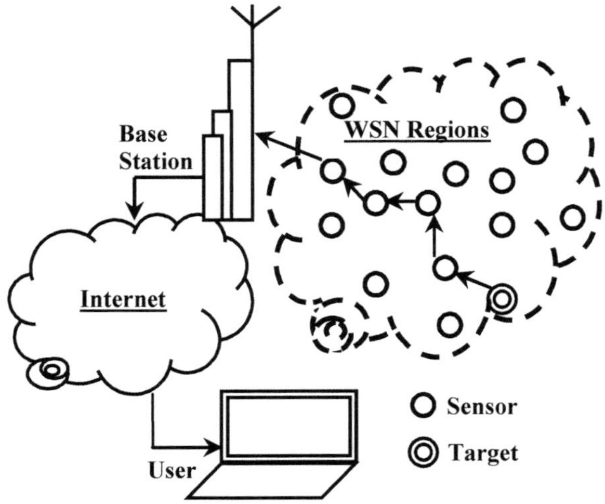

Fig. 1 Typical wireless sensor network environment

toward the base station (BS) (refer Fig. 1). The BS processes data and directs it to other networks, e.g., Internet for further analysis and study.

This article is structured as mentioned here. Section 2 shows the review of the radio models and the related works. We debate our proposed routing approach in Sect. 3. We present mathematical analysis of cluster-based routing protocol and our DACH-based routing approach in Sect. 4. In Sect. 5, the simulation study is presented. And we conclude our work in conclusion section.

2 Existing Related WSN Routing Protocols and Issues

There are four major components in the wireless sensor nodes [3], e.g., sensor, processing, communication, and power. To work these components appropriately, there is a power unit which supplies the power to all other units. The radio model in [4] evaluates the power consumed in both sending and receiving a message of k-bits through a d-unit of distance. Below equations provide the sender's power consumption and the power consumption of receiver node.

$\in_{\mathrm{te}}$ power consumption in transmitter circuitry
$\in_{\mathrm{r}}$ power consumption in receiver circuitry
$\in_{\mathrm{ta}}$ power required for amplifier in transmitter unit

$$E_{\mathrm{tr}}(p_d, d) = \in_{\mathrm{te}} p_d + \in_{\mathrm{ta}} p_d d^2.$$

$$E_{\mathrm{r}}(p_d) = \in_{\mathrm{r}} p_d.$$

2.1 Low-Energy Adaptive Clustering Hierarchy (LEACH)

Low-energy adaptive clustering hierarchy (LEACH) [4] is built on the clustering hierarchy design model. The whole set of WSNs are organized and form number of groups called clusters. There is a head node in each group which is chosen by cluster members. This head node accepts the data from group members, fuses it, and then forwards to the next device (nodes) in the path toward the base station. There is a cluster network setup and election process to elect the new cluster head periodically.

2.2 Power-Efficient Gathering in Sensor Information Systems (PEGASIS)

This protocol [5] is based on a chain mechanism and does not form cluster like LEACH protocol. The sensor node communicates with its nearest nodes. The nearest nodes which receive the information again transfer the message to their nearest nodes and so on.

2.3 LEACH and PEGASIS Issues

LEACH is having a better performance over the flooding-based protocols [6], and the few limitations are as follows.

a. Unbalanced energy load: Single cluster head per cluster.
b. More transmission distance: send to CH instead of nearest neighbor.

PEGASIS is a well-balanced protocol and has better performance with respect to flooding-based protocols, and some of the limitations [5] are as follows.

a. Unawareness of energy status: no awareness of nearest node's energy level.
b. More transmission distance: nearest neighbor chain, no specific route.

3 Our Proposed Dual Alternate Cluster Head Approach

As we see in above section, the devices in group (cluster)-based protocols transfer the gathered information to head node, and cluster head now collects the information from all its cluster member devices and transfers it to BS to complete the data transmission. In multi-hop transmission, the cluster head (CH) forwards the data to BS through intermediate device(s) and their corresponding cluster head(s) (shown in Fig. 2). In our proposed energy-efficient dual alternate cluster head routing (DACH) approach, an additional cluster head within the cluster, which would be a device located in the BS path, shares the data gathering and transmission load of cluster head.

The nodes which are located close to the additional (alternate) group head transmit the information to additional cluster head in place of transmitting data to primary head node, as displayed in Fig. 3. The main idea of introducing an

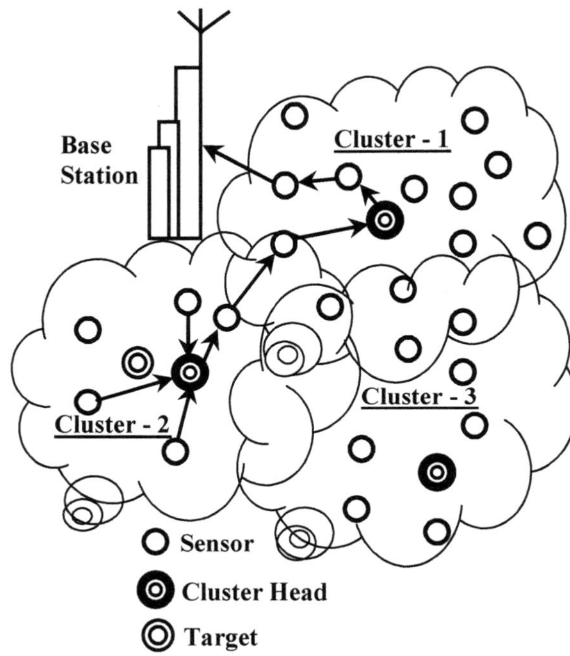

Fig. 2 Conventional single cluster head-based data transmission

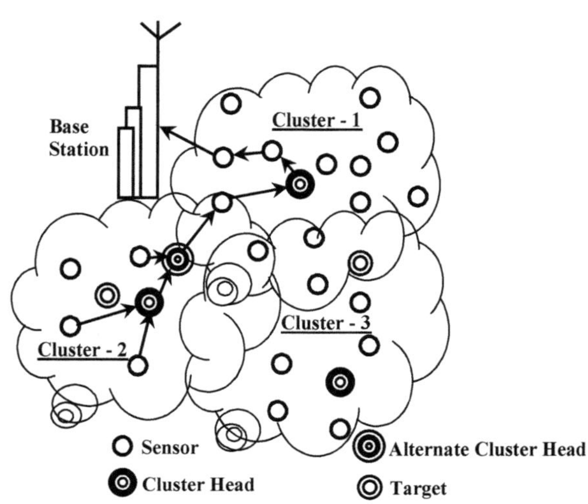

Fig. 3 DACH routing-based data transmission

additional head node (group or cluster head) is to balance the data gathering and transmission load and to elongate the live period of the network.

The proposed approach is being initially applied in a restricted environment where all individual devices in the group are assumed to sense different information of the environment, i.e., the data collected by different nodes present in the cluster are mutually exclusive. The considerations of the data, collected by sensor nodes, which are not mutually exclusive, are left for our future research work. The goal of DACH approach is to reduce the power absorption in information communication period.

Also, it avoids a frequent cluster head selection to some extent, as the additional cluster head can carry forward the data transmission for some more periods. As shown in Fig. 4, the remaining power (often denoted as residual energy, i.e., RE) of the primary cluster head ($E_r^{c_1}$) is compared with the threshold (edge) energy (ξ).

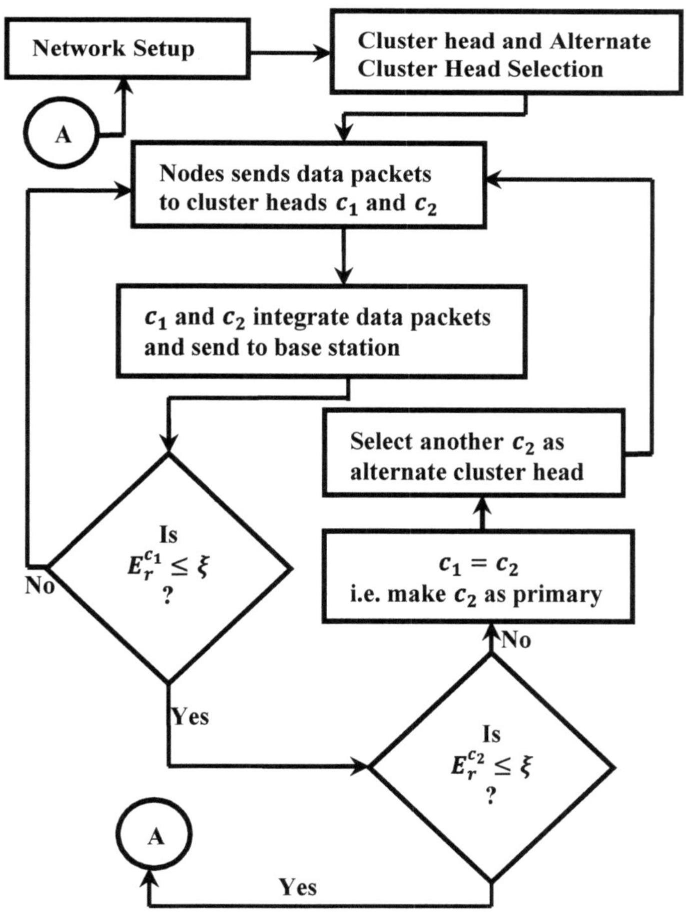

Fig. 4 DACH approach—a typical process flow

The ξ is the minimum energy required to actively play the role of cluster head. If the head node is not able to play the role of a cluster head, then the RE of the additional head node is compared with the edge energy (ξ). Here, the additional cluster head can transmit the data for more time before the network proceeds for network setup phase, where all nodes exchange the information for the cluster head selection.

In general, it is true that out of the two cluster heads (primary and alternate), one will fall in one's routing path to base station, as all devices in a group send their information to base station through the group head. In our approach, we have taken a use case (scenario 1) of keeping the alternate cluster head as an intermediate node in routing path of primary group head to BS. In similar fashion, the other case (scenario 2), i.e., keeping primary cluster head as an intermediate node in the alternate cluster head's routing path, could also be shown. In this paper, we have considered scenario 1 for the mathematical analysis and network simulation.

The proposed approach has an alternate cluster head along with a group head in each group (cluster). This approach is different than two cluster (group) heads in two clusters in the following ways.

1. Alternate cluster head is a backup or secondary cluster head in the same cluster whereas all earlier approach has a single cluster head in a cluster.
2. DACH avoids frequent complex network setup phases. In DACH approach, the secondary cluster will become the primary cluster and select another alternate group head within the group, and the network continues to work till both CHs (primary and alternate) die.
3. Multiple granular clusters (small-sized and many clusters) have network complexity and maintenance of routing tables which reduced in this DACH approach.
4. Target object property sensing (data capture) and data integration within a cluster by alternate CH and primary CH become easy than two different CHs in two different clusters (two different object visibilities).

Lemma 1 *For two mutually exclusive property message sets, the combined property message size is equal to the sum of the individual property message size.*

Proof Let the two property message sets be

$$M_1 = \{p_1, p_2, p_3 \cdots p_m\}$$
$$M_2 = \{p_{m+1}, p_{m+2}, p_{m+3} \cdots p_n\}.$$

As M_1 and M_2 are mutually exclusive, so $M_2 \cap M_2 = \varnothing$, and hence, $|M_2 \cap M_2| = 0$.

So, the combined message size is given by

$$\begin{aligned} |M| &= |M_1 \cup M_2| \\ &= |M_1| + |M_2| \text{ As } |M_2 \cap M_2| = 0. \end{aligned}$$

Lemma 2 *If the additional cluster head* (c_2) *is located on the path to base station from the cluster head* (c)*, then the distance* $(d_{c_2,b})$ *between* c_2 *and BS is less than or equal to the distance* $(d_{c,b})$ *between the cluster head and base station, i.e.,* $d_{c_2,b} \leq d_{c,b}$.

Proof As the additional cluster head is on the path to BS from CH, so

$$d_{c,b} = d_{c,c_2} + d_{c_2,b}.$$

Case I

if $c = c_2$, then $d_{c,c_2} = 0$. So $d_{c,b} = d_{c_2,b}$.

Case II

if $c \neq c_2$, then $d_{c,c_2} > 0$. So $d_{c_2,b} < d_{c,b}$.

So, combining Case I and Case II, we conclude $d_{c_2,b} \leq d_{c,b}$.

The sensor nodes, which perform the role of CH and the additional CH, execute the algorithm 1 (executeDACHForCH). The cluster heads receive the gathered information from the devices, combine the information, and transfer to BS. If the node plays the role of primary cluster head, then it sends the combined information to the additional CH c_2, and c_2 finally sends information to BS. However, in every iteration of data transmission, the residual power of the CH is compared with the threshold (ξ) value, and then, setup phase is called, if required.

```
Algorithm 1: executeDACHForCH
Data:S_A = {n_i | E_r^{n_i} ≥ ξ}/* S_A denotes set of all active nodes */
M_c = Φ /* Integrated message for base station BS*/
Result: Send Integrated message to Base Station
for each n_i ∈ S_A do
    receive(n_i, m_i);
    M_c = M_c ∪ m_i
end
if (self == c_1)  then/* c_1 denotes primary cluster head */
    send(c_2, M_c);       /* c_2 denotes alternate cluster head */
    if E_r^{c_1} ≤ ξ then
        if E_r^{c_1} ≤ ξ then
        call networkSetup()
    end
end
else
    send(path(c_2, BS), M_c);
end
```

Algorithm 2 describes the network setup process to elect cluster head and select alternate cluster head for our proposed DACH approach.

```
Algorithm 2: networkSetup
Data: x_i = longitude of i^th node and y_i = latitude of i^th node
So, location of i^th node is l_i = (x_i, y_i), node's Remaining
Energy denoted as E_r^{n_i}
Result: Network cluster setup with primary and alternate
cluster head.
for (each n_j) do
  | receive (E_r^{n_j'}, l_j');
  | send (E_r^{n_j}, l_j);
  | updateCache (E_r^{n_j'}, d_jj');
end
if (E_r^{n_j} ≥ max(E_r^{n_j'})) then
  | elect self (j^th node) as primary cluster head(ch)
end
else
  | elect other (j'^th node) having max(E_r^{n_j'}) as primary ch
end
if(self == ch)
  | select j'^th node with Min (E_r^{n_self}, max (E_r^{n_j'})) as alternate ch
end
```

4 Mathematical Analysis

Let us consider we have a multiple group formed in a sensor node network which is deployed arbitrarily. The following are some of the mathematical symbols used in our analysis.

d_{cj}	remoteness (distance) between CH and *j*th node.
$d_{c'j}$	remoteness between *j*th device and the next hop cluster head c'.
n	device counts in the group (cluster).
$\in_{\mathrm{te}}, \in_{\mathrm{ta}}$	device's radio transmitter power and transmitter amplifier energy.
$\in_{\mathrm{r}}$	device's radio receiver power.
$p_d, p_{cd}, p_{cd'}$	size of data (information) packets for single, $(n-1)$, and $(n-2)$ aggregated data packet, respectively.
$\Psi_f^t, \Psi_f^r, \Psi_f$	transmitting energy, receiving energy, and combined energy for flooding-based routing.

$\Psi^{t}_{d,c}, \Psi^{r}_{d,c}, \Psi_{d,c}$ energy dissipated for transmitting, receiving, and total energy for DACH approach with cluster head as c.

As per the Dijkstra's algorithm, the minimum path is

$$d_{cj} = \mathrm{Min}\left(d_{cj}, d_{ck} + d_{kj}\right), \quad \forall k.$$

For p_d-sized data packet to send, energy consumption in cluster head is given by

$$\Psi^{t}_{f} = (n-1) \in_{\mathrm{te}} p_d + \in_{\mathrm{ta}} p_d \sum d^{2}_{cj}.$$

Similarly, energy required for $(n-1)$ number of receiving packets with p_d size and again for forwarding combined p_{cd}-sized packets to base station is given by

$$\Psi^{r}_{f} = (n-1) \in_{r} p_d + \in_{\mathrm{te}} p_{cd} + \in_{\mathrm{ta}} p_{cd} d^{2}_{cb}.$$

Again, in case of multi-hop communication

$$d_{cb} = \mathrm{Min}(d_{cb}, d_{ck} + d_{kb}) \quad \forall k.$$

So, in the mth cluster, the overall energy consumed by all nodes is

$$\begin{aligned} \Psi_{f} &= \Psi^{t}_{f} + \Psi^{r}_{f} \\ &= p_d \in'_{d} + p_{cd} \in'_{cd} \end{aligned}$$

where $\in'_{d} = \left[(n-1)(\in_{\mathrm{te}} + \in_{\mathrm{r}}) + \in_{\mathrm{ta}} \sum d^{2}_{cj}\right]$ and $\in'_{cd} = \left[\in_{\mathrm{te}} + \in_{\mathrm{ta}} d^{2}_{cb}\right]$.

Now, in a proposed DACH approach, let $(m-1)$ count of devices are close to the CH c_1 and $(l-1)$ count of devices are close to the CH c_2, i.e., now, the energy dissipated by all $(m-1)$ number of nodes is given by

$$\Psi^{t}_{d,c_1} = (m-1) \in_{\mathrm{te}} p_d + \in_{\mathrm{ta}} p_d \sum d^{2}_{c_1 j}.$$

So, the total energy consumed by the $(m-1)$ number of devices with primary CH c_1 is

$$\begin{aligned} \Psi^{r}_{d,c_1} &= (m-1) \in_{r} p_d + \in_{\mathrm{te}} p_{c_1 d} + \in_{\mathrm{ta}} p_{c_1 d} d^{2}_{c_1 b} \\ \Psi_{d,c_1} &= \Psi^{t}_{d,c_1} + \Psi^{r}_{d,c_1} \\ &= p_d \left[(m-1)(\in_{\mathrm{te}} + \in_{\mathrm{r}}) + \in_{\mathrm{ta}} \sum d^{2}_{c_1 j}\right] + \in_{\mathrm{te}} p_{c_1 d} + \in_{\mathrm{ta}} p_{c_1 d} d^{2}_{c_1 b}. \end{aligned}$$

Similarly, the energy equation for the $(l-1)$ number of devices which are close to the additional CH c_2 is given by

$$\Psi'^{t}_{d,c_2} = (l-1) \in_{\text{te}} p_d + \in_{\text{ta}} p_d \sum d^2_{c_2 j}$$

and

$$\Psi'^{r}_{d,c_2} = (l-1) \in_{\text{r}} p_d + \in_{\text{te}} p_{c_2 d} + \in_{\text{ta}} p_{c_2 d} d^2_{c_2 b}.$$

So, the total energy consumed by the $(l-1)$ number of devices with additional CH c_2 is

$$\begin{aligned}\Psi_{d,c_2} &= \Psi'^{t}_{d,c_2} + \Psi'^{r}_{d,c_2}\\ &= p_d\left[(l-1)(\in_{\text{te}} + \in_{\text{r}}) + \in_{\text{ta}} \sum d^2_{c_2 j}\right] + \in_{\text{te}} p_{c_2 d} + \in_{\text{ta}} p_{c_2 d} d^2_{c_2 b}.\end{aligned}$$

Hence, the energy consumed by the $(m+l)$ number of devices with primary CH c_1 and additional CH c_2 is given by

$$\begin{aligned}\Psi_d &= \Psi_{d,c_1} + \Psi_{d,c_2}\\ &= p_d[(n-1)(\in_{\text{te}} + \in_{\text{r}}) + \in_{\text{ta}} s_d] + \in_{\text{te}} p'_{cd} + \in_{\text{ta}} s^p_b\end{aligned}$$

where

$$s_d = \left[\sum d^2_{c_1 j} + \sum d^2_{c_2 j}\right], p'_{cd} = [p_{c_1 d} + p_{c_2 d}] \quad \text{and} \quad s^p_b = \left[p_{c_1 d} d^2_{c_1 b} + p_{c_2 d} d^2_{c_2 b}\right].$$

So, if we compare the above expression, we have $\Psi_f > \Psi_d$; as from Theorem 3, we have $p_{c_1 d} d^2_{c_1 b} + p_{c_2 d} d^2_{c_2 b} \leq p_{cd} d^2_{cb}$ and from Theorem 4 we have

$$\left[\sum_{j=1, j\neq c_1}^{m-1} d^2_{c_1 j} + \sum_{j=1, j\neq c_2}^{l-1} d^2_{c_2 j}\right] \leq \sum_{j=1, j\neq c}^{n-1} d^2_{cj}.$$

Hence, we mathematically found that the energy dissipated by a single cluster head is more than the energy dissipated in the DACH approach.

Theorem 3 *For two mutually exclusive property messages with size $p_{c_1 d}$ and $p_{c_2 d}$, the sum of the cluster heads and base station distance square multiplied with the respective message size is less than the primary cluster head and base station distance square multiplied with the combined property message with size p_{cd}, i.e., $p_{c_1 d} d^2_{c_1 b} + p_{c_2 d} d^2_{c_2 b} \leq p_{cd} d^2_{cb}$.*

Proof As the CH (c) is assumed as the primary CH in our proposed DACH approach, so $c = c_1$. Hence, $d^2_{c_1 b} = d^2_{cb}$, and from Lemma 2, we have $d_{c_2 b} < d_{cb}$.

So,

$$\begin{aligned} &p_{c_1d}d_{c_1b}^2 + p_{c_2d}d_{c_2b}^2 \le p_{c_1d}d_{cb}^2 + p_{c_2d}d_{cb}^2 \\ &\Rightarrow p_{c_1d}d_{c_1b}^2 + p_{c_2d}d_{c_2b}^2 \le p_{cd}d_{cb}^2 \end{aligned}$$

where from Lemma 1, we have $p_{cd} = p_{c_1d} + p_{c_2d}$.

Theorem 4 *If d_{c_1j} and d_{c_2j} be the distances (remoteness) from the jth node to CH c_1 and c_2, respectively,* (m − 1) *and* (l − 1) *are the number of nodes close to CHs c_1 and c_2, and then,*

$$\left[\sum_{j=1, j\neq c_1}^{m-1} d_{c_1j}^2 + \sum_{j=1, j\neq c_2}^{l-1} d_{c_2j}^2\right] \le \sum_{j=1, j\neq c}^{n-1} d_{cj}^2$$

where $n = (m + l)$.

Proof Let $c = c'$. So,

$$\sum_{j=1, j\neq c}^{m+l-1} d_{cj}^2 = \sum_{j=1, j\neq c}^{m-1} d_{cj}^2 + \sum_{j=1, j\neq c'}^{l-1} d_{cj}^2 + d_{cc'}^2.$$

Now, if we consider $c = c_1$ and $c' = c_2$ as the nodes which are nearer to additional cluster head, c_2 transfers the data to CH c_2 rather than sending to the primary cluster head $c_1 (d_{c_1j} \ge d_{c_2j})$, for all *j*th nodes those are in close proximity of the additional CH (c_2). So

$$\begin{aligned} &\sum_{j=1, j\neq c}^{m+l-1} d_{cj}^2 \ge \sum_{j=1, j\neq c}^{m-1} d_{cj}^2 + \sum_{j=1, j\neq c'}^{l-1} d_{cj}^2, \quad \text{As } d_{cc'}^2 \ge 0 \\ &\sum_{j=1, j\neq c}^{n-1} d_{cj}^2 \ge \sum_{j=1, j\neq c}^{m-1} d_{cj}^2 + \sum_{j=1, j\neq c'}^{l-1} d_{cj}^2. \end{aligned}$$

Hence, the inequality holds true.

5 Simulation

In the simulated environment study, the below limit values have been taken to replicate the flooding-based route protocol, LEACH, and PEGASIS route mechanism versus DACH route mechanism.

a. Every device has initial residual energy 1 J.
b. $\in_r = 50\,\text{nJ}/\text{bit}$ i.e. 50 nJ amount of energy consumed by receiver circuitry for 1-bit data processing.
c. $\in_{ta} = 100\ \text{pJ}/\text{bit}/\text{m}^2$ i.e. 100 pJ amount of energy required to send/broadcast 1 bit of data in 1 m^2 area by the transmit amplifier.
d. $\in_{te} = 100\ \text{pJ}/\text{bit}$ i.e. 100 pJ amount of energy required to receive 1 bit of data by the receiving amplifier.
e. Base station location $(x, y) = (\text{random}, \text{random})$.
f. Size of the data packets $p_d = 1000\,\text{bits}$.
g. Magnitude of the combined packet
    i. $p_{cd} = \text{random} \leq (n-1).1000$.
    ii. $p_{cd} = \text{random} \leq (n-2).1000$.
h. Network coverage area of each device = 7.25 m.

The sensor devices are deployed haphazardly as we have taken the random values for the sensor nodes' coordinates. For cluster (group) formation, we have considered the network coverage and number of nodes present inside the wireless coverage area. To find the cluster head, we select the node which can communicate maximum number of nodes available inside its wireless coverage area. The RE of the devices vs the network live period is displayed below (refer Fig. 5a) (60 nodes deployed over 100 × 100 Sq. unit area) and Fig. 5b (150 nodes deployed over 120 × 120 Sq. unit area). We observed that our proposed DACH scheme is providing approximately 1.2–1.5 times more lifetime to the network in both scenarios.

Similarly, in case of PEGASIS, we have 120 nodes installed on 80 × 80 unit square area (Fig. 6a) and 200 nodes installed on 150 × 150 square unit area (Fig. 6b). In both the scenarios, PEGASIS has a significant more lifetime over the

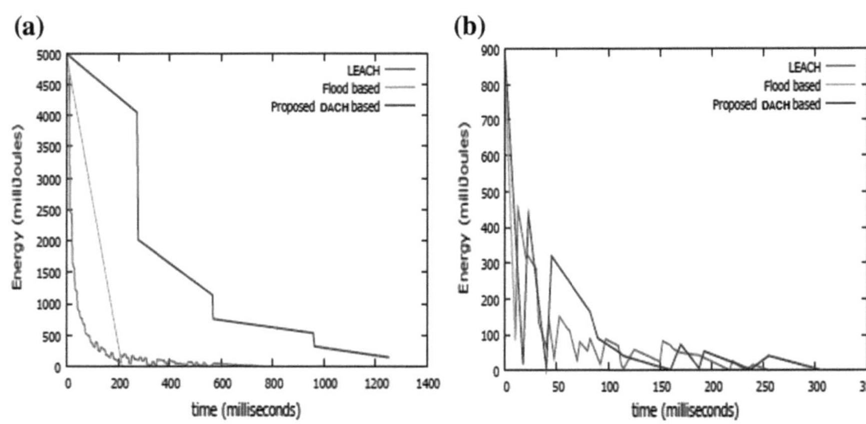

Fig. 5 **a** RE versus life period of network for DACH versus LEACH and flood based for 60 devices on 100 × 100 Sq. area. **b** RE versus life period of network for DACH versus LEACH and flood based for 150 devices on 120 × 120 Sq. unit area

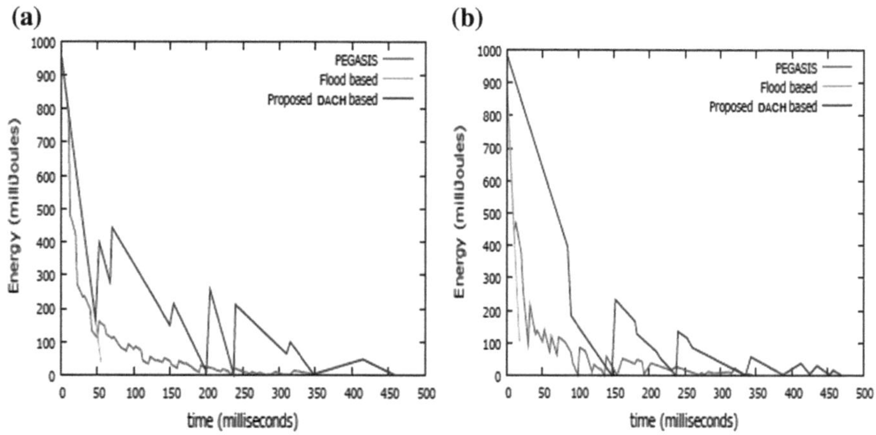

Fig. 6 **a** RE versus life period of network for DACH versus PEGASIS and flood based for 120 devices on 80 × 80 unit Sq. area **b** RE versus life period of network for DACH versus PEGASIS and flood based for 200 devices on 150 × 150 unit Sq. area

flooding-based approach and our proposed DACH approach is adding approximately 0.28–0.38 times more network lifetime over PEGASIS. Though PEGASIS is a chain-based protocol, its nearest neighbor broadcast mechanism consumes more energy. In DACH, devices transmit the information packets to the group heads (primary or alternate) instead of broadcast the data packet, and DACH shares the load between primary and alternate cluster heads. The simulation depicts that though LEACH and PEGASIS have more network lifetime over the flooding-based routing approach, DACH approach is giving an improved performance in conserving the residual energy.

6 Conclusion

Different routing techniques, e.g., flooding, LEACH, PEGASIS, are evolved to minimize the energy consumption. However, there is an energy loss in transmitting the data to a CH which is in a more distance than the alternate cluster head in the same group. The DACH mechanism is an attempt to decrease the transmission energy and to share and balance the information gathering and data transmission load with the primary cluster head. Our mathematical observation shows that DACH approach is conserving more power than the cluster-based routing approach. The simulation results show the effectiveness of the proposed DACH protocol's performance for the energy conservation. This approach could now be applied and integrated into other clustering-based routing protocol to manage the sensor device power effectively and to extend the sensor network lifetime. The more relevant mechanism, e.g., neural network in primary cluster head and alternate cluster head selection process, is our future scope of work.

References

1. Kamila, N.K., Dhal, S.,: Overview of WSN Infrastructure Models, Design & Management, In: International Journal on Recent and Innovation Trends in Computing and Communication (IJRITCC), vol 3, Issue 3, pp. 9–13 (2016).
2. Kamila, N.K., Dhal, S., Samantaray, A.K.,: A Survey of Neural Network Energy Efficiency Management in Wireless Sensor Networks, In: International Journal of Applied Engineering Research (IJAER), vol 10, Number 21, pp. 42023–42036 (2015).
3. Muruganat, S.D.,: A Centralized Energy-E_client Routing Protocol for Wireless Sensor Networks, In: IEEE Radio Communications, vol 43, pp. 8–13 (2005).
4. Heinzelman, W.R., Chandrakasan, A., Balakrishnan, H.: Energy-Efficient Communication Protocol for Wireless Microsensor Networks, In: Proceedings of the 33rd Hawaii International Conference on System Sciences, pp. 1–10 (2000).
5. Lindsey, S., Raghavendra C.S.,: PEGASIS: Power-Efficient Gathering in Sensor Information Systems, In: IEEE Aerospace Conference Proceeding (2002).
6. Intanagonwiwat, C., Govindan, R., Estrin, D.,: Directed diffusion: a scalable and robust communication paradigm for sensor networks, In: Proceedings of ACM MobiCom (Boston, MA), pp. 56–67 (2000).

Horizontal Motion Control of Underactuated Quadrotor Under Disturbed and Noisy Circumstances

M. Kamran Joyo, Syed Faiz Ahmed, Mohd Izhar Abu Bakar and Athar Ali

Abstract In recent years, the area of quadrotor UAV has drawn prominent attention of the researchers, enabling them to develop an immense research area in the field of UAVs. A quadrotor system has a simple and nonlinear architectural design, so it requires a suitable controller to ensure its stability during flight. However, due to its architectural structure several issues were found regarding its controlling such as angular stability, altitude, and position control of quadrotor under the challenging conditions such as wind burst and noisy measurements, and these issues are still not successfully resolved. In this paper, modern control design techniques are discussed and their application in quadrotor control issues is presented. An innovative and more robust control technique, compared to the available renowned control techniques, is proposed for the position controlling quadrotor system. The controller is designed by fusing two distinct control techniques PID and LQR, which is named PID–LQR which deals with two major issues faced by the flying quadrotor, i.e., external disturbance and noise, respectively. Furthermore, the effectiveness of the proposed control technique is also verified by comparing it with auto-tuned PID and optimized LQR techniques under disturbed and noisy conditions. The simulated results indicate that the proposed method yields a better response as compared to the conventional methods.

Keywords Quadrotor · PID · LQR · UAV · PID–LQR

M. Kamran Joyo (✉) · S.F. Ahmed · M.I.A. Bakar · A. Ali
Electrical Section, Universiti Kuala Lumpur, British Malaysian Institute,
53100 Kuala Lumpur, Malaysia
e-mail: muhammad.kamran@s.unikl.edu.my

S.F. Ahmed
e-mail: syedfaiz@unikl.edu.my

M.I.A. Bakar
e-mail: mizhar@unikl.edu.my

A. Ali
e-mail: athar.ali@s.unikl.edu.my

D.K. Mishra et al. (eds.), *Information and Communication Technology*,
Advances in Intelligent Systems and Computing 625,
https://doi.org/10.1007/978-981-10-5508-9_6

1 Introduction

The unmanned aerial vehicles have been the subject of interest to the researchers due to the rapid advancement in technology and demand as they are expected to become a major asset to the aviation industry [1–4]. UAVs are space-traversing remote-controlled flight vehicles that work without any human intervention [5]. There exist numerous types of UAVs, one of them is a quadrotor UAV consisting of four rotors mounted at the four ends of the cross-frame [1, 4, 6, 7], which is prominent among the researchers. The cause of their prominence is due to their light weight, simple mechanical structure that makes its assembly trouble-free [1].

Quadrotor UAV might look simple by design but is nonlinear in nature and has caused considerable annoyance to researchers in the past [1]. The control for stabilization of the system for longitudinal motion under external disturbance such as air turbulence and noisy conditions is still a subject that is concerned to researchers, and this matter requires a reasonable consideration [8]. These issues can affect the performance of quadrotor during flight as they can drift the UAV from its original position [1–10]. Hence, it requires an efficient, robust position control design which is able to react quickly to the wind gusts and overcome the noisy measurements of the GPS sensor. This research work helps to resolve these issues faced by fully autonomous quadrotor UAV during its longitudinal motion under uncertain conditions such as external disturbance and faulty sensor measurements.

The purpose of the research is to develop an efficient, robust control design for longitudinal motion of quadrotor under various uncertainties such as external disturbances and noisy sensor measurements, acting on the system. PID and LQR control techniques are used to develop such control design that can collectively react efficiently over these uncertainties acting on the system. Furthermore, the proposed control technique (PID–LQR) is implemented to quadrotor to analyze the real-time response of the quadrotor under uncertain conditions.

2 Related Work

2.1 Survey of Former Designed Quadrotors UAVs

Since last few decades, UAVs are gaining attention and are continuously developing at a phenomenal rate. At present, there are more than 1000 UAVs developed by more than 50 countries as a major contribution to the military and civil assistance. UAVs can be categorized as shown in Fig. 1 [11].

Among these UAVs, quadrotor is one of the most popular center of research since recent years due to its demand in civil and military applications [4, 8, 12, 13]. Quadrotor idea starts building in early 1900s, an experimental rotorcraft was built by Breguet brothers which flew in the year 1907 for the very first time named Breguet-Richet Gyroplane 1 [11, 14, 15], shown in Fig. 2. The whole body is made

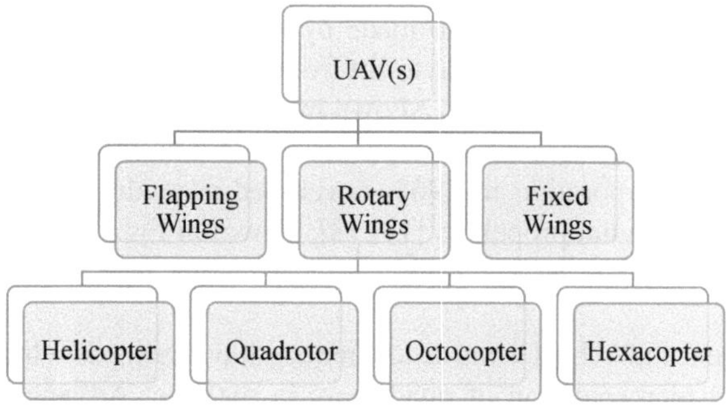

Fig. 1 Classification of UAV(s)

Fig. 2 Breguet-Richet Gyroplane1

Fig. 3 De Bothezat design

up of steel that is why the weight of aircraft is around 500 kg without the pilot; finally, this aircraft did not fly well and was not controllable at all by any means [16].

After Gyroplane 1 another quadrotor was developed by Georges de Bothezat as Flying Octopus in 1922 [11, 15]. This was the first remote control (RC) based quadrotor [14], shown in Fig. 3. The design got the capabilities to fly at low altitude but was very slow in moving toward the horizontal motion because it has no control algorithm for avoiding the uncertainties [17].

In 1922, other attempts were also made by Etienne Oemichen. The first model made by him failed as it was not able to lift from the ground for flight then he made the second model as Oemichen No. 2 shown in Fig. 4 [11], consisting of four rotors and eight propellers, supported by a cruciform steel-tube framework layout [15].

Curtiss-Wright Corporation in 1963, developed a quadrotor Curtiss X-19 that used special type of radial propellers [11, 18], shown in Fig. 5. This quadrotor was destroyed during its first test flight and no further work was done on its development [19].

More complex and robust control techniques have been developed in order to provide detailed representation of quadrotors in real life. Advancement in sensor technology and processors have led to the development of small quadrotor UAVs. Since last few years, the development of micro quadrotors has potentially increased [18]. In 1992, Hoverbot, as shown in Fig. 6, was developed using four radio-controlled helicopters [20].

Fig. 4 Oemichen No. 2

Fig. 5 Curtiss X-19

Fig. 6 Hoverbot

Fig. 7 Mesicopter

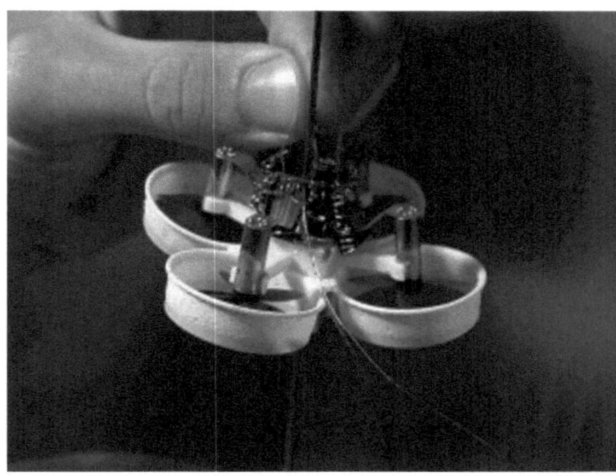

Standford University developed Mesicopter shown in Fig. 7, in late 90s as centimeter-scale quadrotor project. Their research study included aerodynamic design, fabrication, power issues, and stability and control [18, 21].

The indoor micro quadrotor UAV was developed and was named as OS4 [18, 22]. Figure 8 shows the diagram of the model. The mathematical design and modeling were presented and the quadrotor was equipped with the sensors that could measure position, altitude, and orientation [23].

Besides these quadrotors many other quadrotors were also developed like X4–flyer, a CEAs project for operating under the environmental uncertainties [18, 24, 25], a commercially available quadrotor Draganflyer, is designed for autonomous flight [26]. Starmac II [24] is a classical type of quadrotor which is composed of electronic interface board and several processors [27, 28]. The two currently operational quadrotors are the Draganfly Innovations Draganflyer X-Pro and SIM Sky Eye [18].

Fig. 8 OS4 quadrotor

2.2 *Survey of Proposed Position Control Algorithms of Quadrotors UAVs*

In the recent years, position controlling of quadrotor has remained a problem due to the constraints and unstable kinematics and dynamics. However, some of the techniques of control were developed in this field.

2.2.1 Proportional Integral Derivative (PID)

Most of the controller applications utilize PID control technique and are the most commonly used control technique opted by the industry in many cases [6, 9]. PID control technique was utilized in the quadrotors to address the issues of position and orientation of quadrotors. Initially, the model was simulated on MATLAB and then it was implemented [1, 7]. With the appropriate values of control variables, minimum overshoot and steady error minimized to zero were achieved [29]. The research work did not include any situation to overcome sensor noises.

In 2011, another author researched on quadrotor and prepared a model. The analysis of the model was done on MATLAB. PID control was implemented and compiled better results [30]. The study did not contain the idea for minimizing or eliminating the sensor noises.

In Li research work [29], PID control was applied to resolve afore mentioned issues of position and orientation of quadrotors. The response of the PID control shows the stability with almost zero steady-state error [6].

2.2.2 Linear Quadratic Regulator

LQR control technique is used to stabilize the system optimally at minimum cost. The comparison of LQR with the PID algorithm was implemented by applying it on the quadrotor system [31]. The result of the comparison shows that LQR has a better performance than PID because LQR works for the complete dynamic model while PID works for the simplified model [6].

A researcher implemented LQR control technique on a quadrotor for its position, altitude, and attitude control. Author's research concluded that LQR performs well under noisy conditions but does not perform well under disturbance conditions [7].

In 2009, another researcher used LQR for position controlling of a quadrotor. The results were satisfactory but in his research work author did not consider the effect of disturbance on a quadrotor system [32].

In 2012, another researcher surveyed about control algorithms developed for quadrotor discussing their pros and cons. The survey mentioned that LQR controller is a linear controller but it lacks the robustness which is highly required by the quadrotor UAV [33].

In 2013, another researcher implemented LQ control technique on all the dynamics including position controlling. The findings of the research did not consider disturbances [34].

2.2.3 Sliding Mode Control (SMC)

SMC is a nonlinear control algorithm in which a discontinuous control signal is applied to the system and move it in the prescribed path. This control is applied to stabilize the underactuated systems [6, 35].

In 2006, a researcher used an impulse function as disturbance to the quadrotor system. Sliding Mode control was used to overcome the applied disturbance [2, 7, 36].

In 2007, another researcher used sliding mode to observe the disturbance. In the simulation tests step function was applied on the vehicle as a wind gust and external disturbance [37].

2.2.4 (Integrator) Backstepping Control

Backstepping control algorithm is recursive and modular in nature, it breaks the system into subsystems and successively stabilizes the system [6]. In [38], backstepping control technique was applied for stabilizing the quadrotor by using the Lyapunov stability analysis [2, 6, 7].

2.2.5 Adaptive Control Algorithms

Adaptive control algorithms adapt the changing parameters like uncertainties or change in time that occurred in the system [6]. Another researcher introduced a technique called Adaptive Integral Backstepping. The technique was developed for position control of quadrotor. Initially, it was observed that only integral backstepping was not enough, so adaptive technique was introduced to gain better robustness [39]. The performance of controller was better but it did not provide any solution to noisy measurement issues.

In 2011, a researcher developed robust adaptive control for a quadrotor helicopter. The researcher applied a step function as disturbance to its system. The developed controller was able to regain the equilibrium state [16].

In 2016, Adaptive Super Twisting Controller was developed for tracking the position in the presence of uncertainties. This controller deals with all the uncertainties present in the system utilizes the feedforward dynamic inversion (FF) to minimize the discontinuities, and improves the performance. This controller was found effective in reducing the effect of uncertainties and disturbance on the system [40].

2.2.6 Reinforcement Learning

In 2013, reinforcement learning based control technique was developed for controlling the quadrotors. It was applied on the black box quadcopter system along with the learned control. This technique was more preferred for the nonlinear system as it treats the system as black box and the response is stored in the transition matrix [40].

Besides these control algorithms, other control algorithms are also developed and implemented such as robust control algorithms which deals with all the types of uncertainties and disturbances in the system and ensures performances within acceptable ranges, optimal control algorithms which includes LQR, L1, $H\infty$, and Kalman filter, feedback linearization which converts the nonlinear system into its equivalent linear system, and hybrid control algorithms which involves more than one control techniques for controlling the quadrotors [6].

3 Methodology

In this section, a control technique is proposed to solve the various uncertainties using PID and LQR control techniques and control the position of quadrotor. The proposed control design is based on switching technique which is developed by fusing two distinct algorithms called PID–LQR. The motivation to develop this algorithm was to develop a control technique for point-to-point movement of quadrotor under certain uncertainties acting on quadrotor system. These uncertainties acting on a quadrotor can drift the UAV from its desired position. The proposed algorithm is robust enough that it can overcome acting disturbance and noisy sensor data.

Generally, quadrotor can be configured in two different configurations '+' and 'x'. In this study, cross-frame, i.e., 'x' configuration is focused. The detailed kinematics and dynamics and the model of quadrotor are explained in the following research papers [3, 7, 13].

3.1 Proposed Control Design for Point to Point Movement of Quadrotor

The proposed control technique is designed for point to point movement of quadrotor for tackling heavy an external disturbance and overcome the effect of noisy measurement data. The technique involves two control algorithms PID and LQR to overcome the trouble caused by the external disturbances and noisy conditions, respectively.

Two distinct algorithms PID and LQR are fused together to deliver quick and smooth response that may improve flight performance. Figure 9 shows the block diagram of PID–LQR control design for position control of quadrotor.

Figure 10 shows the MATLAB block diagram of the overall diagram of position controlling of quadrotor for single axis.

3.2 Implementation of PID and LQR on Longitudinal Motion of Quadrotor

The horizontal motion control is responsible for the movement of quadrotor UAV in x- and y-axis. The horizontal motion is attained by rolling and pitching the

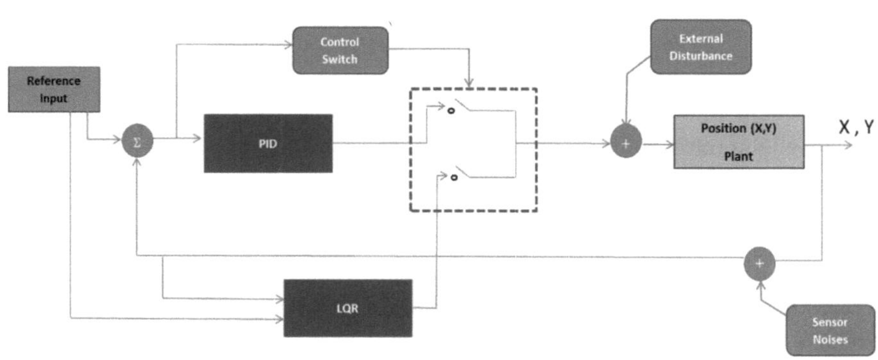

Fig. 9 PID–LQR for position controlling of quadrotor

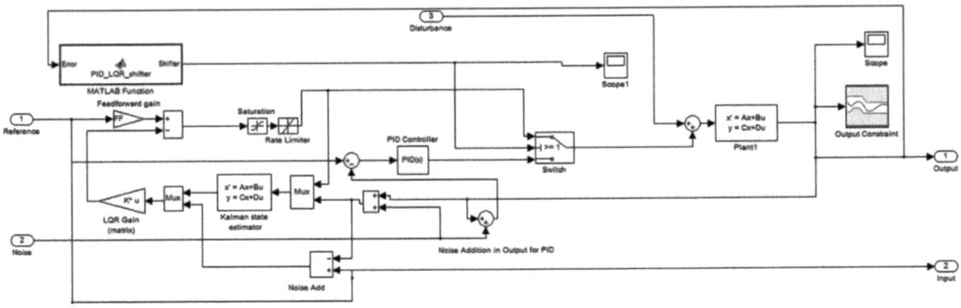

Fig. 10 MATLAB block diagram for position controlling of quadrotor with PID–LQR

quadrotor. The controller is also responsible to recover from the circumstances if the UAV experiences a drift in *x*- and *y*-positions due to external disturbance [24]. For position, only *X* and *Y* equations will be taken into account defined in [1, 2].

The objective of using PID is to express how each K_p, K_i, K_d contributes to obtain Fast rise time, minimum overshoot and illuminate steady-state error in quadrotor system. To set the parameters of PID, auto-tune technique is regulated [2]. The values of these parameters are K_p = 15.8408199084157, K_i = 5.10153277345552, K_d = 4.33434893675906. We consider horizontal equation of motion with rotor dynamics from following equations:

$$\ddot{X} = (\sin\varphi \sin\varphi + \cos\varphi \sin\theta \cos\varphi)\frac{U_1}{m} \tag{1}$$

$$\ddot{Y} = (-\cos\varphi \sin\varphi + \sin\varphi \sin\theta \cos\varphi)\frac{U_1}{m} \tag{2}$$

After applying angle approximation and rotor dynamics to the translational motion equation [22].

$$x(S) = \left(\frac{K_x}{S^2}\right)\left(\frac{0.936}{0.178S+1}\right)^2 \tag{3}$$

$$y(S) = \left(\frac{K_y}{S^2}\right)\left(\frac{0.936}{0.178S+1}\right)^2 \tag{4}$$

the close loop transfer function for the position '*x*' will be:

$$\frac{Y(s)}{R(s)} = \frac{K_a K_d K_j s^2 + K_a K_P K_j s + K_a K_i K_j}{K_b s^5 + K_d s^4 + s^3 + K_a K_d K_j s^2 + K_a K_P K_j s + K_a K_i K_j} \cdot e(s) \tag{5}$$

where 'K_j' can be K_x or K_y for position *x* and *y*, respectively. Equation (5) is the position controller equation for *X* and *Y*-axis, which represents actual, linearized and controlled outputs.

For LQR the value of 'K' is an important term, after optimization using Ricatti equation we get the value [7]:

$$K = \begin{bmatrix} 0.0649 & 0.8929 & -2.0818 & -1.1046 & 0.0227 \end{bmatrix} \quad (6)$$

With the help of Equation $u = -K_x$, the close loop transfer function for x- and y-axis can thus be defined as:

$$G(S) = \frac{K_x K_a K_{\hat{x}}}{K_b S^4 + K_c S^3 + S^2} \cdot e(S) \quad (7)$$

The close loop transfer function can be written as:

$$\frac{Y(S)}{R(S)} = \frac{K_x K_a K_{\hat{x}}}{K_b S^4 + K_c S^3 + S^2 + K_x K_a K_{\hat{x}}} \cdot e(S) \quad (8)$$

4 Results and Discussion

In this section, a detailed analysis of the proposed control design called PID–LQR. The incentive of this research work is to develop a control technique for point to point movement of quadrotor under certain uncertainties acting on quadrotor system. The proposed algorithm is expected to be robust enough to overcome the acting wind gusts or disturbances applied on the system and faulty sensor data.

Figure 11 shows the comparison of the responses of PID and LQR for quadrotor horizontal motion control system under noisy measurements and external disturbance.

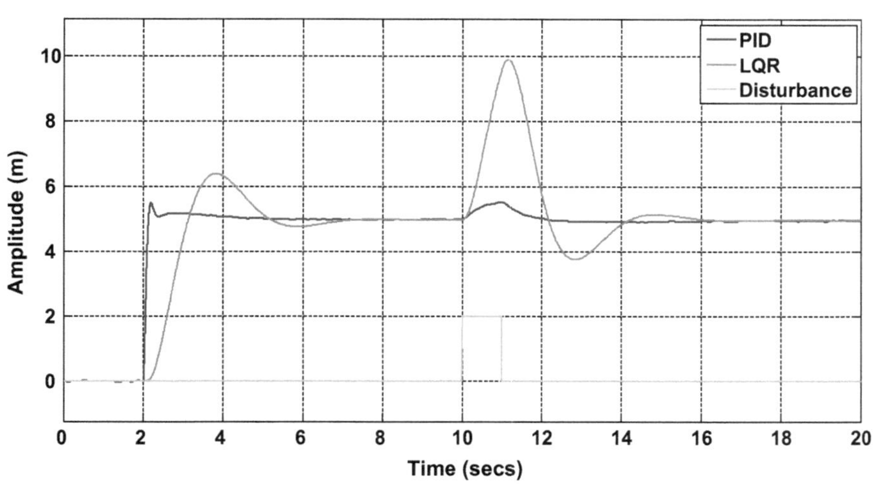

Fig. 11 PID versus LQR

The proposed control algorithm works in such a way that LQR remains active throughout the flight because sensor noises remain all the time. The controller is switched from LQR to PID whenever the disturbance is applied on the system or the reference is changed. Figure 12 shows that 50% external disturbance is applied on the system with respect to input. The sensor noises are also considered which is due to GPS drift. The system reacts effectively from the disturbance applied on the system. PID–LQR overshoot is around 0.586 m and takes 2 s to stabilized and the under the effect of disturbance the drift from the reference is around 0.160 m and stabilizes within 1 s.

In Fig. 13 disturbance level is increased to 200% at the same instance. The overshoot is around 0.585 m and becomes stables in 2 s and the drift in the system

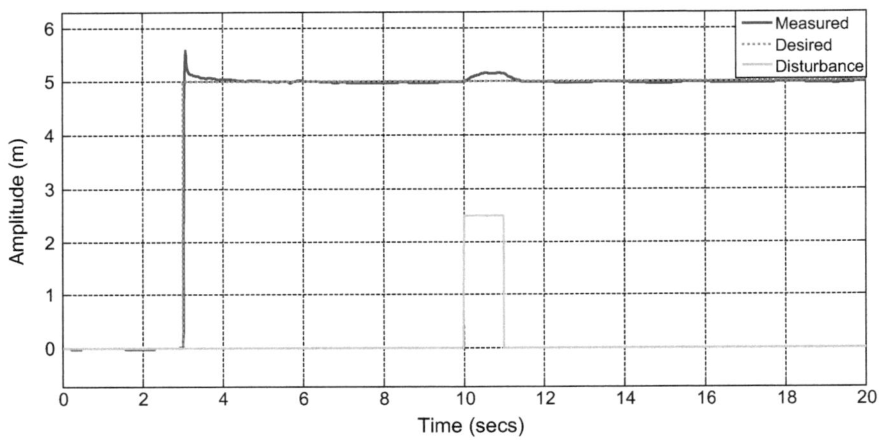

Fig. 12 Effect of 50% disturbance on system with PID–LQR

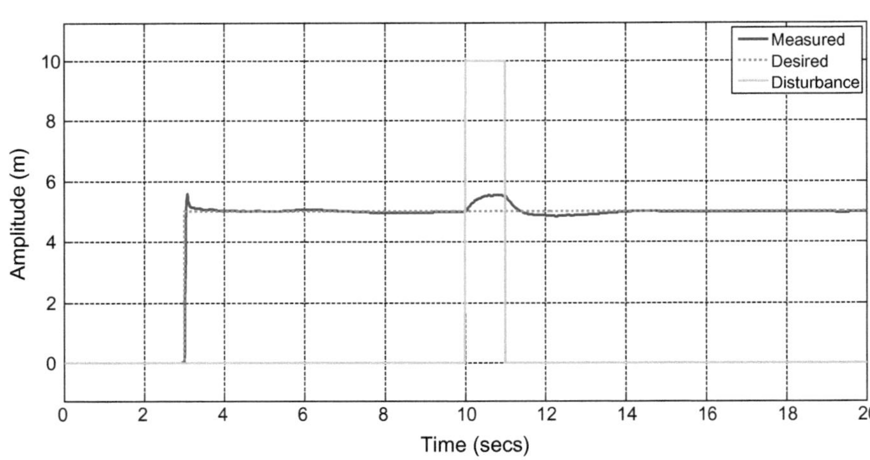

Fig. 13 Effect of 200% disturbance on system with PID–LQR

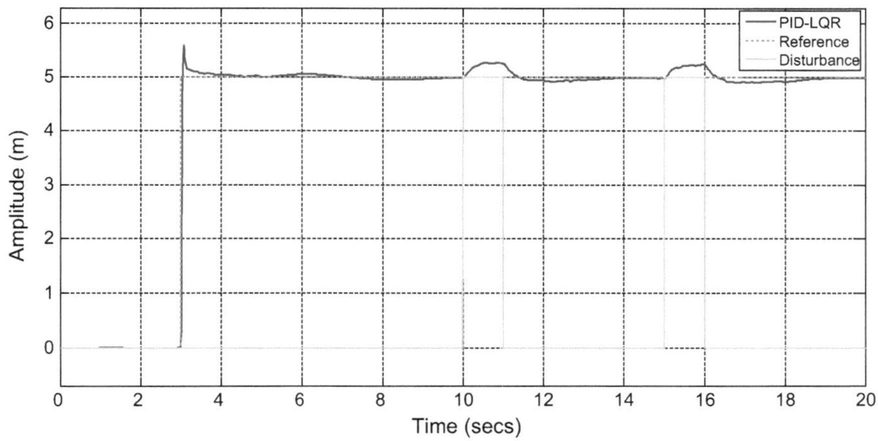

Fig. 14 Effect of 100% disturbance twice applied on system with PID–LQR

due to disturbance is 0.542 m. It can be observed that system reacts quickly to the disturbance and stabilizes it in 3 s.

In Fig. 14 disturbance effect is 100% with respect to the input but it has been applied twice on the system. The overshoot is around 0.585 and takes similar time to 3 s to stabilize. When the disturbance is applied first the drift is around 0.270 m and takes 3 s to stabilize. When the disturbance is applied second time the drift is around 0.239 m and it stabilizes within 3 s. This variance is due to the noise effect.

4.1 Point to Point Movement

A scenario of position controller is simulated and shown in Figs. 15 and 16 in which working of proposed control algorithm is shown. The system is effected by external disturbances and sensor noises. The path defined to quadrotor is such that its present position is considered to be an origin, i.e., $X = 0$, $Y = 0$, the planned path is to maneuver from $X = 0$, $Y = 0$ to $X = 5$, $Y = 5$ and then move to $X = 2$, $Y = 3$. On every change in reference, control switch activates PID control until the response is settled within the range of error defined.

While maneuvering a huge disturbance of 100% of the reference input is applied twice on the system. Once the disturbance is applied in '*Y*' direction and then applied in '*X*' direction. It can be observed that when disturbance acted on the system, the control switched from LQR to PID due to which there is no huge impact on the system as it tackles very quickly to the applied disturbances. Once the error is minimized to the defined range the control is again shifted to LQR to smoothen the response.

Figure 17 shows the point to point movement of quadrotor in 3d. This shows a detailed view of the flight path followed by quadrotor. Even though in very harsh

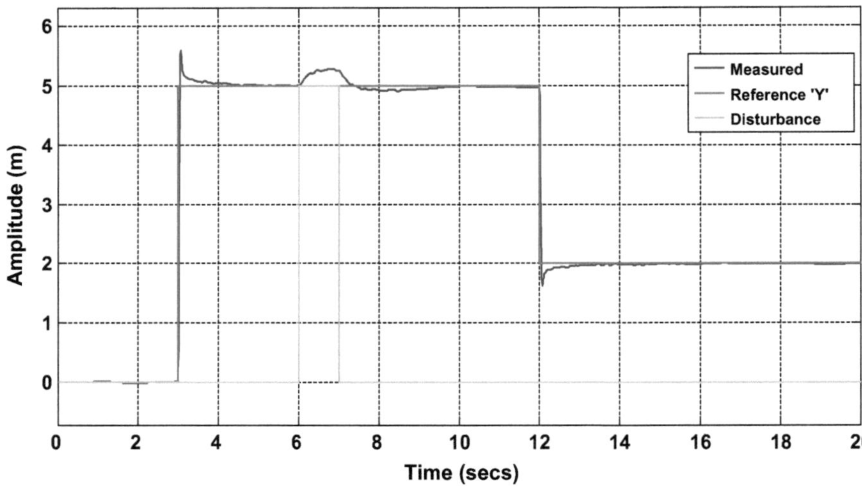

Fig. 15 Movement of quadrotor in *X*-axis direction

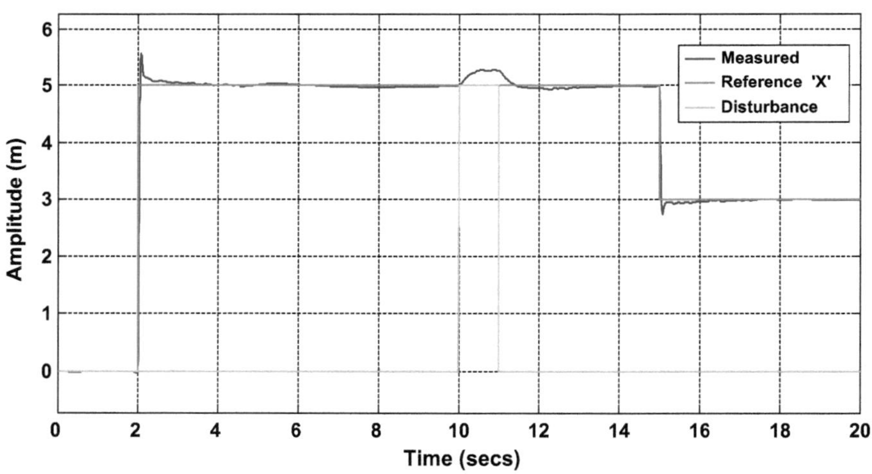

Fig. 16 Movement of quadrotor in *Y*-axis direction

conditions the quadrotor system is able to reach its destination effectively and efficiently. Table 1 indicates the comparison of responses generated by the classical control techniques and proposed control algorithm under impact of external disturbances and sensor and system noisy measurements.

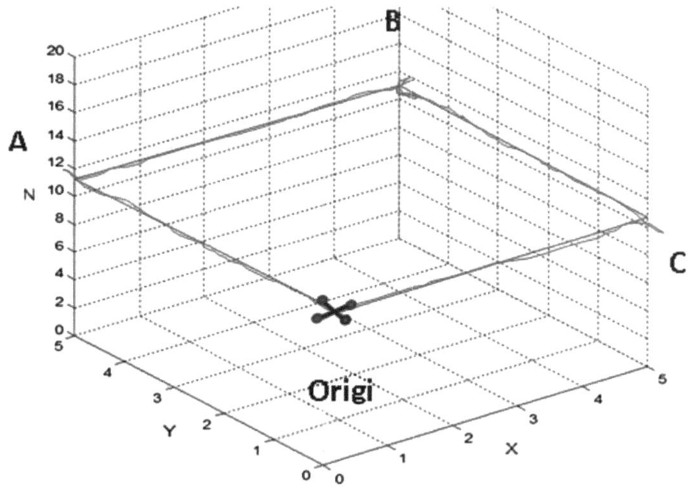

Fig. 17 Point to point movement of quadrotor

Table 1 Individual responses under various uncertainties

Impact	PID	LQR	PID–LQR
Disturbance	Quick	Very slow	Quick
Noise	Unsteady	Smooth	Smooth

5 Conclusion

The research work provided an effective way for autonomous point to point maneuvering of quadrotor UAV under severe flight conditions. The conditions were set tough for quadrotor, sensor noisy measurements and external disturbance were acted on the system to analyze the robustness of the proposed algorithm. Previous researchers worked on algorithms such as PID, LQR and Integral Backstepping, but each of them was not able to tolerate either impact of disturbance or sensor noisy measurements. With these drawbacks, there is a maximum possibility that quadrotor is drifted from its desired position.

The proposed control algorithm uses PID control technique to quickly react to the external disturbance acting on the system. If any drift is occurred due to external disturbance PID recovers the drift very quickly. The external disturbance applied on the system is a step input. The second part of the proposed control technique is LQR which neglects the noisy measurements from the sensors. LQR smoothens the response of the GPS sensor and no much variation is observed. The noise generated from the sensors is known as white noise. The switching between two controllers depends upon the error range.

References

1. Muhammad Hassan, T., Hazry, D., Faiz, S., Muhammad Kamran, J., Faizan, A. W., Zuradzman, M. R., … & Abadal-Salam, T. H. (2014). PID based controller design for attitude stabilization of Quad-rotor.
2. Joyo, M.K. et al., 2013. Position Controller Design for Quad-rotor under Perturbed Condition., 20(7), pp. 178–189.
3. Máthé, K. & Buşoniu, L., 2015. Vision and Control for UAVs: A Survey of General Methods and of Inexpensive Platforms for Infrastructure Inspection. Sensors (Basel, Switzerland), 15 (7), pp. 14887–916. Available at: http://www.scopus.com/inward/record.url?eid=2-s2.0-84933556460&partnerID=tZOtx3y1.
4. Hassan Tanveer, M. et al., 2013. Stabilized controller design for attitude and altitude controlling of quad-rotor under disturbance and noisy conditions. American Journal of Applied Sciences, 10(8), pp. 819–831.
5. Cai, G., Dias, J. & Seneviratne, L., 2014. A Survey of Small-Scale Unmanned Aerial Vehicles: Recent Advances and Future Development Trends. Unmanned Systems, 02(02), pp. 175–199. Available at: http://www.worldscientific.com/doi/abs/10.1142/S2301385014300017.
6. Zulu, A. & John, S., 2014. A Review of Control Algorithms for Autonomous Quadrotors. Open Journal of Applied Sciences, 4(December), pp. 547–556. Available at: http://www.scirp.org/journal/ojapps.
7. Joyo, M.K. et al., 2013. Horizontal plane motion controller design for Quadrotor under noisy conditions., 63(9), pp. 239–246.
8. Hassan Tanveer, M. et al., 2013. Disturbance And Noise Rejection Controller Design For Smooth Takeoff/Landing And Altitude Stabilization Of Quad-rotor. Journal of Applied Sciences Research, 9(5), pp. 3316–3327.
9. John, S. (2013) Artificial Intelligent-Based Feedforward Optimized PID Wheel Slip Controller. AFRICON, 12 September 2013, Pointe-Aux-Piments, 1–6.
10. Tanveer, H. & Hazry, D., 2014. Model Predictive Control based reference point tracking of quad-rotor UAV in prevalence of disturbance 1., 8(4), pp. 428–431.
11. Norouzi Ghazbi, S. et al., 2016. Quadrotors unmanned aerial vehicles: A review. International Journal on Smart Sensing and Intelligent Systems, 9(1), pp. 309–333.
12. Mustapa, M.Z., 2015. Altitude controller design for quadcopter UAV. Jurnal Teknologi, 74 (1), pp. 187–194.
13. Form, O.A., 2016. International Journal of Unmanned Systems Engineering., pp. 5–7.
14. Leishman. (2006). Principles of Helicopter Aerodynamics (2nd ed.): Cambridge university press.
15. Domingues, J., 2009. Quadrotor prototype. Instituto superior tecnico, universidade tecnica de …, (October), p. 129. Available at: http://scholar.google.com/scholar?hl=en&btnG=Search&q=intitle:Quadrotor+prototype#4.
16. Nicol, C., Macnab, C., & Ramirez-Serrano, A. (2011). Robust adaptive control of a quadrotor helicopter. Mechatronics, 21(6), 927–938.
17. Orsag, M., Poropat, M., & Bogdan, S. (2010). Hybrid fly-by-wire quadrotor controller. AUTOMATIKA: časopis za automatiku, mjerenje, elektroniku, računarstvo i komunikacije, 51(1), 19–32.
18. Pounds, P.E.I., 2007. Design, Construction and Control of a Large Quadrotor Micro Air Vehicle., (September), p. 193. Available at: http://www.eng.yale.edu/pep5/P_Pounds_Thesis_2008.pdf\nhttp://www.eng.yale.edu/pep5/publications.html.
19. J. Winchester, X-Planes and Prototypes, Amber Books, London, United Kingdom, 2005.
20. J. Borenstein, The Hoverbot—An Electrically Powered Flying Robot, Unpublished paper, University of Michigan, 1992, ftp://ftp.eecs.umich.edu/people/johannb/paper99.pdf (2004).
21. Kroo, I., Prinz, F., Shantz, M., Kunz, P., Fay, G., Cheng, S., … & Partridge, C. (2000). The Mesicopter: A miniature rotorcraft concept–phase ii interim report. Stanford university, USA.

22. Bouabdallah, S., Becker, M., & Siegwart, R. (2007a). Autonomous miniature flying robots: coming soon! - Research, Development, and Results. Robotics & Automation Magazine, IEEE, 14(3), 88–98. doi: 10.1109/MRA.2007.901323.
23. Bouabdallah, S., Murrieri, P., & Siegwart, R. (2004a). Design and control of an indoor micro quadrotor. Paper presented at the Robotics and Automation, 2004. Proceedings. ICRA'04. 2004 IEEE International Conference on.
24. Joyo, M.K. et al., 2013. Altitude and horizontal motion control of quadrotor UAV in the presence of air turbulence. Proceedings - 2013 IEEE Conference on Systems, Process and Control, ICSPC 2013, (December 2013), pp. 16–20.
25. N. Guenard, T. Hamel and V. Moreau, Dynamic Modeling and Intuitive Control Strategy for an "X4-Flyer". In Proc. International Conference on Control and Automation, Budapest, Hungary, 2005.
26. Lacher, A., & Maroney, D. (2012). A new paradigm for small UAS. http://wwwmitre.org/sites/default/files/pdf/12_2840.Pdf.
27. Hoffmann, G. M., Huang, H., Waslander, S. L., & Tomlin, C. J. (2007b). Quadrotor helicopter flight dynamics and control: Theory and experiment. Paper presented at the Proc. of the AIAA Guidance, Navigation, and Control Conference.
28. Hoffmann, G. M., Huang, H., Waslander, S. L., & Tomlin, C. J. (2011a). Precision flight control for a multi-vehicle quadrotor helicopter testbed. Control engineering practice, 19(9), 1023–1036.
29. Li, J., & Li, Y. (2011). Dynamic analysis and PID control for a quadrotor. Paper presented at the Mechatronics and Automation (ICMA), 2011 International Conference.
30. Salih, A. L., Moghavvemi, M., Mohamed, H. A., & Gaeid, K. S. (2010). Flight PID controller design for a UAV quadrotor. Scientific Research and Essays, 5(23), 3660–3667.
31. Bouabdallah, S., Noth, A., & Siegwart, R. (2004b). PID vs LQ control techniques applied to an indoor micro quadrotor. Paper presented at the Intelligent Robots and Systems, 2004. (IROS 2004). Proceedings. 2004 IEEE/RSJ International Conference.
32. Öner, K. T., Çetinsoy, E., Sırımoğlu, E., Hancer, C., Ayken, T., & Ünel, M. (2009). LQR and SMC stabilization of a new unmanned aerial vehicle.
33. Yibo, L. (2012, 18-20 Oct. 2012). A survey of control algorithms for Quadrotor Unmanned Helicopter. Paper presented at the Advanced Computational Intelligence (ICACI), 2012 IEEE Fifth International Conference.
34. Rinaldi, F., Chiesa, S., & Quagliotti, F. (2013). Linear quadratic control for quadrotors UAVs dynamics and formation flight. Journal of Intelligent & Robotic Systems, 70(1–4), 203–220.
35. Xu, R. and Ozguner, U. (2006) Sliding Mode Control of a Quadrotor Helicopter. Proceedings of the 45th IEEE Conference on Decision and Control, San Diego, 13–15 December 2006, 4957–4962.
36. Benallegue, A., Mokhtari, A., & Fridman, L. (2006). Feedback linearization and high order sliding mode observer for a quadrotor UAV. Paper presented at the Variable Structure Systems, 2006. VSS'06. International Workshop on.
37. Besnard, L., Shtessel, Y. B., & Landrum, B. (2007, 9–13 July 2007). Control of a Quadrotor Vehicle Using sliding mode disturbance observer.
38. Huo, X., Huo, M. and Karimi, H.R. (2014) Attitude Stabilization Control of a Quadrotor UAV by Using Backstepping Approach. Mathematical Problems in Engineering, 2014, 1–9.
39. Fang, Z., & Gao, W. (2011). Adaptive integral backstepping control of a Micro-Quadrotor. Paper presented at the Intelligent Control and Information Processing (ICICIP), 2011 2nd International Conference.
40. Rajappa, S. et al., 2016. Adaptive Super Twisting Controller for a Quadrotor UAV., (1), pp. 2971–2977.

Detecting Concentration Condition by Analysis System of Bio-signals for Effective Learning

Kuniaki Yajima, Yoshihiro Takeichi and Jun Sato

Abstract E-learning system is comfortable for us to learn the new knowledge, because people have tablet, mobile phone, and note PC nowadays, and they always can connect to the Internet. So we can learn by e-learning system using learning management system (LMS), anywhere and anytime. But e-learning style are passive, we watch the contents. Sometime students are boring, they are not easy to concentration. Students are interested in good contents, but it is not easy to recognize which part of contents is good. Usually, we research these, we take questionnaire for students, but it is not true, because students estimate an average of the lecture of the teacher. So we use bio-signal for concentration of the contents and analysis and detective form the measured bio-signal. If we get students condition, we become possible to teach effectively.

Keywords Bio-signal · Detecting concentration · Galvanic skin resistance

1 Introduction

Currently, most people as well as students are experiencing learning by e-learning system. This background has improvement on the network environment (the quality and the rate) and performance improvement on our information terminal (mobile phone, tablet). The merit of e-learning is below. We can learn anytime, anyplace

---

K. Yajima (✉)
National Institute of Technology, Sendai College, 4-16-1, Ayashi-chuo, Aokba-ku, Sendai, Miyagi 9893128, Japan
e-mail: yajima@sendai-nct.ac.jp

Y. Takeichi · J. Sato
National Institute of Technology, Tsuruoka College, 104 Sawada, Inooka, Tsuruoka, Yamagata 997-8511, Japan
e-mail: takeichi@tsuruoka-nct.ac.jp

J. Sato
e-mail: jun@tsuruoka-nct.ac.jp

D.K. Mishra et al. (eds.), *Information and Communication Technology*, Advances in Intelligent Systems and Computing 625,
https://doi.org/10.1007/978-981-10-5508-9_7

when we want to learn. A large number of contents exist in the world. Those contents are made by teacher, customer, and so on. Some contents are developed to concentrate on learning, to attract interest, and to learn for one thing deeply. It looks like school or training. Therefore, a participant may not easy to interest and not easy to concentrate on any contents. The state that a participant can't concentrate with concentration during watching of contents is repeated. We have continued the study which judges a concentration ratio to learning contents so far using bio-signals. By measuring and analyzing skin impedance (GSR) from the former study results. It was confirmed to judge a concentration ratio objectively [1–5].

In this paper, we report on development of a measurement system of GSR and detect of a concentration from measured data. In Chapter "Extracting Hidden Patterns within Road Accident Data Using Machine Learning Techniques", we discuss measurement of a concentration ratio from bio-signals, and development of the professional type system is described. We show the result by which GSR was measured using BIOPAC system which measures the system that it has been developed and bio-signals. We describe analysis results of their concentration ratio using our developed measurement system. In Chapter "Implementation of Smart Job First Dynamic Round Robin (SJFDRR) Scheduling Algorithm with Smart Time Quantum in Multi-core Processing System", we describe down the size of the system and a measurement system. We are explained about an analysis of a concentration ratio from the body surface temperature and the heart rate. In Chapter "Security Enhancement in MANETs Using Fuzzy Based Trust Computation Against Black Hole Attacks", we describe an upgrade to the wireless system that the usability was considered. Finally, we tell this system about expansion to the value of the activity of active learning. We are aiming to support the efficient active learning by using this result.

2 Bio-signals Measuring System

2.1 Bio-signals

The concentration of human is able to be measured by brain wave. Under state of tension, beta waves are generated; by contrast, under state of relaxing, alpha waves are generated. Alpha waves are able to be measured while the concentration is long-lasting. The parasympathetic nerve is measured to be predominant between 8 and 13 Hz, and so symptoms, such as lowering of the heartbreak number and the body temperature, a constricted pupil, decreasing the time of breathing, are occurred. In addition, skin response on the wrist and brain wave indicates similar movement.

GSR measures the change of skin electric resistance by sweating. There are two kinds of sweating: thermal sweating and mental sweating. In measuring of a concentration degree, the change of GSR by mental thermal is used.

The GSR is declining by mental sweating. Also, the concentration is able to be measured by using a blood or saliva, but it takes time to analyze that result. Therefore, we measure the body temperature, GSR, and the pulse. Moreover, we research this validity with concentration and consider whether the concentration is possible to measure more easily. Measuring is conducted on the wrist to relieve stress.

2.2 Measuring System

A measuring system of bio-signals is necessary to measure the bio-signals. BIOPAC, which is the measuring system of bio-signals, is able to measure the brain wave, an electrocardiogram and an electromyogram. However, the measuring system is so expensive; in addition, it may provide a stress to people. Therefore, it is necessary to develop the measuring system which does not provide a stress, does not restrict a person's activities, and is cheap. In the previous study, Fig. 1 shows that bio-signals are measured by incorporating sensor in head phone. GSR is measured by putting sensor on a forehead. However, in this study, measurement is conducted by integration of wristband and sensor, because restraint of forehead may provide a stress to people.

Figure 2 shows simple measuring system which is made in the previous study. It adopted serial communications which are using a USB cable. It consists of mbed and some sensor driver circuits. We are aiming for a cordless, since a cord also provides a stress during AL. Simple measuring system in this study is codeless; moreover, the data are stored in USB memory, but they are not able to measure in real time. Data transfer from a plurality of measuring system will be available by a one-to-multiple which is using Bluetooth. However, there are problems such as a lost data between communication devices. To solve this problem, we aim to protect the measuring data and preserve the dropping data by using the data transfer to a current USB memory.

Fig. 1 A part of sensor

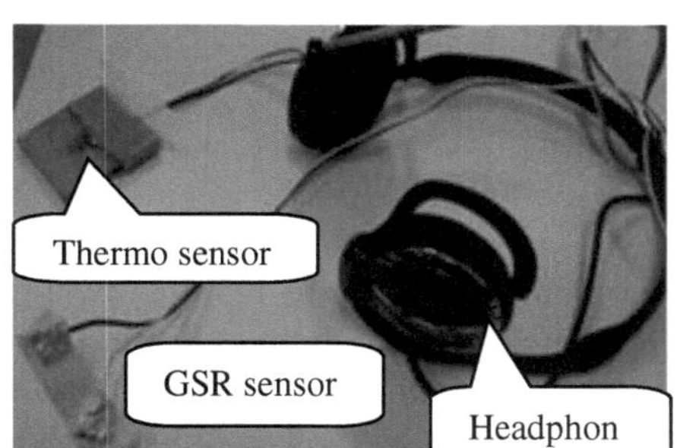

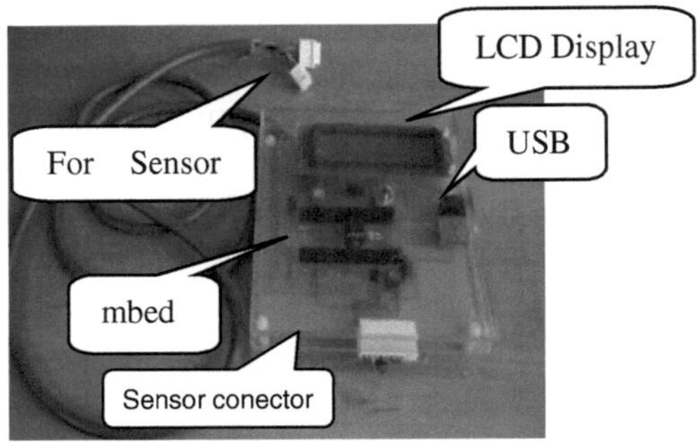

Fig. 2 Measuring system

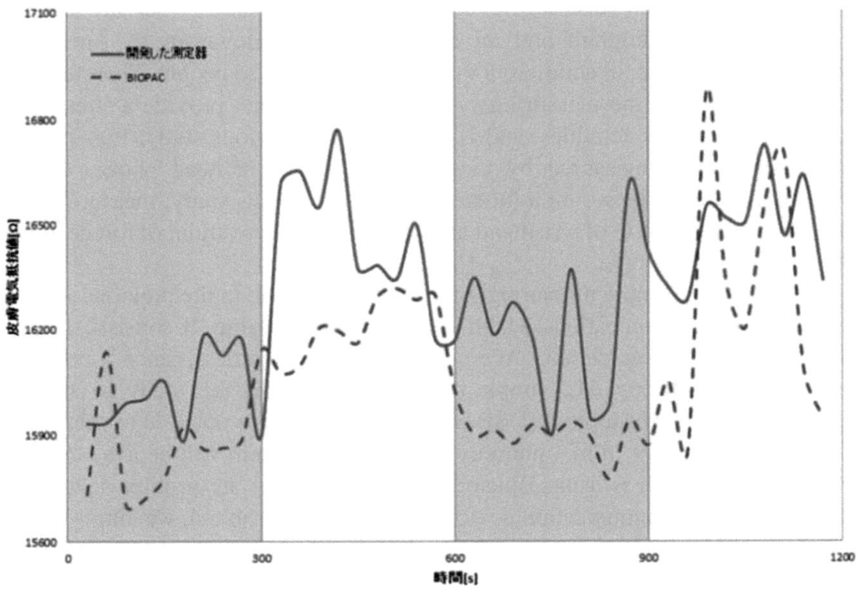

Fig. 3 Comparison with simple measuring system and BIOPAC

2.3 *Detection of Degree of Concentration*

In the previous study, student's bio-signals are measured by using BIOPAC and Finometor, which are commercially available; moreover, a causal relationship between the degree of concentration and bio-signals is proved. The measuring of GSR and body temperature is conducted for 20 min, in which each mathematical question and break was repeated twice in experiments. Figure 3 shows the result that compares the simple measuring system and BIOPAC. As a result, a skin electrical resistance was conjugated in the same way. Therefore, it suggests that GSR is becoming lower during solving the questions (during concentrating);

conversely, it is becoming higher during break or sleeping (during cannot concentrating). It is a symptom of mental sweating; additionally, it is useful to decide the degree of concentrations.

3 Development New System

3.1 Hardware

Figure 4 shows the measuring system, which is miniaturized and optimized than the previous circuit, that size could be reduced from 11 × 9 (cm) to 4 × 9.5 (cm). Skin response on the wrist and brain wave indicates similar movement; hence, integrated wristband with sensor is used as shown in Fig. 5.

3.2 Principle of Pulse Measuring

Pulse wave is a wave motion when the pressure changes in the blood vessel, which is occurred when the blood is pushed into aorta by the contraction of the heart, is

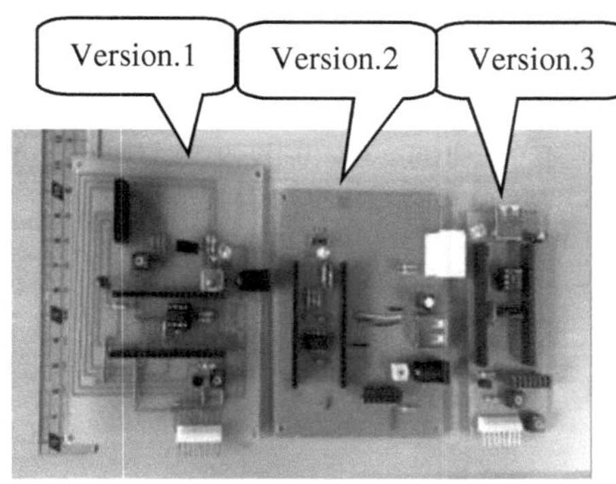

Fig. 4 Measuring system

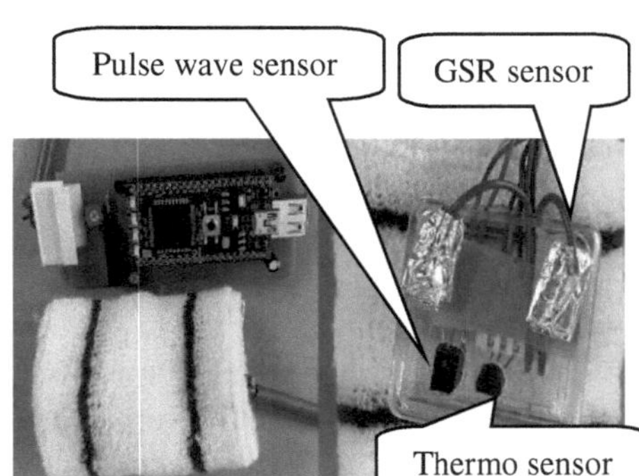

Fig. 5 Integration of wristband and sensor

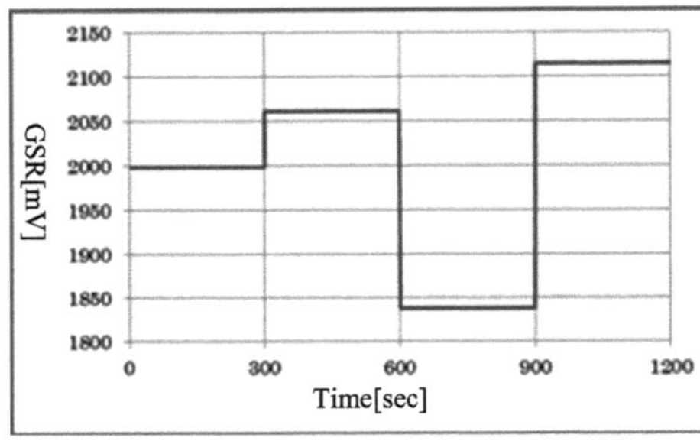

Fig. 6 Average value of GSR

transmitted to peripheral direction. In this study, we develop the measuring system of pulse wave (Fig. 6).

The following is principle of pulse measuring in this study. Pulse wave detecting is classified as photoelectric and piezoelectric. Moreover, detection of photoelectric pulse wave is classified as permeable and reflecting. Hemoglobin in blood has a high absorbing spectrum to light of specific wavelength range. The light of this wavelength range is changed in accordance with the amount of hemoglobin. In permeable type, the measurement part (such as fingertip section) is put between light-emitting part (such as infrared LED) and the light-receiving part (such as phototransistor); pulse wave is detected by changing penetrating light to an electrical signal. In reflection type, a method which is sticking to a measurement point to the light-receiving part and light-emitting part is able to choose any measurement point. We adopt the reflection type in this study because there are many measurement points. Photoreflector RPR-220 which is combined with phototransistor and infrared LED was used.

3.3 Results

The measuring is conducted for 20 min, in which each mathematical question and break was repeated twice in experiments. Figure 7 shows result in this study that is same change of GSR with result in the previous study. It suggests that GSR is becoming lower during solving the questions (during concentrating); conversely, it is becoming higher during break or sleeping (during cannot be concentrating). By contrast, Fig. 8 shows that temperature is becoming higher. It seems that the cause of temperature rise is accumulation of heat in a wristband. Waveform both temperature and GSR is an average value for 5 min, because fluctuation of value from sensor is large.

Figure 8 shows that peripheral skin temperature is changeable, thus measurement is conducted on fingertip. In pulse measurement, a result of it was unstable, because the sensor sticks on skin. Accordingly, we will use elastomer resin to

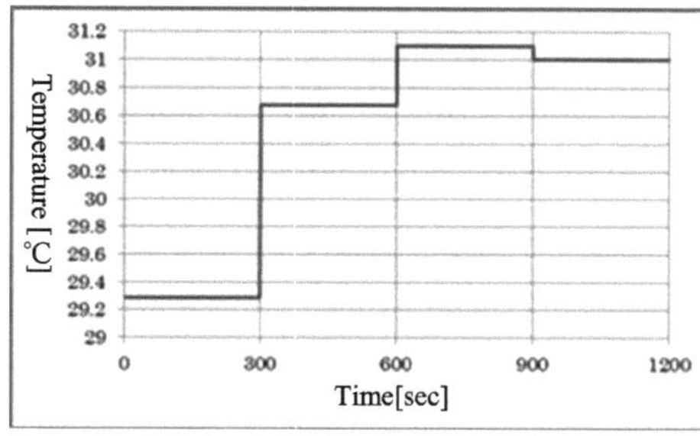

Fig. 7 Average value of temperature

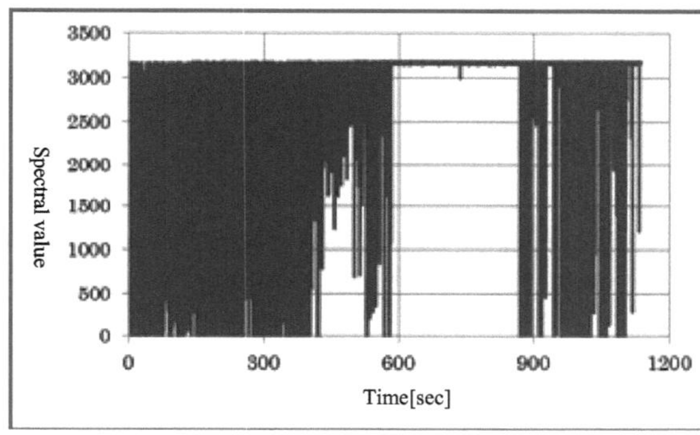

Fig. 8 Pulse waveform

measuring element, so that we will heighten adhesion on skin. Elastomer resin has elasticity, a smooth texture, and lower conductivity.

4 Future Work

We will improve the sensor part. Measurement of GSR will be conducted on wrist. Measurement of temperature is obtained by peripheral skin temperature. As for the measurement of pulse, we will select the site which is possible to reduce the stress in order to aim to adhere to the finger.

We will analyze the result of actual measurement. We will select the useful bio-signals by comparing the data of the previous study; in addition, we will develop the simple measuring system which is able to measure several data. Simultaneously, we will improve the application based on the evaluation.

We reduced the size of the system. So student would not be conscious of measurement equipment and could attend e-learning. That is effective to analyze a concentration ratio to objective contents from bio-signals. We are considering detection of a concentration ratio from bio-signal except GSR, for example information on the number of the eye blink and the point of view of the head. In recent years, JIN CO., LTD., which develops the glasses brand “JINS,” announces the glasses “JINS MEME” which is able to measure the fatigue degree of eye at work, the drowsiness during driving, and the amount of activity. It has three types of sensors, three-point electrooculography sensors, three-axis accelerometer sensors, and three-axis gyroscope sensors. Communication mode is Bluetooth Low Energy. JINS MEME can obtain the data by connecting smartphone applications. We will verify the validity of relationship between concentration and bio-signals by using the results, which are measured by the simple measuring system and a commercial wearable appliance such as JINS MEME.

We need to another improvement that is wireless communication.

If measure system and PC for analyses are wired communication, student’s movement is restricted. We need to develop to communication by a Wi-Fi or Bluetooth is needed.

5 Conclusions

In order to evaluate the degree of concentration during AL, we suggested evaluating it by measuring the GSR, temperature, and pulse. GSR was changed in proportion to the degree of concentration. Regarding the temperature and pulse, further work is needed. Therefore, we propose a system which is able to measure the degree of concentration by measuring the temperature and pulse on the fingertips. It seems that student’s concentration can be evaluated by using three bio-signals. We will verify the validity of relationship between concentration and bio-signals by using the results, which are measured by the simple measuring system and a commercial wearable appliance such as JINS MEME. That is glasses had built in the sensors which is marketed in Japan. That can measure the posses myoelectric potential around the eyes and some axis positions.

References

1. Kuniaki Yajima, Shusaku Nomura, Nobuyuki Ogawa and Yoshimi Fukumura, “Objective Evaluation of e-learning Contents Based on Biological Signals”, International Symposium on Technology for Sustainability, pp. 140–143, (2012.11), Bangkok.
2. Kazunori Nishino, Tetsuo Mayumi, Yurie Iribe, Shinji Mizuno, Nobuyuki Ogawa, Kuniaki Yajima, Kumiko Aoki, Yoshimi Fukumura: Consideration on the Relationship between Changes in Learners’ Learning Preferences and the Differences in e-Learning Modes of a Course, KES-2012 16th Annual KES Conference, San Sebastian, Spain, 10–12 September 2012.
3. Santoso Handri, Shusaku Nomura, Yoshimasa Kurosawa, Kuniaki Yajima, Nobuyuki Ogawa and Yoshimi Fukumura: Evaluating hemodynamic responses of student in e-learning courses by using MPCA and SVM,” Proc. Joint 5th International Conference on Soft Computing and Intelligent Systems and 11th International Symposium on advanced Intelligent Systems, pp. 1459–1464, 2010.12.8, Okayama.

4. Shusaku Nomura, Masako Hasegawa Ohira, Yoshimasa Kurosawa, Yasushi Hanasaka, Kuniaki Yajima, and Yoshimi Fukumura: SKIN TEMPRETURE AS A POSSIBLE INDICATOR OF STUDENT'S INVOLVEMENT IN ELEARNING SESSIONS, 2011 NETs International Conference on Internet Studies, pp.-pp, 2011.9.8-10. Kuala Lumpur.
5. Santoso Handri, Shusaku Nomura, Kuniaki Yajima, Nobuyuki Ogawa, Yoshimi Fukumura and Kazuo Nakamura: STUDENT'S ATTITUDE IDENTIFICATION TOWARDS ELEARNING COURSE BASED ON BIOSENSORS INFORMATION, 2011 NETs International Conference on Internet Studies, pp.-pp, 2011.9.8-10. Kuala Lumpur.

Image Fusion Using Uniformity in HT Domain

Kilari Veera Swamy, Vadhi Radhika and Samayamantula Srinivas Kumar

Abstract The intent of the digital image fusion is a process to obtain important information from acquired images and then form as a distinct fused image. Image fusion algorithms are popular in transform domain than spatial domain methods. The usual methods in transform domain are block-based and multi-resolution transforms. Commonly used orthogonal transforms for image processing are SVD, DCT, KLT, CT, and DWT, but hardware implementation of these transforms is difficult because of the floating-point arithmetic operations. Hadamard transform (HT) is preferred, where the computational speed is the criterion for real-time implementation. In general, block-based methods suffer from blocking artifacts. It influences the features of the fused image. To reduce these problems, statistical measures like mean, contrast, and variance are applied. In the current proposal, statistical measures like entropy and uniformity are explored in HT domain. Further, all statistical measures in HT domain are compared and analyzed. Application of statistical measures in HT domain gives better image fusion results than conventional HT domain fused techniques. Dominance of the uniformity measure in HT domain is observed based on the experimental results.

Keywords Image fusion · Hadamard transform · Entropy · Uniformity

K.V. Swamy (✉)
ECE Department, QIS College of Engineering and Technology, Ongole, AP, India
e-mail: kilarivs@yahoo.com

V. Radhika · S.S. Kumar
ECE Department, JNTUK, Kakinada, AP, India
e-mail: radhikav139@gmail.com

S.S. Kumar
e-mail: samay_ssk2@yahoo.com

D.K. Mishra et al. (eds.), *Information and Communication Technology*,
Advances in Intelligent Systems and Computing 625,
https://doi.org/10.1007/978-981-10-5508-9_8

1 Introduction

Image fusion plays a vital role in image processing. Image fusion is a process of combining the different acquired images into a fused image. The fundamentals of image operations are explained by Gonzalez et al. [1]. In general, the source image should be registered. Generally, image fusion is executed in two domains: spatial domain [2] and transform domain [3, 4]. Stathaki [5] briefly discussed different fusion methods. Yang and Li [6] proposed an image fusion on patches instead of complete image. Amina et al. [7] discussed the contrast enhancement of image with fusion rules. Tang [8] proposed contrast-based fusion method in discrete cosine transform. Mohammad et al. [9] proposed variance-based fusion method in DCT domain. HT and its variations have been greatly used for video applications. The elements of the basis vectors of HT take only the plus or minus one (fixed point). Hardware implementation is easy with the HT. HT has better edge because of simple integer manipulations (other transforms need floating-point operations). So, HT is computationally effective than other transforms. In this work, entropy and uniformity in HT domain are introduced in extension to the existing statistical measures. Results are verified with the various quantitative measures available in the literature.

The organization of this paper is as follows: In Sect. 2, the basics about HT and mean, contrast, and variance in HT domain are discussed. Entropy and uniformity in HT domain are explained in Sect. 3. The proposed fusion algorithm is discussed in Sect. 4. Results are elaborated in Sect. 5. Conclusions are presented in Sect. 6.

2 Hadamard Transform

2.1 Basics of HT

HT is used for various applications, viz. image compression, watermarking, and video coding. Let [g] be the original one and [G] be the transformed one. HT [10] is given by

$$[G] = \frac{A_H[g]A_H}{n}, \tag{1}$$

where A_H is an $n \times n$ Hadamard block. The inverse DHT is given as

$$[g] = \frac{A_H[G]A_H}{n}. \tag{2}$$

The advantages of HT are given below:

1. HT takes plus or minus one. No multiplications are required calculations.
2. HT is a fast transform.
3. HT represents high-frequency content of the images effectively.

The disadvantages of HT are given below:

1. Energy packing efficiency is less than DCT.
2. HT fails to represent low-frequency content of the images effectively.

2.2 Mean Calculation in HT

Mean is the basic measure in image processing. In Eq. (1), $G(0,0)$ is the DC term and others are AC. Mean value [1] in HT domain is given in Eq. (3).

$$G(0,0) = \frac{1}{N}\sum_{m=0}^{N-1}\sum_{n=0}^{N-1} g(m,n). \tag{3}$$

2.3 Contrast Calculation in HT

Contrast-based image processing plays a vital role in image fusion. All coefficients are arranged as various frequency bands. The contrast [8] at every coefficient of nth band in z block is

$$C(i,j) = \frac{G(i,j)}{\sum_{t=0}^{t=n-1} E_t}, \tag{4}$$

where E_t is the normal value of a particular band and i,j are the positions in z block.

2.4 Variance Calculation in HT

Variance is spread among pixel values. This measure is low when changes of transformed coefficient values are low. It is high when changes of transformed coefficient values are high. Variance of the z block is computed as given below [9]:

$$\sigma^2(z) = \frac{1}{N^2}\sum_{k=0}^{N-1}\sum_{l=0}^{N-1} G^2(k,l) - G^2(0,0). \tag{5}$$

In finish, the $\sigma^2(z)$ of block is calculated from transform coefficients.

3 Proposed Statistical Approaches in HT Domain

3.1 Entropy Calculation

Entropy is an arithmetical determination of arbitrariness which can be used to distinguish the consistency of the image [1]. Entropy is more when variations of transformed coefficients are high. Entropy is less when variations of transformed coefficients are low. Consider L as the number of possible coefficient levels after transform and G_i as a random variable indicating coefficient value. Probability of the coefficient values in a region is denoted as $p(G)$. Entropy of the z block is computed as given below:

$$E(z) = -\sum_{i=ll}^{L} p(G_i) \log_2 p(G_i). \tag{6}$$

where '*ll*' is the least value in a region after transform.

3.2 Uniformity Calculation

Uniformity is high when changes in AC values are low. It is low when changes in AC values are high. In the transform equation, $G(0,0)$ is the DC coefficient and all other coefficients are the AC coefficients. To compute uniformity, consider only AC coefficients. Uniformity of the z block is computed as given below:

$$U(z) = \sum_{k,l} |F(k,l)| \quad \text{where} \quad k \neq 0, \;\; l \neq 0. \tag{7}$$

In finish, the $U(z)$ is estimated from HT values. It is the fixed addition of the AC values of the HT block.

4 Proposed Fusion Algorithm

Proposed fusion algorithm is explained below:

- Two or more acquired images are taken.
- Each acquired image is partitioned into sub-matrix.
- Compute 2D-HT for each sub-matrix.

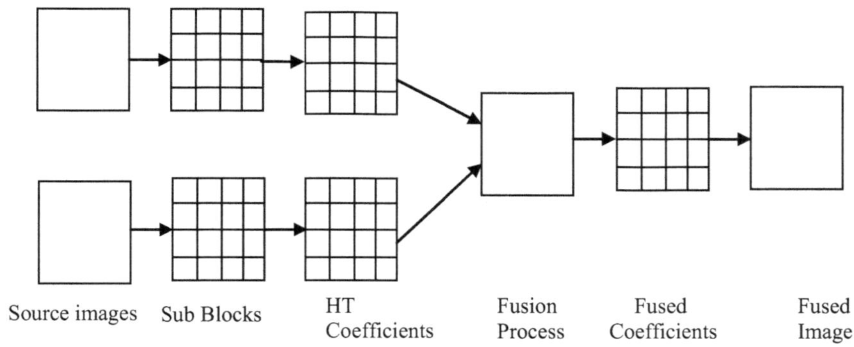

Fig. 1 Details of the fusion process

- The proposed statistical measures like entropy/uniformity of each HT sub-matrix are measured by considering Eqs. (6)/(7), respectively.
- Sub-matrix with higher arithmetical value is identified for fusion technique.
- Apply 2D-IHT for each higher arithmetical valued sub-block.

Details of the fusion are presented in Fig. 1.

The general procedure of fusion is discussed here. The acquired images are partitioned into sub-matrices. HT is applied for each sub-matrix. Arithmetical measures (entropy and uniformity) are calculated for the all the sub-blocks using Eqs. (6) and (7). The highest value of statistical measure is chosen as the suitable one. Inverse is applied to get the image.

5 Experimental Results and Discussion

Experimental findings of the proposed algorithms for image fusion are presented and analyzed. Images used to verify the proposed methods are given in Fig. 2.

5.1 Measuring Parameters

Many metrics are available to judge the performance of image fusion. Some of the qualitative measures are considered for performance evaluation; those are mutual information (MI) [11], edge strength and orientation preservation ($Q^{f1f2/fs}$) [12], feature similarity (FSIM) [13], normalized cross-correlation (NCC) [14].

Fig. 2 Images used for fusion process

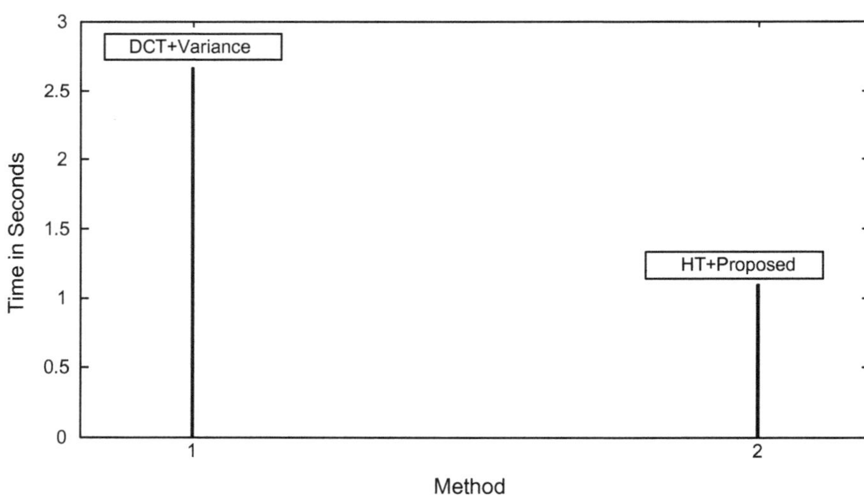

Fig. 3 Graphical analysis

5.2 *Experimental Analysis*

The proposed methods are implemented on six images as shown in Fig. 3. Sub-block size of 8 × 8 is considered for experimentation. MI judges the information quantity present in the image. The ESOP ($Q^{f1f2/fs}$) evaluates the orientations and edge information. FSIM is a metric for local structure. NCC is a computation for correlation between original images and fused image. Experimental values are given in Table 1.

Table 1 Results of the image fusion algorithms

Image	Fusion rule	MI	$Q^{f_1 f_2/f_S}$	FSIM	NCC
Clock	HT + avg	4.3933	0.9050	0.9997	0.9991
	HT + contrast	4.4911	0.9165	0.9997	0.9992
	HT + variance	4.5347	0.9149	0.9997	0.9992
	HT + entropy	4.4910	**0.9166**	0.9997	0.9992
	HT + uniformity	**4.5693**	0.9149	**0.9997**	**0.9992**
Toy	HT + avg	3.2979	0.8613	0.9998	0.9979
	HT + contrast	3.2802	0.8691	0.9996	0.9960
	HT + variance	3.5601	0.8748	0.9998	0.9987
	HT + entropy	3.2766	0.8691	0.9996	0.9960
	HT +uniformity	**3.6076**	**0.8780**	**0.9999**	**0.9990**
Disk	HT + avg	3.3177	0.8798	0.9996	0.9974
	HT + contrast	3.9217	0.8959	0.9996	0.9978
	HT + variance	4.1189	0.9012	0.9997	0.9983
	HT + entropy	3.9216	0.8958	0.9995	0.9978
	HT + uniformity	**4.1688**	**0.9030**	**0.9998**	**0.9986**
Pepsi	HT + avg	3.8730	0.8813	0.9998	0.9988
	HT + contrast	3.9476	0.8994	0.9998	0.9986
	HT + variance	4.4757	0.9138	0.9999	0.9994
	HT + entropy	3.9474	0.8993	0.9998	0.9986
	HT + uniformity	**4.5097**	**0.9143**	**0.9999**	**0.9995**
Paper	HT + avg	2.9544	0.8626	0.9994	0.9843
	HT + contrast	3.1566	0.8498	0.9988	0.9688
	HT + variance	3.9129	0.8947	0.9997	0.9928
	HT + entropy	3.1565	0.8497	0.9989	0.9688
	HT + uniformity	**3.9275**	**0.8950**	**0.9997**	**0.9928**
Lena	HT + avg	3.7464	0.8709	0.9997	0.9973
	HT + contrast	3.9497	0.8925	0.9997	0.9979
	HT + variance	4.2997	0.8919	0.9998	0.9984
	HT + entropy	3.9450	0.8923	0.9997	0.9979
	HT + uniformity	**4.3311**	**0.8933**	**0.9999**	**0.9987**

MI, FSIM, and NCC are better with HT + uniformity approach as per the results presented in Table 1. However, ESOP is better in HT + entropy-based fusion algorithm for the clock image. It can be observed that the HT + uniformity is competent than the other HT-based methods. HT + avg, HT + contrast, HT + variance are existed algorithms. HT + entropy and HT + uniformity algorithms are proposed in this paper. Uniformity is a statistical method to diminish the Gaussian noise. It is included in the image acquisition process. Hence, uniformity algorithm performs better than existing algorithms. Results are also compared with DCT-based image fusion. The comparative results of DCT + variance and

Table 2 Results with run-time values

Image	Fusion rule	MI	$Q^{f_1f_2/fs}$	FSIM	NCC	Run time (s)
Disk	DCT + variance [9]	4.1189	0.9012	0.9997	0.9983	2.662359
	HT + uniformity	**4.1688**	**0.9030**	**0.9998**	**0.9986**	**1.089344**

HT + proposed methods are presented in Table 2. Time taken in seconds is also shown in Table 2 for both the methods. It is implemented on Pentium 4 processor with 3 and 2.99 GHz, 500 MB of RAM. The computer has a system of Microsoft Windows XP Professional, and its version is 2002.

Graphical analyses of DCT + variance and HT + proposed methods are shown in Fig. 3. Time taken for the proposed method is lesser when compared with that of DCT-based method.

6 Conclusions

In this work, two parameters (entropy and uniformity) are implemented. Uniformity is a statistical method which eliminates the Gaussian noise. Further, all statistical measures in HT domain for image fusion are compared. Abundant experiments on assessing the fusion performance are performed, and the outcomes exhibit that the uniformity in HT dominates the earlier techniques. Difficulty of the considered methods is less. Computational time is lesser in uniformity method than other statistical measures. HT shows a number of advantages, i.e., it does not blur at edges, it converges fast and with insignificant error, it requires only integer manipulation. Run time for the proposed method is 50% lesser when compared with DCT-based fusion. These factors allow for ease of implementation and high computational efficiency. Speed algorithms are easily implemented when considering VLSI creation using HT. In all real-time applications, the proposed fusion scheme is the preferred choice. Hence, it is appropriate for practical applications.

References

1. Rafael C. Gonzalez, Richard E. Woods and Steven L. Eddins: Digital Image Processing using MATLAB. Pearson Education (2005).
2. A.A. Goshtasby, S. Nikolov.: Image fusion: advances in the state of the art. Information Fusion, vol. 8, no. 2, pp. 114–118 (2007).
3. G. Piella. A general framework for multiresolution image fusion: from pixels to regions. Information Fusion. 4 (4), pp. 259–280 (2003).
4. Tanish Zaveri and Mukhesh Zaveri.: A Novel Two Step Region Based Multi-focus Image Fusion Method. International journal of Computer and Electrical Engineering, Vol. 2. No. 1. pp. 86–91 (2010).
5. Tania. Stathaki. Image fusion algorithms and applications. Academic Press (2008).

6. Bin Yang and Shutao Li.: Pixel-level image fusion with simultaneous orthogonal matching pursuit. Information Fusion 13, pp. 10–19 (2012).
7. Amina Saleem, Azeddine Beghdadi and Boualem Boashash. Image fusion-based contrast enhancement. EURASIP Journal on Image and Video Processing, doi.10.1186/1687-5281-2012-10 (2012).
8. Jinshan Tang.: A contrast based image fusion technique in the DCT domain. Digital Signal Processing, vol. 14, pp. 218–226 (2004).
9. Mohammad Bagher Akbari Haghighat, Ali Aghagolzadeh, and Hadi Seyedarabi.: Multi focus image fusion for visual sensor networks in DCT domain. Computers & Electrical Engineering, vol. 37, no. 5, pp. 789–797 (2011).
10. Bendalam Vijay and Jallu Swathi.: An efficient fast hadamard oriented digital image in image watermarking. International Journal for Research in Applied Science and Engineering Technology, vol. 3, no. V, pp. 570–585 (2015).
11. G.H. Qu, and D.L. Zhang. Information measure for performance of image. Electronic Letters, Volume 38. No. 7. pp. 313–315 (2002).
12. CS Xydeas and V. Petrovic. Objective image fusion performance measure, Electronic Letters, Volume 36, No. 4, pp. 308–309 (2000).
13. Lin Zhang and Xuanqin Mou. FSIM: A Feature Similarity Index for Image Quality Assesment. IEEE transactions on Image Processing. Vol. 20. No. 8. pp. 2378–2386 (2011).
14. K Veeraswamy, B Chandramohan and S Srinivas Kumar. HVS based robust Image Watermarking scheme using Slant Transform. Second International Conference on Digital Image Processing. SPIE proceedings. Vol. 7546. Singapore (2010).

Crowd Density as Dynamic Texture: Behavior Estimation and Classification

Neeta A. Nemade and V.V. Gohokar

Abstract Extracting crowd feature is a key step for crowd density estimation. This paper proposes a simple and novel approach of preprocessing and extraction of crowd feature. A 5×5 mask is defined for finding isolated components in the image, which proved very efficient for classification of crowd density. SVM classifier is used for classifying the crowd in five different levels. The proposed method is powerful to understand crowd behavior such as crowd coming toward camera and exiting from the camera site. The results are analyzed for PETS dataset and are very promising for images that have bright sunlight and shadow frames too. This method can be used for intelligent surveillance system in public places.

Keywords Convolution of mask · Isolated components · Statistical features · Thresholding

1 Introduction

Computer vision-based surveillance systems for crowd density analysis have tremendous demand these days. Tragedies involving large crowds occur, especially during religious, musical, and political events. To prevent the problems caused by large crowds, proper control and management of crowd is essential. To know the crowd distribution from the crowd density estimation, crowd density is one of the basic features of the crowd status. It is required to give different level of attention to the crowd of different density. Crowd feature extraction is one of the key aspects of crowd analysis. Research of crowded scene analysis could lead to a lot of grave

N.A. Nemade (✉)
Shri Sant Gadge Baba College of Engineering & Technology, Bhusawal, India
e-mail: neeta.nemade@gmail.com

V.V. Gohokar
Maharashtra Institute of Technology, Kothrud, Pune, India
e-mail: vinaya.gohokar@mitpune.edu.in

D.K. Mishra et al. (eds.), *Information and Communication Technology*, Advances in Intelligent Systems and Computing 625,
https://doi.org/10.1007/978-981-10-5508-9_9

applications of crowd analysis such as crowd density estimation, people tracking and detection, seclusion preservation, and crowd activities analysis.

Crowd counting based on features can be analyzed using texture, dynamic texture, and pixel-based algorithms. Texture and dynamic texture analysis methods are used by various researchers [1–8], where the features extracted are gray-level co-occurrence matrix (GLCM) and local binary pattern (LBP). Pixel analysis method used in [9–12] is based on background removal, foreground picture elements, edge detection, Minkowski fractal dimension (MFD), corner detection, edge orientation, blob size histograms, foreground area, perimeter area ratio, shape appearance, etc. Hajer Fradi, Xuran Zhao, Jean-Luc Dugelay put forward subspace of the high-dimensional LBP as raw feature vector and obtained efficiency 89.75% [9]. Wang et al. proposed texture descriptor based on local binary pattern co-occurrence matrix with 94.25% accuracy [12]. Yang et al. proposed sparse spatiotemporal local binary pattern (SST-LBP) as a dynamic texture descriptor and accuracy results got are 90–95% [11].

This paper proposes a feature-based method, in which various statistical features are extracted from image. The correspondence between features extracted and classification of the crowd density is learned. Using this relationship, crowd density is categorized into five distinct levels using SVM classifier.

The remainder of this paper is organized as per the following queue: Sect. 2 presents framework of proposed methodology with detailed analysis of feature extraction process. Datasets used for analysis and performance are discussed in detail in Sect. 3. Results and analysis for different datasets are presented in Sect. 4. Conclusion is discussed in Sect. 5.

2 Proposed Methodology

Figure 1 explains the crowd density estimation flow used in proposed methodology.

2.1 Image Preprocessing

The image acquired by camera is first converted into grayscale for size reduction. The image is then filtered using average filter of 5×5 mask. This helps in removal of noise as well as blurs the edges. The filtered image is then given to feature extraction block.

2.2 Feature Extraction

Two types of features are used for classification.

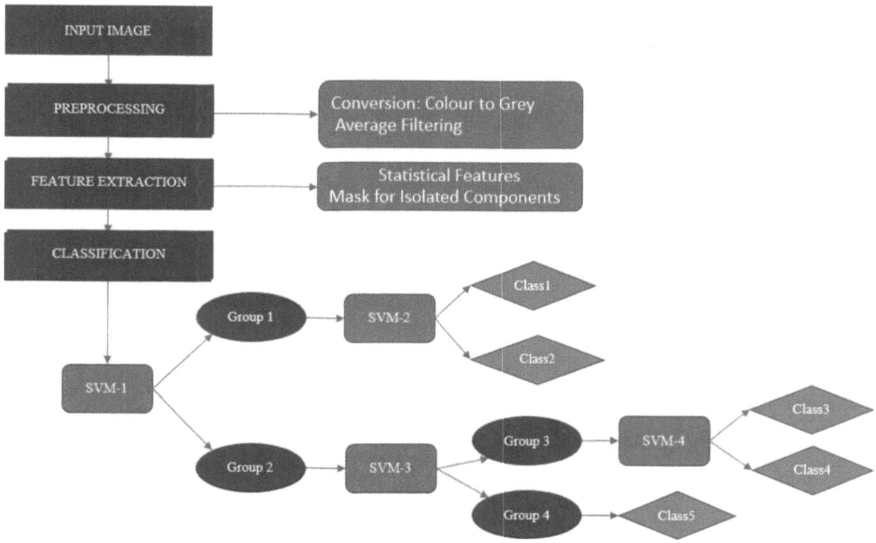

Fig. 1 Crowd density estimation flow

Fig. 2 Mask for isolated nodes

-2	-2	-2	-2	-2
-2	-2	8	-2	-2
-2	8	8	8	-2
-2	-2	8	-2	-2
-2	-2	-2	-2	-2

- Statistical features
- Number of isolated components

Various statistical features are studied including average, standard deviation, variance, kurtosis. It was observed from the study on various datasets that variance is the most suitable parameter. For detection of crowd density, a mask is proposed which segments the crowd into number of isolated components. Using the principle of point detection, a mask called K-mask is proposed as shown in Fig. 2.

This mask is modified version of mask for point detection. Center pixel and the nearest four neighbors are boosted, while the remaining pixels of the 5×5 mask are multiplied by -2. Convolution of this mask with the crowd image segments the image into individual element. The output of the convolution is threshold to get binary image. The threshold is found by averaging the minimum and maximum value of the output. The number of isolated components can be found which will give an approximate value of crowd density. Figure 3 shows the effect of convolution of this mask with image and thresholding.

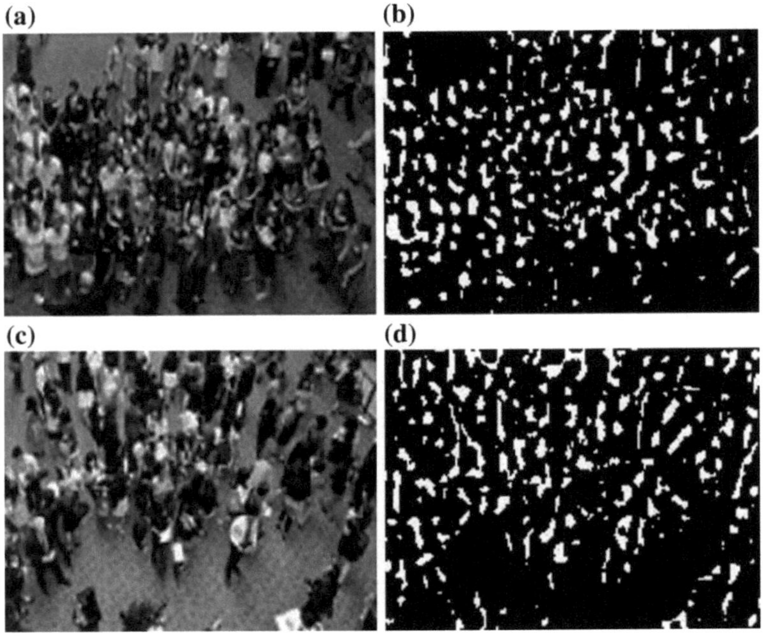

Fig. 3 Effect of convolution of mask **a**, **c** original images and **b**, **d** are corresponding segmented images

Table 1 SVM classification techniques

Method	One- to-one technique	One-to-all technique
Approach	SVM two-class	SVM multiclass
Involves	Builds one SVM for each pair of classes	Builds one SVM per class. It is trained to differentiate the samples in a single class from the samples in all residual classes
Advantage	The training process is quicker in the decision-making process and high classification results, so more practical to get used	With few classes, strategy seems to be significantly more accurate
Disadvantage	More computationally thorough	Due to unbalanced training datasets, recital can be compromised

2.3 Classification

Support vector machine (SVM) is used for classification. It is very useful tool because of its high generalization performance. It does not require prior knowledge. Dimension of input space is also not a problem, while using SVM it is a machine learning tool that is based on the idea of large margin data classification. SVM multiclass strategies is that it is a binary (two-class) classification technique, which has to be adapted to handle the multiclass tasks. Two of the common methods to enable this variation include the one- to-one and one-to-all techniques as shown in Table 1 [13, 14].

3 Dataset

For testing the performance of algorithm, PETS (Performance Evaluation of Tracking and Surveillance) dataset (http://www.cvg.rdg.ac.uk/PETS2009/a.html) is used with the two different scenarios as given in Table 2. Highlights, recording details, and density details are advantages of this dataset. Total 480 frames including different crowd density levels, camera views and crowd directions are examined and results are summarized in Table 3. It also had variation in lighting conditions and shadow.

4 SVM Classification

Crowd density is defined for different levels as per the pedestrian per m^2, number of pedestrian or area occupied by the pedestrians. Person counting is not always necessary for density analysis. The crowd frames are clustered according to congestion degree of the crowds. Clear idea of the problem of level of services for a pedestrian flow is provided in the literature according to which the crowd density is categorized into five levels as jammed flow, very dense, dense, restricted, free flow [15].

SVM is designed for binary classification, but the crowd density estimation is a multiclass problem. Therefore, it needs to be extended for the multiclass problem. Considering the computation complexity and the feature vector property, we used the one-against-one method. In order to reduce the number of SVM classifiers, the classes are divided into groups and at next level group is divided into class. The MaxWins strategy is utilized to decide the density level of the crowd [13, 14].

As shown in Fig. 1, SVM 1 is used to classify Group 1 (defined for class 1 and class 2 crowd density levels) and Group 2 (class 3, 4, and 5 crowd density levels). SVM 2 classifies the frame into class 1 or class 2 density levels. SVM 3 classifies between Group 3 (defined for class 3 and class 4 density level) and Group 4 (defined for class 5 density level). SVM 4 classifies the frame into class 3 or class 4 crowd density levels. Thus, four SVMs are sufficient to classify the five crowd density levels.

Table 2 Two different scenarios from PETS dataset

Data	Scenes	Elements	Time/view	No. of frames	Highlights	Recording details	Density details	Resolution	Compression method/disadvantage
S1	Person count and density estimation for applications as pedestrian tracking, behavior analysis and event detection	L2 walking: medium and high density crowd	14_06, 001	120	Normal flow and the background for the training and density estimation data	Multiple cameras to film the scenarios and involve multiple actors	Multisensory sequences containing different crowd activities	PAL standard (i.e., full color, 768×576 pixels, and 25 frames per second)	JPEG hard to reach crowd level five, i.e., jammed flow [10]
		L3 Running: medium density crowd	14_17, 001 002 003 004	90 for each view					

Table 3 Crowd density frames from different scenarios, graph between isolated nodes and five different crowd density levels and remarks from the graphs

PETS dataset	Number of isolated nodes for five different crowd density levels for PETS dataset
S1: L2: Time_14_06: View_001 frames	    
Graph between isolated nodes and five different crowd density levels	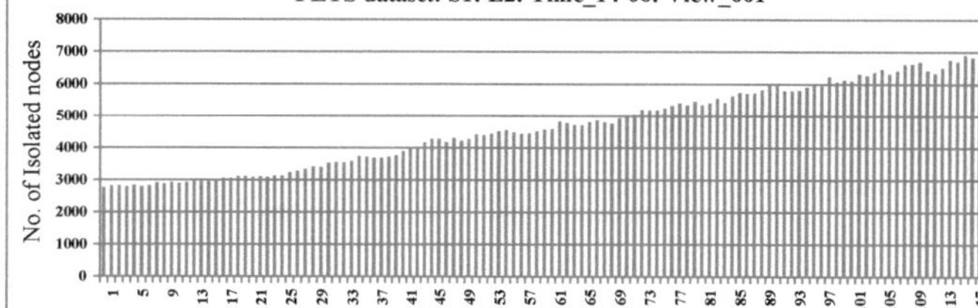
Remark	For S1: L2: walking frames, the number of isolated nodes increases with the crowd density level
S1: L3: Time_14_17: View_001 frames	    

(continued)

Table 3 (continued)

PETS dataset	Number of isolated nodes for five different crowd density levels for PETS dataset
Graph between isolated nodes and five different crowd density levels	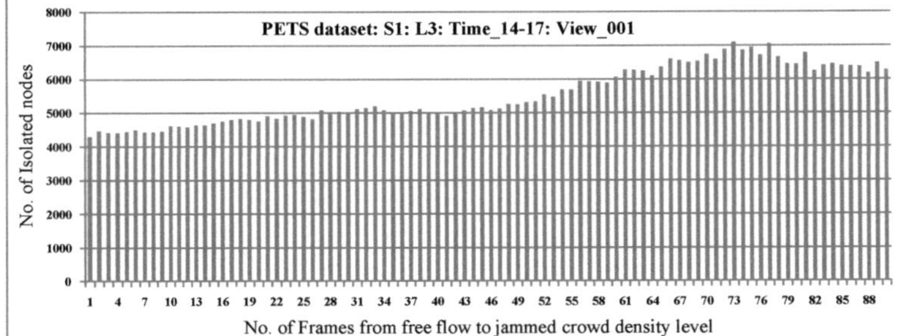
Remark	For S1: L3: walking (side view camera) Time_14_17: View_001 frames, though there is bright sunlight and shadows, the number of isolated nodes increases with crowd density level. As the persons from crowd exit from the camera view, the number of isolated nodes decreases
S1: L3: Time_14_17: View_002 frames	
Graph between isolated nodes and five different crowd density levels	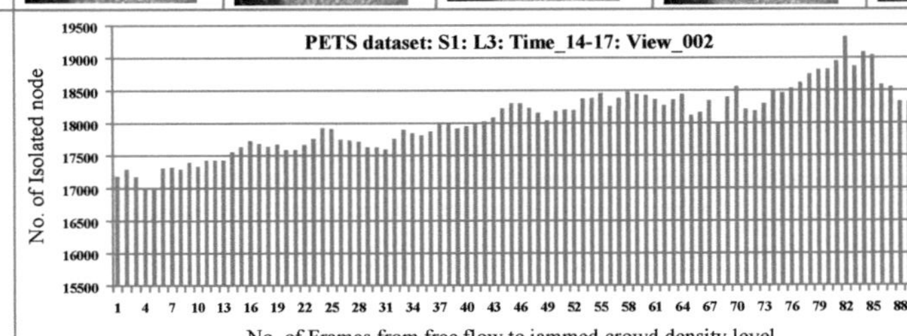

(continued)

Table 3 (continued)

PETS dataset	Number of isolated nodes for five different crowd density levels for PETS dataset
Remark	For S1: L3: walking (front view) Time_14_17: View_002 frames, the crowd is coming toward the camera location. Crowd density increases
S1: L3: Time_14_17: View_003	
Graph between isolated nodes and five different crowd density levels	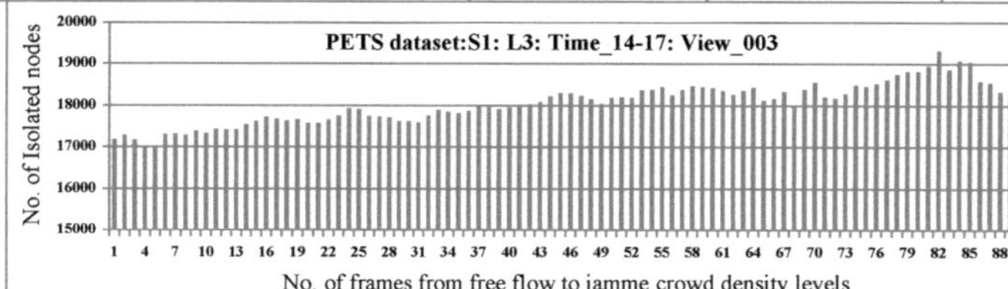
Remark	For S1: L3: walking (side view with occlusions) Time_14_17: View_003 frames, with the bright sunshine and object (tree) between the crowd and camera view, higher the value of number of isolated nodes for close crowd to camera view
S1: L3: Time_14_17: View_004 frames	

(continued)

Table 3 (continued)

PETS dataset	Number of isolated nodes for five different crowd density levels for PETS dataset
Graph between isolated nodes and five different crowd density levels	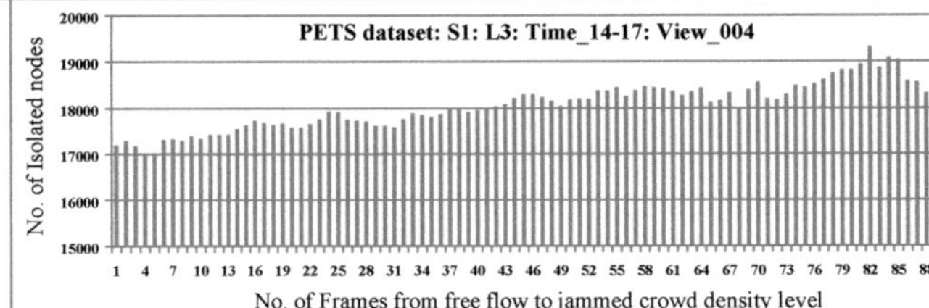
Remark	For S1: L3: walking (back view and Time_14_17: View_004 frames, crowd is going away from camera view, bright sunshine, the number of isolated nodes is proportional to crowd density

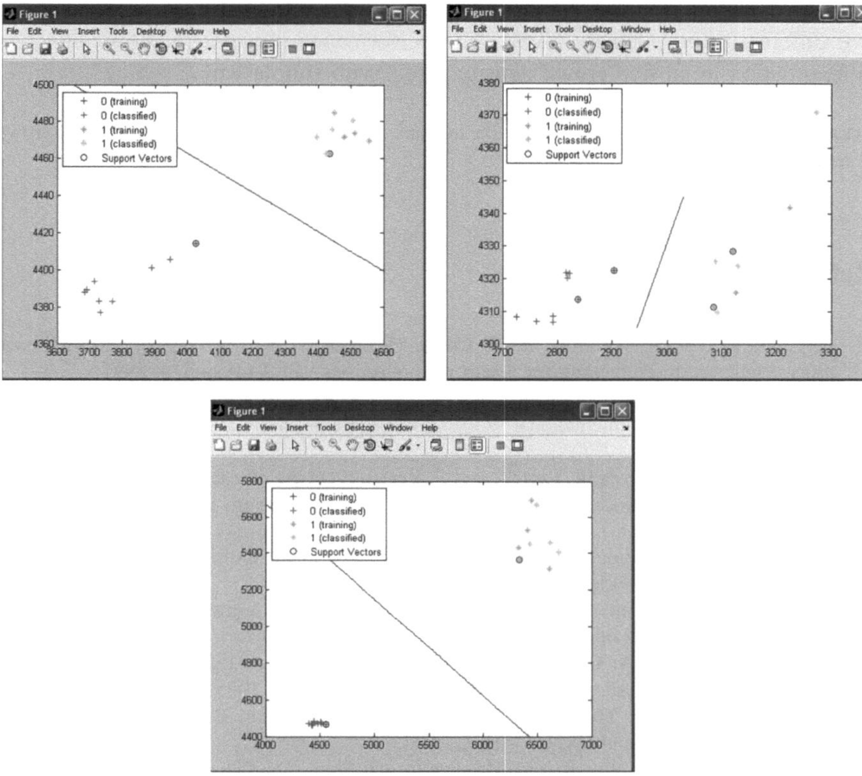

Fig. 4 SVM 4 classification results for group 3 as dense (class 3) and very dense (class 4), group 1 as free flow (class 1) and restricted (class 2) and group 3 (class 4—very dense) and group 4 (class 5—jammed) crowd density levels for PETS dataset

4.1 Classification Results

Figure 4 shows classification between two different crowd density levels for PETS dataset: S1:L2: Time_14_06: View_001. Number of isolated nodes and maximum variance both parameters are considered for the classification.

5 Conclusion

For different frames from PETS dataset, the crowd behavior is examined. The results are promising even under influence of bright sunshine, shadows, direction away and toward camera view. SVM classifier, one-versus-one is used as the classification algorithm with a linear Kernel function that separates a set of objects into their respective groups. Nearly 100% accuracy is achieved with zero error rate when tested on PETS dataset. The time taken is very less for the classification of five different crowd density levels. The performance parameters shows effective differentiation in all five crowd density levels with simple way.

Acknowledgements The authors would like to thank to the PETS dataset for providing the crowd counting datasets.

References

1. A.N. Marana, S. A. Velastin, L. F. Costa, 'Estimation of crowd density using image processing', Image Processing for Security Applications, IEE Colloquium, 1997.
2. A.N. Marana, S.A. Velastin, L.F. Costa, R. A. Lotufo. "Automatic estimation of crowd density using texture", International Workshop on Systems and Image Processing, IWSIP'97, May 28–30, Poland.
3. A.N. Marana, L. da Costa, R. Lotufo, and S. Velastin. On the efficacy of texture analysis for crowd monitoring', Proceedings of the International Symposium on Computer Graphics, pages 354–361, 1998.
4. A.N. Marana, S. Velastin, L.F. Costa and R. Lotufo, Automatic Estimation of Crowd Density Using Texture. Safety Science, Vol. 28, pp: 165–175, 1998.
5. Marana A. N. Verona V, "Wavelet packet analysis for crowd density estimation," International Symposia on Applied Informatics, Inns-bruck, Austria, pp. 535–540, 2001.
6. A.N. Marana and M.A. Cavenaghi, Real-Time Crowd Density Estimation Using Images. International Symposium on Visual Computing, pp: 355–362, 2005.
7. Haibin Yu, Zhiwei He, Yuanyuan Liu, Li Zhang, "A Crowd Flow Estimation Method Based On Dynamic Texture and GRNN", IEEE, 2011.
8. Wei Li and Xiaojuan Wu, Koichi Matsumoto and Hua -An Zhao, "A New Approach of Crowd Density Estimation", TENCON 2010, IEEE.
9. Hajer Fradi, Xuran Zhao and Jean-Luc Dugelay, "Crowd Density Analysis Using Subspace Learning On Local Binary Pattern", July 2013.
10. Hajer Fradi, Jean-Luc Dugelay, A New Multiclass SVM Algorithm and its Application to Crowd Density Analysis Using LBP Features, ICIP 2013, IEEE.
11. Hua Yang, Hang Su, Shibao Zheng, Sha Wei and Yawen Fan, "The Large-Scale Crowd Density Estimation Based on Sparse Spatiotemporal Local Binary Pattern", IEEE, 2011.
12. Zhe Wang, Hong Liu, Yueliang Qian and Tao Xu, "Crowd Density Estimation Based on Local Binary Pattern Co-Occurrence Matrix," 2012 IEEE International Conference on Multimedia and Expo Workshop.
13. Gidudu Anthony, Hulley Gregg and Marwala Tshilidzi, 'Image Classification Using SVMs: One-against-One Vs One-against-All', 2007.
14. Jonathan Milgram, Mohamed Cheriet, Robert Sabourin, ""One Against One" or "One Against All": Which One is Better for Handwriting Recognition with SVMs?, Tenth International Workshop on Frontiers in Handwriting Recognition, Oct 5 2006.
15. A. Polus, J. Schofer and A. Ushpiz, "Pedestrian Flow and Level of Service," J. Transportation Eng. Volume 19, Issue 5–6, September 2008. Pages 345–357.

Restricted Turn Model Fault Tolerant Routing Techniques for 3D Mesh Network-on-Chip: An Evaluation

Ravindra Kumar Saini and Mushtaq Ahmed

Abstract Communication plays a crucial role in design and performance of multi-core system-on-chips (SoCs). Recent development in nanoscale has opened an alternative option to conventional on-chip communication network with uniform stackable multi-chip modules in three dimensions. As the feature size continues to shrink, transient failures or permanent physical damages of on-chip network links are becoming a critical issue. To overcome these failures, network-on-chip (NoC) routing scheme can be enhanced by adding fault tolerant capabilities. In this paper, we analyze the performance of restricted turn model-based routing for the 3D mesh NoC, namely partially adaptive fault tolerant odd even (FTOE3D) routing, fault tolerant negative first (FTNF3D) routing, and fault tolerant XYZ (FTXYZ) routing. As compared to other two routing algorithms, FTOE3D gives the promising results. This document is in the required format.

Keywords Reliability · Fault tolerant routing · 3D mesh

1 Introduction

System-on-chip (SoC) consists of multiple shareable resources that need to communicate at very high speed. An alternate reliable, scalable, and efficient communication infrastructure is required with the shrinking of feature size and higher scaling of cores in SoC. In terms of simplicity and ease of implementation, the bus-based architectures have obvious advantage, but lack scalability and higher bandwidth required. To overcome these limitations, a new communication infras-

R.K. Saini (✉) · M. Ahmed
Department of Computer Science and Engineering,
Malaviya National Institute of Technology Jaipur, Jaipur, India
e-mail: ravisaini.bits@gmail.com

M. Ahmed
e-mail: mahmed.cse@mnit.ac.in

D.K. Mishra et al. (eds.), *Information and Communication Technology*,
Advances in Intelligent Systems and Computing 625,
https://doi.org/10.1007/978-981-10-5508-9_10

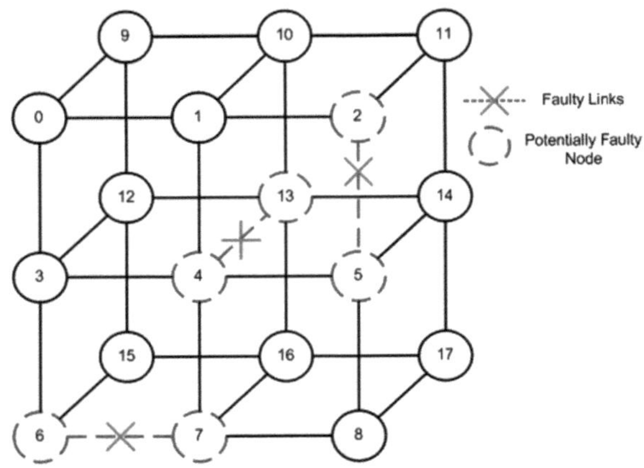

Fig. 1 3 × 3 × 2 3D mesh showing faulty links and potentially faulty nodes

tructure is needed for multi-core chips. Network-on-chip is emerging as an alternate solution to deal with complex system design in SoC [1, 2].

However, increasing the number of cores over a 2D plane is not efficient due to long network diameter and overall communication distance [3]. Three-dimensional (3D) integration is a viable design paradigm to overcome the existing interconnect bottleneck in integrated systems to enhance system performance characteristics [4]. 3D mesh NoC architecture comprises multiple homogeneous and heterogeneous cores interconnected through routers as shown in Fig. 1. The complex 3D NoC architecture is more vulnerable to the faults such as link failure or processing element (PE) or router failure. Every router must be aware of the faults in channels or adjacent nodes. A higher degree of tolerance is desirable in a routing algorithm without any deadlock or livelock condition. An efficient algorithm with fault tolerance exhibits higher degree of tolerance while exploring multiple paths to deliver the packet to the destination.

This paper is organized as follows: Sect. 2 presents related work in 3D NoC routing. Section 3 explores the restricted turn model routing for 3D NoC. Experimental setup, results, and performance analysis of fault tolerant algorithm is discussed in Sect. 4, and finally we end the paper with conclusion and future work in Sect. 5.

2 Related Work

In [5], a fault resilient routing algorithm for vertically partially connected 3D NoC is discussed where each node in network does not have a vertical link in order to deliver a packet to the destination layer. The routing algorithm requires two virtual channels along the Y dimension while one each in X and Z dimensions. In [6], authors address a low-cost solution to improve fault tolerance in horizontal interconnections only, in order to minimize the fault susceptibility in 3D NoCs.

Whereas, reliability issues are discussed in [7] on the aggregated faults that affect through silicon vias (TSV) links in 3D NoC.

A traffic distributing routing algorithm for the 3D mesh network by limiting bandwidth in the vertical dimension is discussed in [8]. In this algorithm, routing decision is taken on the basis of distance between current and the destination node along with the congestion information from the neighboring nodes. However, algorithm focuses solely on congestion avoidance.

Another fully adaptive routing algorithm 3D FAR for homogeneous networks is presented in [9]. This algorithm requires two virtual channels along the *X*, *Y*, and four along *Z* direction that affects area and cost. In their proposed architecture, the network is divided into four disjoint networks and packets are routed using shortest paths between the source and destination nodes as long as there is no fault. In case of any fault non-minimal routes are used. An efficient router structure is suggested for 3D NoC called true NoC architecture by [10]. The architecture consists of vertical links that are embedded in the crossbar and extend to all vertical layers.

3 FTOE3D and FTNF3D Routing Algorithms

Minimal path routing algorithms are advantageous as delays are smaller compared to other adaptive routing algorithms. FTOE3D and FTXYZ are minimal, whereas FTNF3D is non-minimal. The restricted turns of FTOE3D and FTNF3D are shown in Fig. 2a, b, respectively.

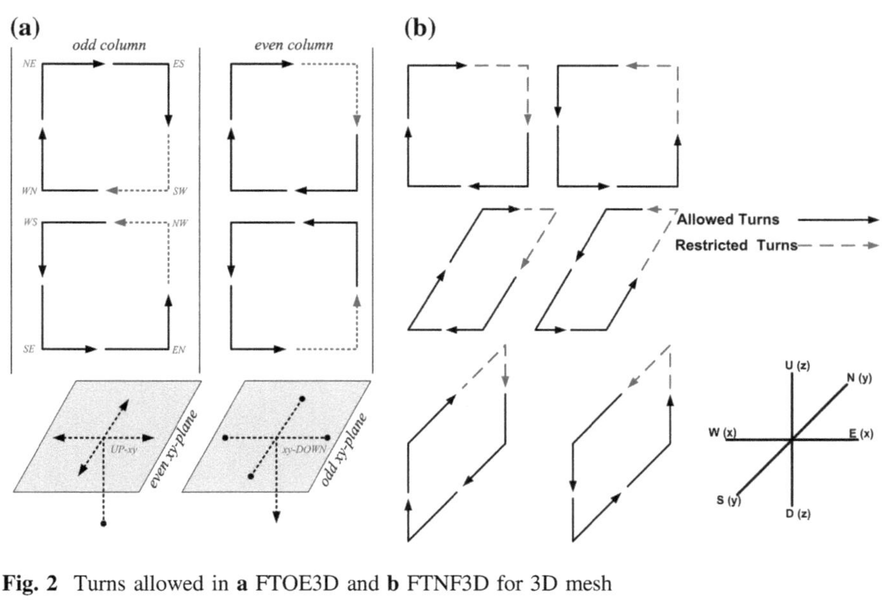

Fig. 2 Turns allowed in **a** FTOE3D and **b** FTNF3D for 3D mesh

Odd-even turn model for adaptive routing is proposed by Chiu [11], which is further extended by Nizar et al. [12]. Turn restrictions are made on the basis whether the current node is in even column or odd column and even slice or odd slice as listed below and shown in Fig. 2a.

1. In odd column, packets are not allowed to take North-West and South-West turns.
2. In even column, packets are not allowed to take East-North and East-South turns.
3. Up-*XY* turns are not allowed in an even *XY*-plane, and *XY*-down turns are not allowed in an odd *XY*-plane. Also, down-*XY* turns are not allowed in an even *XY*-plane, and *XY*-up turns are not allowed in an odd *XY*-plane.

FTOE3D routing algorithm is partially adaptive, and even in the presence of multiple faults it always follows minimal path from source to destination node. As FTOE3D routing prohibits turns to break any waiting cycles hence, prevent deadlocks. Further, it adheres to shortest path in cuboid of interest from source to destination making it free from livelock also. The algorithm is shown in Algorithm 1.

Algorithm 1. Fault Tolerant Odd Even (FTOE3D) Routing Algorithm for 3D Mesh NoC

procedure ► (FTOE3D Routing Algorithm)

Case 1: If current node is having Even column and Even plane.
  If input direction of current node is East.
    If links attached except North and South are failed, drop the packet.
    Else, send in any one direction depending on congestion.
  If input direction of current node is Rear
    If minimal direction contains Rear and link not failed, send packet in Rear direction
    Else, drop the packet.
  Otherwise
  If all links failed, drop the packet.
  Else, send packet in any one direction depending on congestion.
Case 2: If current mode is having Even column and Odd plane:
  If input direction of current node is East:
    If links attached except North, South and Front failed, drop the packet.
    Else send in any one direction depending on congestion.
  If input direction of current node is West or North or South:
    If links attached except Front failed, drop the packet.
    Else, send in any one direction depending on congestion.
  Otherwise
  If all links failed, drop the packet.
  Else, send packet in any one direction depending on congestion.
Case 3: If current node is having Odd column and Even plane:
  If input direction of current node is North or South:
    If links attached except West failed, drop the packet.
    Else, send in any one direction depending on congestion.
  If input direction of current node is Rear:
    If minimal direction contains Rear and link not failed, send packet in Rear direction.
    Else, does not contain Rear or contains Rear but Rear link is failed, drop the packet.
  Otherwise
  If all links failed, drop the packet.
  Else, send packet in any one direction depending on congestion.
Case 4: If current node is having Odd column and Odd plane:
  If input direction of current node is North or South:
    If links attached except West and Front failed, drop the packet.
    Else, send in any one direction depending on congestion.
  If input direction of current node is West or East:
    If links attached except Front failed, drop the packet.
    Else, send in any one direction depending on congestion.
  Otherwise
  If all links failed, drop the packet.
  Else, send packet in any one direction depending on congestion.
end procedure

Algorithm 2. Fault Tolerant Negative First (FTNF3D) Routing Algorithm for 3D Mesh NoC

Procedure ▶ (FTNF3D Routing Algorithm)

Case 1: If all the three minimal directions of current node are negative and input direction at current node is west or south or front or current node is source node.

If all the links are failed, drop the packet.

Else, send the packet in any of these directions depending on congestion.

Case 2: Else if the minimal direction of current node has any two negative directions and input direction at current node is west or south or front or current node is source node.

If Link in minimal negative direction fail, check for non-minimal negative direction.

If this link fails, drop the packet.

Else, send the packet in that direction.

Else, send in any of those directions depending on congestion.

Case 3: Else if the current node contains only one minimal negative directions and input direction at current node is west or south or front or current node is source node

If link in minimal negative fails, check for the remaining two non-minimal negative directions.

If all the links fail, drop the packet.

Else, send the packet in these directions depending on congestion.

Else, send the packet in that minimal negative direction.

Case 4: Else

If all links attached with the current node in minimal negative directions fail

Send the packet in any of non-minimal negative directions, if not failed.

Else drop the packet.

Else

If some or all links attached with the current node in minimal negative directions are not failed, send the packet in any of these directions depending on congestion.

end procedure

FTNF3D is partially adaptive, and non-minimal fault tolerant routing algorithm based on turn restricted routing algorithms is proposed by Glass [13]. Six turns are restricted in this algorithm as shown in Fig. 2b. This algorithm implies that all turns from positive direction to negative direction, i.e., West, South, and Front are prohibited. Forwarding packet is first routed toward West or South or Front until offset is zero and then, turned toward East or North or Rear direction. The FTNF3D is described in Algorithm 2, where packets from source to destination node are routed through minimal path until there is no fault in network. If all links in minimal negative direction are failed, then packets are routed through non-minimal negative direction paths.

4 Experimental Setup and Results Analysis

4.1 Experimental Setup

In order to evaluate the efficiency of the proposed routing algorithms, cycle accurate simulator written in SystemC [14] is used. Simulation parameters are taken as listed in Table 1. We have evaluated performance of proposed routing method with random and transpose traffic patterns using bursty data, with the burst length of 4 at an interval of 3 cycles. To evaluate effectiveness of FTOE3D, FTNF3D, and FTXYZ, we have introduced 10 link failures across the network.

4.2 Results Analysis

For 64 nodes (4 × 4 × 4) uniform 3D mesh NoC, load ranging from 5 to 50% with the increase of 5% each time is applied for initial 2000 cycles and simulated for 10,000 cycles. Every 20 bytes packet is fragmented by wormhole routing in 4 bytes flits. Experiment is executed 10 times with a different load values to achieve a better level of confidence. Table 2 shows the path taken by routing algorithms for single source and destination pair. Every time, a new fault is injected in the current path to see the degree of tolerance.

The graph is plotted for the latency at different load conditions for FTOE3D, FTNF3D, and FTXYZ shown in Fig. 3a, b, respectively. FTOE3D performs better as it tends to find the minimal path from source to destination in multilink failure environment. After 20%, load congestion starts building, causing the higher latency and more failure of packet delivery. Both FTOE3D and FTXYZ routing prohibit

Table 1 Simulation parameters

Parameter	Values
Mesh size	4 × 4 × 4; total 64 nodes
Packet size	20
Buffer size	8
Flit size	4
Simulation cycles	10,000
Test gen. number	2000
Traffic patterns	Random and transpose
Load in %	5–50 with 5% steps
Data pattern	Data pattern bursty data with burst length 4
No. of simulation	10 times with different load and traffic pattern for FTOE3D, FTNFD, and FTXYZ

Table 2 Path for single source to destination pair in the presence of multiple faults

No of links fail	Failed links for FTOE3D	FTOE3D path taken from source (3) to destination (60)	Failed links for FTNF3D	FTNF3D path taken from source (3) to destination (60)
0	Nil	3 → 19 → 18 → 22 → 26 → 30 → 29 → 28 → 44 → 60	Nil	3 → 7 → 11 → 15 → 14 → 13 → 12
1	3 → 19	3 → 2 → 6 → 10 → 14 → 13 → 12 → 28 → 44 → 60	3 → 7	3 → 2 → 6 → 10 → 14 → 13 → 12
2	3 → 19, 2 → 3	3 → 7 → 23 → 22 → 26 → 30 → 29 → 28 → 44 → 60	3 → 7, 2 → 3	3 → 19 → 18 → 17 → 16 → 0 → 4 → 8 → 12
3	3 → 19, 2 → 3, 22 → 23	3 → 7 → 23 → 39 → 55 → 54 → 58 → 62 → 61 → 60	3 → 7, 2 → 3, 16 → 17	3 → 19 → 18 → 17 → 21 → 20 → 4 → 8 → 12
4	3 → 19, 2 → 3, 22 → 23, 54 → 58	3 → 7 → 23 → 39 → 55 → 54 → 53 → 52 → 56 → 60	3 → 7, 2 → 3, 16 → 17, 4 → 8	3 → 19 → 18 → 17 → 21 → 20 → 4 → 5 → 9 → 13 → 12
5	3 → 19, 2 → 3, 22 → 23, 54 → 58, 23 → 7	3 → 7 → 11 → 27 → 26 → 30 → 29 → 28 → 44 → 60	3 → 7, 2 → 3, 16 → 17, 4 → 8, 20 → 21	3 → 19 → 18 → 17 → 21 → 25 → 24 → 8 → 12
6	3 → 19, 2 → 3, 22 → 23, 54 → 58, 23 → 7, 26 → 30	3 → 7 → 11 → 27 → 26 → 25 → 24 → 28 → 44 → 60	3 → 7, 2 → 3, 16 → 17, 4 → 8, 20 → 21, 24 → 25	3 → 19 → 18 → 17 → 21 → 25 → 29 → 28 → 12

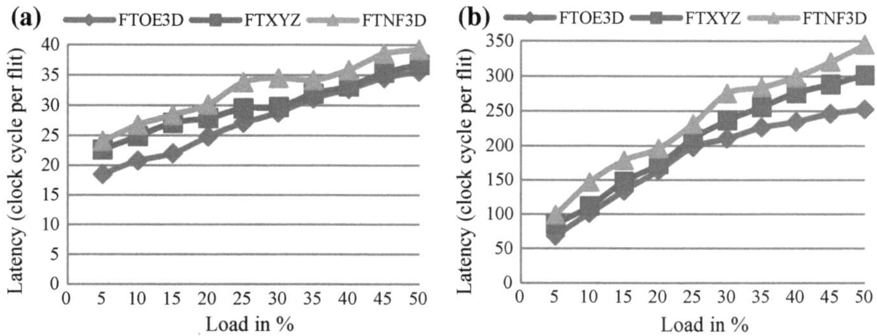

Fig. 3 Latency of **a** random traffic and **b** transpose traffic in the presence of 10 faulty links with load ranging from 5 to 50%

turns to break any waiting cycles and prevent deadlocks as a result virtual channels are not required (Fig. 3).

Average latency in FTOE3D under random and transpose traffic is observed to be lower than FTNF3D and FTXYZ under different load conditions. FTNF3D and FTXYZ use non-minimal path to route the packet form source to destination, whereas FTOE3D uses minimal path to route the packets to surrounding cuboid of interest. FTNF3D selects non-minimal path but less congested. By comparative analysis, we can observe that FTOE3D is giving the best latency results in comparison with FTNF3D and FTXYZ. This may be owing to the facts that turn restrictions in the FTOE3D results in uniform distribution of load. More realistic pattern is observed in transpose traffic pattern, and it is clear that FTOE3D outperforms compared to other routing algorithms making it more reliable for the 3D mesh NoC.

5 Conclusion and Future Work

In this paper, we explored FTOE3D, FTNF3D, and FTXYZ routing algorithms with the fault tolerant under the different traffic with bursty data patterns. Results are indicative that FTOE3D exhibits better performance in terms of latency than FTNF3D and FTXYZ. This is because routing always following a minimal path that does not include faults. In future, robustness of these algorithms is evaluated under more complex and realistic traffic and data patterns like bit shuffle, NED, and multimedia or MPEG4, respectively.

References

1. L. Benini and G. De Micheli, "Networks on **Chips: A New SoC Paradigm," Com**puter, pages 70–78, Jan. 2002.
2. W.J. Dally and B. Towles, "Route Packets, Not Wires: On-Chip Interconnection Networks," Proc. 38th Design Automation Conf. (DAC, '01), pages 683–689, June 2001.
3. V.F. Pavlidis and E.G. Friedman, "3-D Topologies for Networks-on-Chip," IEEE Trans. Very Large Scale Integration (VLSI, '07), pages 1081–1090, Oct. 2007.
4. B. S. Feero and P. P. Pande, "Networks-on-Chip in a Three-Dimensional Environment: A Performance Evaluation," IEEE Transactions on Computers, pages 32–45, 2009.
5. Ebrahimi, Masoumeh, Ronak Salamat, Nader Bagherzadeh, and Masoud Daneshtalab. "A Fault Resilient Routing Algorithm for Sparsely Connected 3D NoCs." 4th MEDIAN Workshop, 2015.
6. A. Kologeski, C. Concatto, D. Matos, D. Grehs, T. Motta, F. Almeida, F. Lima Kastensmidt, A. Susin, R. Reis. "Combining fault tolerance and serialization effort to improve yield in 3D Networks-on-Chip," IEEE 20th International Conference on Electronics, Circuits, and Systems (ICECS), pages 125–128, 8–11 Dec. 2013.
7. A. Eghbal, P. M. Yaghini, N. Bagherzadeh, M. Khayambashi. "TSV Analytical Fault Tolerance Assessment for 3D Network-on-Chip." IEEE Transactions on Computers, vol. Page 99, 06 February 2015.
8. M. Zhu, J. Lee and K. Choi, "An Adaptive Routing Algorithm for 3D Mesh NOC with Limited Vertical Bandwidth, "International Conference on VLSI and System-on-Chip, pages 18–23, 2012.
9. M. Ebrahimi, M. Daneshtalab, P. Liljeberg and H. Tenhunen, "Fault-Tolerant Method with Distributed Monitoring and Management Technique for 3D Stacked Meshes," International Symposium on Computer Architecture and Digital Systems, Pages. 93–98, 2013.
10. J. Kim, C. Nicopoulos, D. Park, R. Das, Y. Xie, V. Narayanan, M. S. Yousif, and C. R. Das, "A novel dimensionally-decomposed router for on-chip communication in 3D architectures," In Proceedings of the International Symposium on Computer Architecture, pages 138–149, 2007.
11. G. M. Chiu, "The odd-even turn model for adaptive routing", IEEE Transaction on Parallel and Distributed Systems, 11:729–738, 2000.
12. N. Dahir, T. Mak, R. Al-Dujaily, A. Yakovlev, "Highly adaptive and deadlock-free routing for three-dimensional network-on-chip," In Computers and Digital Techniques, IET, vol. 7, no. 6, pages 255–263, November 2013.
13. Christopher J. Glass, and Lionel M. Ni. "The turn model for adaptive routing. In ACM SIGARCH Computer Architecture News, 20(2):278–287, 1992.
14. Jain L., Al-Hashimi B., M.S. Gaur, Laxmi V., Ahmed M., and Narayanan A. Nirgam: A simulator for noc interconnect routing and application modeling. In DATE, pages 44–47, 2007.

Power Issues of MANET

Prathmesh Singh, Suruchi Gupta, Lakshita Sejwal and Amrita Mohan

Abstract Mobile ad hoc network (MANET) is a group of wireless and mobile hosts which work effectively and efficiently without any central administration. It has become the most convincing source for future implementations in wireless networks to increase the popularity of mobile and wireless nodes. MANET has several issues such as power efficiency and multicast congestion control. Since MANET uses battery power for functioning, energy turns out to be deficient or one of the major drawbacks for its operation. Our goal for this paper is to find an energy-conserving architecture so that the idle use of power can be reduced in MANET without disrupting the functioning of current MANET algorithms.

Keywords Energy conserving · Mobile ad hoc network (MANET) · Multicast congestion control · Mobile hosts · Power efficiency

1 Introduction

The dependency on technology in today's modern era has increased the need for computational devices and communication irrespective of location and time. Therefore, the usage of MANET has also increased to a great extent in fields such as military operations, emergency areas, and disaster-prone areas as it is wireless communication which does not require any base stations or central infrastructure to operate. Since MANET operates on power derived from a mobile battery, con-

P. Singh (✉) · S. Gupta · L. Sejwal · A. Mohan
G.L. Bajaj Institute of Technology & Management, Greater Noida, Uttar Pradesh, India
e-mail: prathmeshofficial@gmail.com

S. Gupta
e-mail: suruchi.gupta55@gmail.com

L. Sejwal
e-mail: lakshitasejwal@live.com

A. Mohan
e-mail: er.amritacs@gmail.com

D.K. Mishra et al. (eds.), *Information and Communication Technology*,
Advances in Intelligent Systems and Computing 625,
https://doi.org/10.1007/978-981-10-5508-9_11

servation of energy becomes a great issue for MANET networking. Since mobile battery is limited and enhancement in battery life is not as fast as the enhancement of wireless communication, we discuss of the protocols or algorithms of using MANET in such a way that the output is maximum with minimum usage of energy. A protocol's behavior does have a significant impact on power consumption [1]. In this paper, we discuss the issues in energy consumption in the working of MANET. We have mentioned some of the already derived algorithms for increasing the energy efficiency or say reducing the power consumption of MANET. Few such protocols mentioned are:

- Dominating-awake-interval protocol [2]
- Periodically fully awake protocol [2]
- Quorum-based protocol [2]
- Proposed efficient power-aware broadcasting algorithm [3]
- Energy efficiency matrix [4]
- Energy efficient tree construction [5]
- Min-power consumption and Max-life algorithm [6].

We will be discussing in detail about dominating-awake-interval protocol and minimum-power consumption and maximum-life algorithm.

2 Issues in the Energy Consumption of Existing MANET Networking

As mobile ad hoc network (MANET) is widely used in today's modern era due to its easy mobility, its efficiency needs to be increased in terms of power consumption. As MANET works on battery power, its lifetime is very short because a battery is a limited source of energy. The MANET working takes place as follows:

The source nodes send packets to the receiver, and the receiver receives the packet to complete the transmission. Since there are no intermediate stations, MANET works on node-to-node transmission of packets. Ample amount of energy is used in this transmission as the sender as well as the receiver needs to stay awake for detection. There are few MANET algorithms which follow the shortest path routing technique but do not have the energy conservation technique [4, 7]. The energy is consumed during these processes:

- Transmission of message packets between the nodes.
- Idle mode of source.
- Irregular awakening and sleeping of hosts.

In this paper, we have discussed some of the protocols and algorithms which help in making MANET more efficient by reducing energy consumption and increasing its transmission efficiency.

3 Protocols and Algorithms for Lowering Energy Usage in MANET

In this section, we present some of the already derived power consumption protocols which help in increasing efficiency of MANET.

3.1 *Dominating-Awake-Interval Protocol [2]*

The basic idea of this protocol is to place as PS host to remain awake for such long time that the nearby users can know each other and deliver buffer packets if required. In this protocol, the power-saving host must remain awake for at least half of BI in each signal interval [2] (*BI: length of each signal interval*).

This protocol works on the following procedure:

When it is required by the host to go to power-saving mode, it breaks its time into short interludes, each of measure BI. The length of all AW, BW, and MW is constant within each beacon. To '*dominating-awake-interval protocol*,' let us assume that **AW >= ((BI/2) + BW)** [2]. The alternating signals are named as even and odd interludes, having different structure (Fig. 1) defined below:

- Odd signal interludes start by active window followed by signal windows and thereafter proceeded by an MTIM window.
- Even signal interludes are also started by the active window, but the active window is enrooted by an MTIM window followed by a signal window.

We can be optimistic that active window of two hosts will always have some overlapping by imposing the active window covering at least half of each signal interval [2].

Earlier we assumed that AW >= ((BI/2) + BW). The below theorem will prove the statement (Proof in Appendix A of [2]).

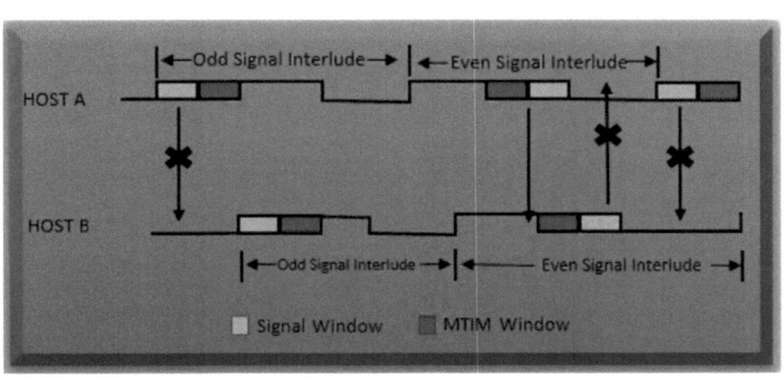

Fig. 1 Architecture of interludes in the protocol

Theorem *The dominating-awake-interval protocol guarantees that when AW >= ((BI/2) + BW), a PS host's entire beacon window always overlaps with any neighboring PS host's active window in every other beacon interval, no matter how much time their clocks drift away* [2].

This theorem assures that a host which is in power-saving mode will receive all its nearby packets in every alternative signal interlude if there is no collision. This reduces the response time for nearby devices discovery.

3.2 Min-Power Consumption and Max-Life Algorithm [6]

This algorithm is based on network's lifetime and minimum power usage. This algorithm was tested on a dynamic stipulating architecture somewhat like the shown in Fig. 2.

The energy used up by every access router during message sending is tracked down, which helps in allocating the efficient host. The proposed algorithm is used in every access router. The threshold value for each node is calculated as min $\{0.15E_i\}$¥ nodes in the transmission path [6]. Every access router collects the data from topology monitors and switches to another route if the residual energy is found to be less than threshold energy.

The following is necessary for designing an algorithm:

No Routing Delay: The goal of the selected algorithm is to choose a path from the beginning node to end node, in such a way that during the transmission of signals the remaining power in each selected node in the path remains greater than the threshold power. The nodes which do not follow this rule are discarded, and new path is selected to decrease the user power and increase the efficiency of MANET.

Optimal Power Consumption: This states that nodes receiving an equal number of packets should consume the same amount of energy. Regular checking of nodes is done at regular intervals to check that remaining power in each node is not less than needed energy level. If the remaining power is less than the threshold

Fig. 2 Spread-out network architecture

energy then the node is sent to sleep mode, and another node is selected for transmission of signals.

The main purpose of this process is to use minimum power to generate maximum efficiency of MANET. The process is mentioned as follows:

- From the starting node, break down the signal packet into same length of small packets and pick a node '*i*' where $min(E_i > Th_i)$ from all the neighboring nodes [6].
- Generate a path to the end where each node has greater power than threshold power.
- Perform the further mentioned steps of the algoithm in harmonic time gap '*t*.'
- Determine the residual energy of every node lying in the selected path by:

$$E_{\text{Res}} = E - E_{c(t)}$$

  where E is power of node at beginning level, $E_{c(t)}$ is used energy in time interval '*t*,' and E_{Res} is remaining power of node [6].
- Power used by a node in time interval '*t*' is defined by: $E_{c(t)} = Nt*a + Nr * b$ [6] where $E_{c(t)}$ is power consumed by node after time '*t*,' Nt is count of signals received by node after time '*t*' 0 <= '*a*', '*b*' => 1.
- If remaining power in each node is greater than the threshold power then transmission continues through similar node.
- If the above statement is not satisfied, find another route to reach the end node which would fulfill the algorithm's criteria.

4 Conclusion

This paper concludes issues related to energy consumption in MANET and the ways through which energy consumption can be reduced to such a level that its lifetime can be increased. The two algorithms/protocols compared here are *dominating-awake-interval protocol* and *Min-Power consumption and Max-Life algorithm.* According to the first algorithm, i.e., *dominating-awake-interval protocol*, the receiver becomes idle when there is no transmission. The second algorithm depicts that if the remaining power in each node is lower than threshold power then the node will get discarded and the neighboring node will be preferred for transmitting the packet. These protocols/algorithms will make MANET more efficient in the near future.

Acknowledgements I would like to acknowledge my fellow partner Suruchi Gupta who cooperated with me in order to complete this paper, and I would like to greet my mentors Ms. Lakshita Sejwal and Ms. Amrita Mohan who gave me an opportunity to do such work and guided in every possible way for the completion of this paper. Last but not least, I would be grateful to the institutional library which helped us in getting all the necessary resources and data for the paper completion.

References

1. B. Chen, K. Jamieson, H. Balakrishnan, R. Morris, Span: an energy-efficient coordination algorithm for topology maintenance in ad hoc wireless networks, ACM MOBICOM (2001) 85–96.
2. Yu-Chee Tseng, Chih-Shun Hsu, Ten-Yueng Hsieh "Power-saving protocols for IEEE 802.11-based multi-hop ad hoc networks" Computer Networks 43 (2003) 317–337.
3. P. Vijayakumar, T. Poongkuzhali " Efficient power aware broadcasting technique for mobile ad hoc network." Procedia Engineering 30 (2012) 782–789.
4. K. ARULANANDAM and Dr. B. PARTHASARATHY "A NEW ENERGY LEVEL EFFICIENCY ISSUES IN MANET" E-ISSN: **2076-3336.**
5. K. Srinivasa Rao, R. Sudhistna Kumar, P. Venkatesh, R.V. Sivaram Naidu, A. Ramesh/ International Journal of Engineering Research and Applications (IJERA) " Development of Energy Efficient and Reliable Congestion Control Protocol for Multicasting in Mobile Adhoc Networks compare with AODV Based on Receivers" ISSN: 2248-9622 www.ijera.com Vol. 2, Issue 2, Mar–Apr 2012, pp. 631–634.
6. Ms. S. Suganya, Dr. S. Palaniammal "An Optimized Energy Consumption for MANET" Procedia Engineering 38 (2012) 903–910.
7. Y. Gadallah and T. Kunz, "Energy consumption in ad-hoc routing protocols: Comparing DSR, AODV and TORA", Proceedings of the 1st International Conference on Ad-Hoc Networks and Wireless, Toronto, Canada, September 2002, pages 161–176.

Social Media for Enhanced e-Education at Namibian Schools

Nobert Rangarirai Jere, Tlou Boikhutso and Pardon Blessings Maoneke

Abstract The continuous technological changes have influenced service delivery in various sectors. Within the education sector, both learners and teachers use different technological devices in teaching and learning. Devices such as computers, mobile phones, tablets, and e-readers are used to access information. Technology has power to transform the teaching and learning paradigm. One should not ignore a variety of e-Learning platforms which are available including social media. This book chapter focuses on the effects of social media in education. A quantitative case study research approach is used. This was supported by document review on current emerging teaching techniques. Approximately 800 participants from high schools and tertiary institutions were engaged. Results from the engaged participants indicate that Facebook is distracting students from studying as they spend a lot of time in online. This book chapter proposes a modern teaching approach that could be used and explains the role of social media.

Keywords e-Learning · Social media · e-Education · Modern teaching

---

N.R. Jere (✉) · P.B. Maoneke
Informatics Department, Namibia University of Science and Technology,
13 Storch Street, Windhoek, Namibia
e-mail: njere@nust.na

P.B. Maoneke
e-mail: pmaoneke@nust.na

T. Boikhutso
Discipline of Public Health Medicine, School of Nursing and Public Health,
University of Kwazulu Natal, Oliver Tambo Building, University Road, Westville,
Durban, South Africa
e-mail: tlou@ukzn.ac.za

D.K. Mishra et al. (eds.), *Information and Communication Technology*,
Advances in Intelligent Systems and Computing 625,
https://doi.org/10.1007/978-981-10-5508-9_12

1 Introduction

This chapter explains modern teaching approaches and how social media could be used to improve teaching. An overview of existing social media applications, i.e., WhatsApp, Facebook, and Twitter, is given. There is no doubt that students at high school and at tertiary institutions are using social media every day. In this research, Facebook is used as the testing platform. The focus of this chapter is to show how social media could be incorporated into teaching and learning within Namibia education sector. We argue that in a situation where learners are having access to technological devices, teachers have to be very innovative and come up with new teaching approaches. Studies on WhatsApp, Internet, and Facebook usage among Namibian teenagers were conducted. However, this chapter explains findings on the use of Facebook only. It is clear that not only social media is used as an emerging approach in classrooms. There are so many other technologies available and becoming popular in many areas. The challenge is to find the most effective approach that ensures learners and teachers do benefit from the emerging technology.

The main focus is to explain how social media platforms such as Facebook could be used to improve teaching and learning. We attempt to demonstrate how Facebook can play a role in the emerging teaching and learning approaches. Our aim is to enable both teachers and learners in developing nations to fully benefit from the emerging technologies and social media. The following sections explain the current approaches that are available and used in different areas within the education sector.

1.1 Emerging Technologies in Teaching and Learning

The emergence of new technologies in the teaching and learning sector has prompted for a review of the way teaching and learning is conducted. The aim is to harness these technologies in such a way that they facilitate the conduct of teaching and learning. For instance [1], recently studied the applicability of modern teaching techniques in the educational process focusing on flipped teaching, whole brain teaching, social media, among other technologies [1]. The research by Tretinjak et al. [1] conclude that the use of different modern teaching techniques such as flipped teaching, whole brain teaching, gamification, and social media is to make learning interesting, improve the students' comprehension, enhance the student–teacher interaction, and improve critical thinking [1].

In particular, to flipped teaching technique, students can use the technology to study course material online before classes. These can include Web 2.0 technologies such as blogs, Twitter, podcasts, wikis, social network sites, virtual worlds, video sharing, and photograph sharing [2]. On the other hand, class time will focus on interactive activities, exercises, illustrating concepts, projects, and discussions under the guidance of the teacher or lecturer [1]. On the other hand, whole brain teaching technique stimulates both sides of the brain and emphasizes active learning in which students repeat core information and practice basic skills through humor and games [1]. The common modern teaching methods include flipped teaching and gamification.

1.2 Social Media as a Learning and Teaching Technology

Among the emerging learning and teaching technologies, social networking or social media is one of the technologies that are slowly gaining popularity. While some of the early research discourages the use of social networks for teaching and learning purposes, recent studies suggest it is practically possible and good practice to engage social networks in education. For instance [2], noted concerns raised in some of the early researches between 2006 and 2007 relating to privacy and anxiety in interacting with professors in social networks motivating a belief that it does not serve an academic purpose and the opinion that faculty should simply avoid "educationally appropriating" these "backstage" social spaces [2]. Nevertheless, more case studies have demonstrated the successful use of social networks in facilitating teaching and learning [3–5]. The use of social networks is motivated by the fact that while students consider teachers as the main source of information, they depend on each other to complete the learning process [3]. Accordingly, teaching and learning methods seek to incorporate the use of already popular social networks or media in facilitating learning. Among other challenges is how to include social media in teaching and learning.

Different approaches to incorporating social networks in education have been suggested. For instance [3], through their study: "Learning by Challenging: a Social Network and Privacy Based Approach" demonstrate the use of social media in education. They indicated how a classroom simulation (Intelligent Tutoring Systems (ITS)) can be achieved in a virtual learning environment using human peers, rather than software, as learning companions. This was achieved through a virtual system (Learning by Challenging strategy) whereby a learner is challenged by another learner/friend or the system to perform a task and obtain a better score. It was found that learners who learn by challenging their friend's scores perform better than those who try to beat the score predefined by the system and that participants who received recommendations from LBC made better progress in their learning [3].

2 Social Media and Facebook as a Case

Social networking applications and sites are now in place to help people connect and communicate with each other [6]. The devices used to access these social networking sites range from personal computers to mobile phones, which are well accessible to learners especially in urban schools. This means that learners have access to mobile phones during school time, thus the intervention by the ministry of education to implement the cell phone policy in public schools. It is perceived that cell phones are a distraction to learners since in the current curriculum everything they need to learn is in the textbooks and thus the introduction of the cell phone policy. In this age, there are a lot of things learners can get on social sites either to their advantage to advance in school or to their detriment.

Faliagka et al. [6] further state that teenagers are at the forefront when it comes to using social networks as a way to share their concerns and socialize without any critique or anyone trying to put them down [6]. This also result in parents coming forth with concerns that their children spend a lot of time on social networking sites and disconnecting from family and friends. It is of utmost importance that the authors chose to investigate the issue of social networking among teenagers due to the concerns raised by parents as well as stakeholders in education.

In addition, Stewart use a case study approach to demonstrate the use of Facebook in building a virtual community of readers [7]. His study reports Jessie's experiment that involves the use of a social literacy activity on Facebook in stimulating critical thinking and animated discussion. The whole process starts with the group leader allocating a book to students to read and then ask related questions. For this to be done on the Internet, Jessie created a group on Facebook named Virtual Literature Circle on which the group leader would post tasks related to the book that the students (group members) have been asked to read. For any questions, students would use the wall and use Facebook's chat system for interacting with each other as they share opinions. Jessie's idea of using Facebook as a platform for encouraging studies was aligned to the theoretical basis of social learning by Vygotsky which argues that social scholarship is important for in cognitive development because students are able to interact, share experiences, and learn from one another [7]. In addition, Bandura's social learning theory highlights the fact that people learn from one another via observation, imitation, and modeling (Learning Theories Knowledgebase 2008a in Steward, 2009). As such, Facebook is herein used as the platform to encourage the interaction and sharing of information among students.

Based on the experiment, Jessie concluded that the Facebook virtual literature circle was an excellent teaching environment for social and group work [7]. It was noted that the development of group dynamics and the application of cooperative structures encouraged equal and shared responsibility from all members, including even the quietest students [7].

While Jessie reports success using her virtual learning group on Facebook [5], argue that a group is one of the weakest Facebook communication methods. Miller and Jensen [5] noted that students may join a group to express an opinion and very little of these students participate in these groups [5]. As such, they focused on amassing friendship and making news feeds as a way of promoting library information. Introduced in fall 2006, the Facebook news feed is now a core feature that is automatically generated for all users every time they sign in [5]. Accordingly [5], report Lauren's experiment on Facebook, courtesy of news feeds, to communicate library news to students. Lauren took advantage of the fact that students keep coming back to Facebook because information is constantly changing and the fact that students read what Facebook puts in front of them not what they seek out on their own.

2.1 Research Approach

A quantitative research approach was used in this study. A questionnaire survey of seven hundred and fifty (750) learners was engaged. Two high schools in Khomas region (Windhoek) and one tertiary institution were engaged. The data was analysed using SPSS version 23 (Statistical Packages for the Social Sciences). A p value <0.05 was considered as statistically significant. Data analysis was initiated with a check of the outliers, missing data, and normality through skewness and kurtosis values that could affect relations between variables. A descriptive statistical analysis of the data (means, standard deviations, ranges, frequencies, and percentages) was initially conducted. Frequency distribution tables and bar graphs were used to summarize categorical variables. An independent samples T-test was used to compare the average number of hours per week on the two groups (High School and Tertiary). This chapter presents the results that were obtained from Facebook survey. The survey engaged Namibian schools and a tertiary institution. The aim of the study was to find out the view and usage on Facebook among the learners at high schools and at tertiary institution. By studying the implications of social networks on teenagers, it helps educators to be aware of the learners' needs. This will enable the integration aspects in the curriculum so that the engagement of learners with social networks does not detriment academic achievement.

Summary of the findings (Table 1).

Results show that tertiary participants had a statistically significantly higher number of hours spent on Facebook (20.93 ± 27.14 h) compared to high school participants (15.43 ± 18.04 h), t (659) = −3.221, p = 0.001. The values of the standard deviation were a bit too high reflecting the high margin on the hours spent on Facebook by the participants.

To ensure that the reproduction of your illustrations is of a reasonable quality, we advise against the use of shading. The contrast should be as pronounced as possible. If screenshots are necessary, please make sure that you are happy with the print quality before you send the files.

This paper gives the results on the Facebook usage among the participants and explains how the results could influence the modern teaching approaches (Fig. 1).

Table 1 Average hours spent on Facebook per week

Category	*N*	Mean
High school	338	15.4294
Tertiary	377	20.9310

The results show that just over 50% of the high school learners agree that they feel Facebook is part of their life. We interpreted this as a feeling which was supported by the hours spent on Facebook. Of course, about 30% of the tertiary students disagree that Facebook is part of their life. This could be due to the fact that tertiary students are matured. This means in as much as they use Facebook, they also have other priorities (Fig. 2).

Results indicate that learners are feeling that Facebook is taking most of their time and this is affecting the schoolwork. The other interesting finding is the variations of the high school learners' results. This could be attributed to the fact that some learners do not have access to Facebook all the time. Some either access Facebook at schools or at home. As for the tertiary learners, we could argue that a greater percentage of over 70% disagree that Facebook is affecting their schoolwork. This could also be attributed to maturity and the ability to manage the Facebook.

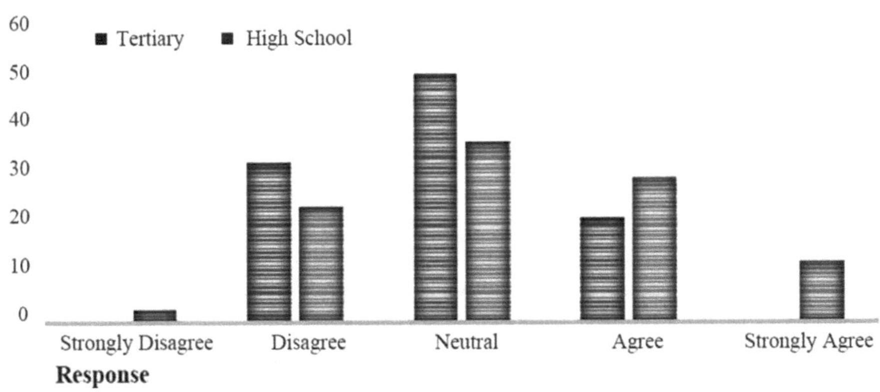

Fig. 1 Facebook as part of life

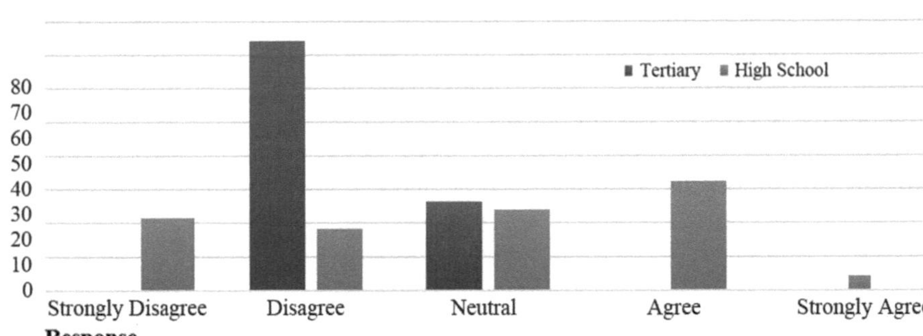

Fig. 2 Has your schoolwork been disturbed because of Facebook

This could show that possibly for high school learners, social media education could be required (Fig. 3).

There were more numbers on high school learners who sleep late using Facebook. At the same time, a significant percentage of almost 15% strongly disagree to that. This could be those who do not have access to Facebook at home. Or those whose parents ask them to sleep early. Still, a greater percentage of tertiary learners indicate that they manage Facebook well and know when to sleep. An almost 50% of the tertiary learners were neutral on this question. This may be due to the definition of sleeping late. So some learners may not be sure on what late hours are (Fig. 4).

On the participation on Facebook discussion, higher percentages of participants were neutral. This indicates that some do participate or at times they initiate discussions so in the end they were not sure whether that could be participating. Of course, a greater percentage of over 35% of high school learners agree that they participate in discussions (Fig. 5).

Much more improved percentages agree that they browse videos, photographs, and images on Facebook. This could support the idea that if Facebook is to be used

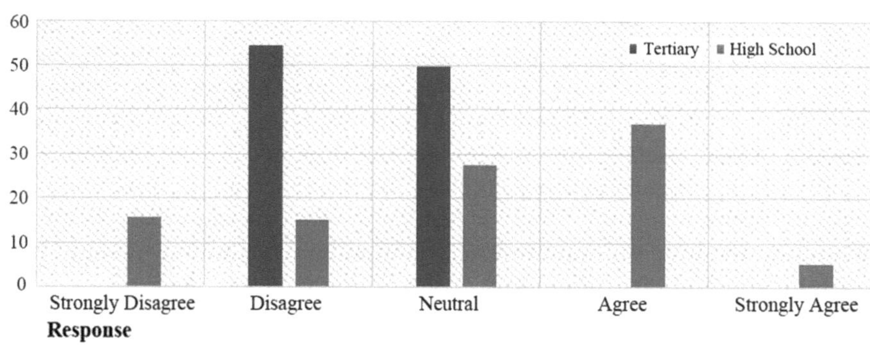

Fig. 3 Facebook affects sleeping time

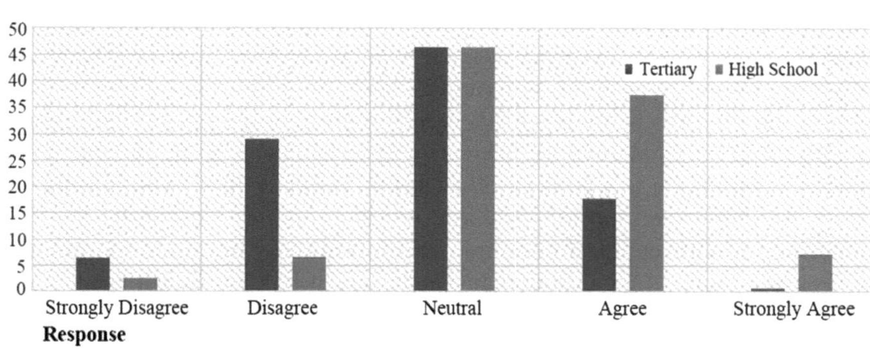

Fig. 4 Participation on facebook disucssion

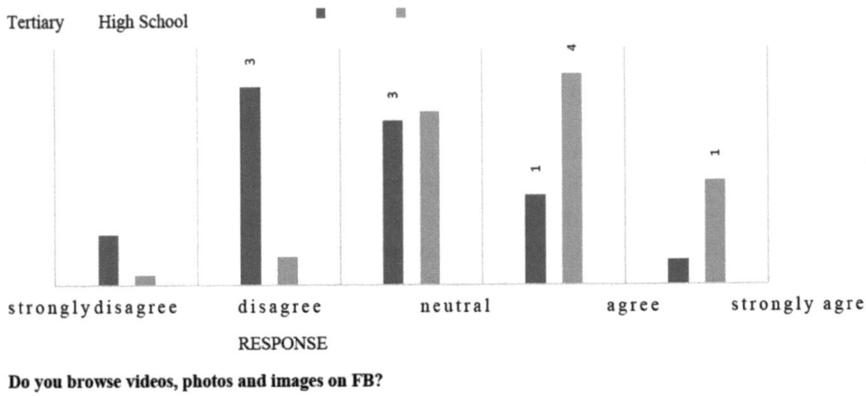

Fig. 5 Facebook usage

in teaching and learning, then it has to be interactive and be in different graphical forms.

Summary of results
Results clearly show variations in the usage and understanding of Facebook among the two groups of participants. However, the notable observation was that Facebook is popular and has been used even during class time. Learners also mentioned during the study that some lecturers and teachers are boring, so they end up using Facebook. If the teaching approach is one and does not engage students, then majority end up sleeping or using Facebook especially tertiary learners who have access to Facebook anytime. We agree that the power of social media to transform the teaching and learning is undebatable. And possibly as technology changes, new ways of accommodating social media in the classroom are required.

3 Achievements and Recommendations

This chapter has exposed the current usage and views of learners on Facebook. There is a clear evidence to support that the Namibian learners from the schools and the tertiary institution engaged are using Facebook. For a much benefit for both the learners and the teachers, we argue that Facebook could be used in the teaching and learning. This could improve the integration of technologies in the classroom and lecturer rooms. There is a need for all education stakeholders to work together in raising the curriculum and accommodate social media. We understand that this cannot happen without proper planning and involvement of the major stakeholders. For example, some of the major recommendations include the following:

- Revisiting of the curriculums to incorporate social media into teaching and learning;
- Coming up with supporting social media policies;
- Equip schools and teachers with the ICT skills;
- Educate learners and teachers on effective ways of using social media;
- Ensure that teachers are involved and not let out.

4 Conclusion

Experience from the current teaching and learning in Namibia shows that the use of different modern teaching techniques such as flipped teaching, whole brain teaching, gamification, and social media could be used to make learning interesting, improve the students' comprehension, enhance the student–teacher interaction, and improve critical thinking. In addition to this, social media, i.e., Facebook, could play a significant role in ensuring the success of emerging technologies. This chapter highlights the views on learners using Facebook. Our proposal is for social media and education experts to engage in discussions in which the incorporation of the social media in teaching and learning could be proposed and implemented. We agree that Facebook has many features that could be used in teaching and learning.

References

1. Tretinjak, M. F., Bednjanec, A., Tretinjak, M.: Application of Modern Teaching Techniques in the Educational Process. Proceedings of the MIPRO Conference, 26–30 May 2014. (2014).
2. Kadry, M. A., Fadl. A. R.: A proposed model for assessment of social networking supported learning and its influence on learner behavior. Proceedings of the International Conference on Interactive Mobile and Computer Aided Learning (IMCL). (2012).
3. Allognon, O., Touré, T., Aïmeur, E.: Learning by Challenging: a Social Network and Privacy Based Approach. Proceedings of the International Conference on Education and e-Learning Innovations. (2012).
4. Greenwood, B.: Facebook: The Next Great Vetting Tool? Information Today. 26(8). (2009).
5. Miller, S. E., Jensen, L. A.: Connecting and Communicating with students on Facebook. Computers in libraries. (2007).
6. Faliagka, E., Tsakalidis, A., Vaikousi, D.: Teenagers' use of social network websites and privacy concerns: A survey. Panhellenic Conference on Informatics, 978-0-7695-4389-5/11. DOI. (2011).10.1109/PCI.2011.17.
7. Stewart, P.: Facebook and virtual literature circle partnership. Building a community of readers. Future. 37(4). (2009).

Fifth-Level Second-Generation Wavelet-Based Image Fusion Algorithm for Visual Quality Enhancement of Digital Image Data

Meenakshi S. Arya and Pratishtha Jain

Abstract Image fusion is a technique that combines the complementary information from multiple images such that the fused image contains possibly the maximum information pertaining to both the constituent images. The objective of the proposed work is to review the existing techniques presented by various researchers for image fusion and devise a more efficient solution. The proposed technique involves the use of lifting wavelet transform technique at fifth level of decomposition and 9 different max–min–mean fusion rule combinations to achieve image fusion. The performance evaluation has been done on the basis of performance parameters considering PSNR, entropy, E-RMS, correlation coefficient, and structural similarity index. The fused images obtained were nearly identical to the ideal images since the correlation coefficient is 0.9947 which is quite close to 1. The observed values of the evaluation parameters show that the proposed scheme shows better performance as compared to others.

Keywords Image fusion · Lifting wavelet transform · SSIM · CR

1 Introduction

In recent years, image fusion has emerged as a new and a promising research area. A significant amount of research is being carried out in the field of image fusion which has offered many benefits to various applications such as medical imaging and multispectral sensor image fusing. Image fusion is the procedure of amalgamating two or more images having some common characteristics or taken from the same scene to produce a single image that has the maximum and the finest information content of the original image, without producing any inconsistencies in the image and simultaneously ignoring the noise and irrelevant features to the

---

M.S. Arya (✉) · P. Jain
Department of Computer Science Engineering,
Baba Farid College of Engineering and Technology, Bathinda, Punjab, India
e-mail: raina.arya@gmail.com

D.K. Mishra et al. (eds.), *Information and Communication Technology*,
Advances in Intelligent Systems and Computing 625,
https://doi.org/10.1007/978-981-10-5508-9_13

maximum extent. Enhanced superior image is thus provided by the fused image than the original image with low details.

As focusing on all the objects present in a scene at times is not possible in all the situations, multifocus image fusion technique is used in those situations. It fuses several images of scene captured with focus on various objects using different sensors, and then, these images are fused to form a resulting image which focuses all the objects in the scene.

The application areas benefited from image fusion include military, remote sensing, machine vision, robotic, and medical imaging. For medical image fusion, the fusion of images can often lead to additional clinical information (such as CT scan, X-ray, diagnostic sonography, PET scan, MRI.) for diagnosis of medical problems. Another important advantage is that it can reduce the storage cost by storing just the single fused image instead of multisource images.

The fusion is fundamentally classified at three levels: pixel level, feature level, and decision level. Pixel level is the lowest fusion level which produces a fused image with each pixel determined from a set of pixels in each input image. Feature level is the medium level which involves feature extraction on the input data to employ them jointly. Decision level is the highest level fusion where each image is processed individually for information extraction.

This paper uses the lifting wavelet transform to perform image fusion. The proposed approach involves 9 max–mean combinations, and the wavelet analysis is done at fifth level of decomposition using Haar transform.

2 Related Work

There are many algorithms that have been proposed for the fusion of images.

Petrovic and Xydis [1] proposed a novel approach that is based on fuse and then decompose idea for multiresolution signal-level fusion where a single fused image is produced without any data loss or distortion. This paper involves multiresolution gradient map representation which potentially reduces the distortion and the loss of contrast information that is usually observed in fused images resulted from the conventional multiresolution fusion schemes.

Wang and Lohmann [2] presented a novel wavelet-based approach for medical image fusion developed by taking into consideration the characteristics of human visual system (HVS) and the physical meaning of the wavelet coefficients. With the decomposition of the medical images by the wavelet transform, different diffusion schemes are proposed for combining the coefficients.

Swathi et al. [3] presented a multimodal image fusion algorithm for medical images based on lifting wavelet transform and neurofuzzy. This paper aims at enrichment of the image content by the amalgamation of the images.

Joseph and Barhatte [4] analyzed image fusion based on three transforms, namely discrete wavelet transform, fast curvelet transform, and discrete fast curvelet transform. The combination of DWT and FCT techniques leads to a new technique named as discrete fast curvelet transform providing better fusion results.

Manna et al. [5] proposed an enhanced block-based feature level image fusion technique using a combination of lifting wavelet transform and neural network to fuse a pair of images. The hybrid algorithm is referred as BFLN method which combines the concepts of both neural network and lifting wavelet transform for fusion of medical images.

Sun et al. [6] proposed an image fusion algorithm based on wavelet transform and the second-generation curvelet transform. This paper involves pixel level fusion. Three fusion algorithms are considered: discrete wavelet transform (DWT), the second-generation curvelet transform that is fast curvelet transform (FCT), and discrete fast curvelet transform (DFCT). This paper gives an idea about the selection principles for low- and high-frequency coefficients according to variable frequency domain after wavelet and the second-generation curvelet transforms.

3 Image Fusion Techniques

There are mainly two types of image fusion techniques. Spatial domain fusion: The spatial domain techniques fuse source images using local spatial features, such as gradient, spatial frequency, and local standard derivation. It further includes the following methods:

- Intensity-hue-saturation (IHS) transform-based fusion;
- Principal component analysis (PCA)-based fusion;
- High-pass filtering method.

Transform domain fusion: For the transform domain methods, source images are projected onto localized bases which are usually designed to represent the sharpness and edges of an image. It is the most well-known type of image fusion technique because of its effortlessness and its capacity to protect the time and recurrence subtle elements of the pictures to be melded. It includes the following kinds of wavelet methods:

- Discrete wavelet transforms (DWTs);
- Lifting wavelet transforms (LWTs);
- Discrete cosine transforms (DCTs).

This paper aims to study and use the second-generation wavelet transform for the image fusion. Wavelet transforms are multiresolution image decomposition schemes that provide a variety of channels representing the image feature using different frequency subbands. The decomposition in the proposed work is carried out using Haar wavelet. The Haar wavelet obtains the lower-frequency components by taking the average of the two pixel values and obtains the higher-frequency components by taking half of the difference of the two pixels. In other words, the

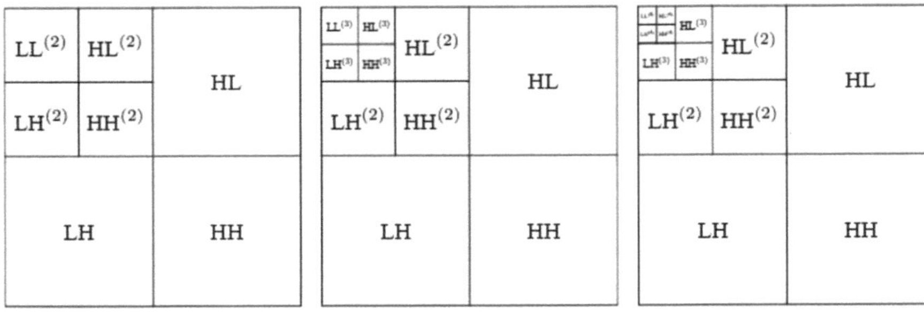

Fig. 1 Decomposition at second, third, and fourth levels

source image is decomposed into rows and columns by low-pass (L) and high-pass (H) filtering and then downsampled to create the coefficient matrices LL and LH. The coefficient matrices LL and LH are low-pass and high-pass filtered in vertical direction and downsampled further to create subbands LL1, LH1, $HL_{1,}$ and HH_1. This type of decomposition can be done at different levels. After the first level of decomposition, four bands, viz LL, LH, HL, and HH, are achieved. Out of these, the LL band, the lower-frequency band, contains most of the information, while the other three higher-frequency bands are having only less information like edge details of the image. The further level of decomposition occurs in the current obtained LL band (Fig. 1).

4 Lifting Wavelet Transform

Sweldens [7–10] proposed lifting wavelet transform using the lifting scheme in time domain. LWT promises to offer a spatial domain interpretation of the transform. Unlike other traditional transforms, it does not demand much memory space, avoids large complex computation, and is able to yield integer-to-integer wavelet transform. These advantages make it appropriate to be applied to image fusion to obtain better fusion effect and improve fusion speed. The LWT comprises of three stages (Fig. 2).

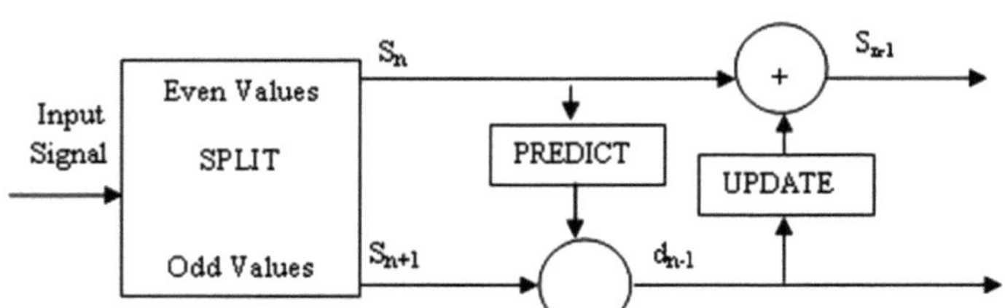

Fig. 2 Lifting wavelet transform

Split: The original signal is divided into two disjoint subsets. Although any disjoint split is possible, the original data set is split into two smaller subsets that are even indexed samples (S_n), odd indexed samples (S_{n+1}).

Predict: The wavelet coefficients dn_{-1} are generated as an error in predicting (S_{n+1}) from (S_n) based on the correlation present in the original data using the prediction operator P. The prediction of the odd subset from the even subset illustrates the lifting of the high-pass subband with the low-pass subband. The difference between the approximation and the actual data replaces the odd elements of the data set. The even elements are kept as their actual form and subsequently become the input for the next step in the transform.

$$(d_{n-1}) = (S_{n+1}) - \mathrm{P}(S_n) \tag{1}$$

Update: (S_n) and (d_{n-1}) are combined to obtain scaling coefficients (S_{n-1}) that represent a coarse approximation to the original signal. This is accomplished by applying an update operator U to the wavelet coefficients and adding the result to (S_n): In this stage, the global properties of the original set in the subsets are maintained and are performed as follows:

$$(S_{n-1}) = (S_n) + \mathrm{U}(d_{n-1}) \tag{2}$$

This step lifts the low-pass subband with the high-pass subband to maintain some properties of the input stream of the low-pass subband. The original value of the odd elements has been overwritten by the difference between the odd element and its even "predictor." Hence, the update phase operates on the differences stored in the odd elements for calculating an average [4].

5 Proposed Technique

The wide availability of multi sensor data in numerous fields such as remote sensing, medical imaging, and machine vision has brought the multisensor data fusion into attention. Various techniques have been proposed till date to address this important research issue. Lifting wavelet theory is a new approach for constructing wavelet, which is also called as second-generation wavelet. The main aim of LWT is to transform a coarser signal (S_{n-1}) into a detailed signal (d_{n-1}). Hence, LWT can be considered as an efficient method for calculating the filtering operations.

The proposed LWT technique is applied to image using fifth level of decomposition and with nine different combination such as min–min, max–min, max–mean using Haar wavelet. The original image is decomposed into subbands, known as subband decomposition level. After the principal level of decomposition, there are 4 subgroups: LL1, LH1, HL1, and HH1. For each progressive level of decay, the LL subband of the past level is utilized as the input. The decomposition of the subsequent approximation bands is done till fifth level, and the same has been depicted in the figure above (till fourth level).

Most of the information from the source images is kept in the low-frequency band as it is a smoothed and subsampled version of original input image. Higher value of wavelet coefficients carries salient information about images such as corners and edges and hence maximum selection rule, gradient, and contrast.

Max fusion rule: The higher value wavelet coefficients contain most important information about images such as edges and corners. Hence, all the smaller magnitude complex wavelet coefficients are replaced by higher magnitude complex wavelet coefficients in maximum selection rule. For every corresponding pixel in input images, the pixel with the maximum intensity is chosen and used as the resultant pixel of the fused image.

Mean fusion rule: This method is a simple one where fusion is achieved by calculating the average of corresponding pixel in each input image. Low-frequency components are fused by averaging method.

Min fusion rule: In this rule, higher magnitude complex wavelet coefficients are replaced by means of smaller magnitude complex wavelet coefficients. For every corresponding pixel in input images, the pixel with the smallest intensity is chosen and used as the resultant pixel of the fused image. The following figure depicts the proposed algorithm (Fig. 3).

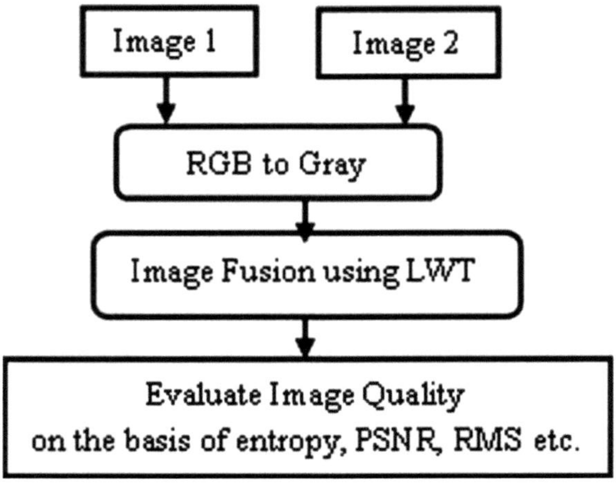

Fig. 3 Flowchart of proposed methodology

6 Simulation Analysis

The proposed approach has been simulated in MATLAB with 9 Max–Min–Mean combinations. Based on the heuristic assessment, from all the 9 combinations, the best one is selected according to the performance parameters. The results of the proposed scheme have been compared with the existing techniques reviewed in the literature on the basis of visualization and qualitative aspect.

The visual evaluation considers the natural appearance, brilliance contrast, presence of complementary features, enhancement of common features, etc.

The quantitative criterion includes five parameters, namely entropy, PSNR, correlation coefficient (CC), structural similarity index (SSIM), and RMS error.

Table 1 is indicative of the fact that as compared to the other techniques, LWT technique performs better using max–mean combination.
Table 2 is indicative of the fact that the proposed algorithm at fifth level of LWT performs better than other image fusion techniques (Figs. 4, 5, 6, 7, 8 and 9).

Table 1 Image quality metrics readings for LWT for various combinations

LWT combination	PSNR	MSE	SSIM	Entropy	CC	RMS
mean–max	29.5150	4.5562	0.9527	7.0589	0.9875	2.1345
mean–mean	33.3915	1.1566	0.9760	6.9196	0.9947	1.0754
mean–min	31.9853	1.9812	0.9506	6.9498	0.9929	1.4076
min–max	29.3939	4.6637	0.9522	7.0576	0.9872	2.1596
min–mean	33.1838	1.2623	0.9760	6.9526	0.9945	1.1235
min–min	31.8365	2.0851	0.9507	6.9496	0.9929	1.4440
max–max	29.5649	4.5407	0.9531	7.0491	0.9877	2.1309
max–mean	33.4194	2.1428	0.9760	6.9374	0.9947	1.0690
max–min	32.0028	1.1428	0.9505	6.9372	0.9929	1.4033

Table 2 Evaluation of the parameters of fused image

Image fusion techniques	Entropy	CC	ERMS
DWT	3.5548	0.9891	2.1564
FCT	3.6455	0.9928	2.0213
DFCT	3.7721	0.9937	1.8678
Proposed LWT	6.9374	0.9947	1.0690

Fig. 4 Input image 1

Fig. 5 Input image 2

Fig. 6 Fused image

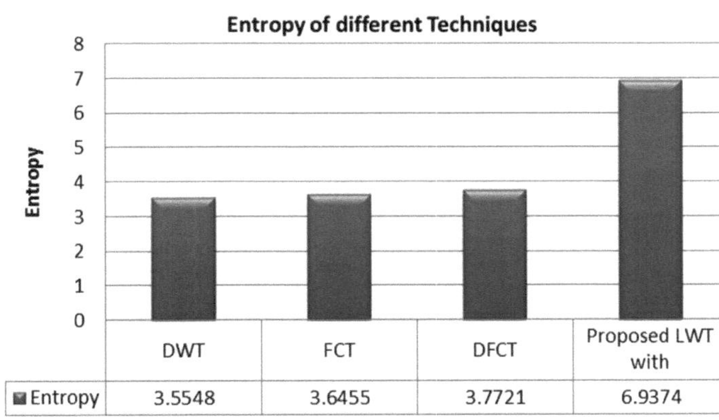

Fig. 7 Entropy comparison of different techniques with the proposed LWT technique

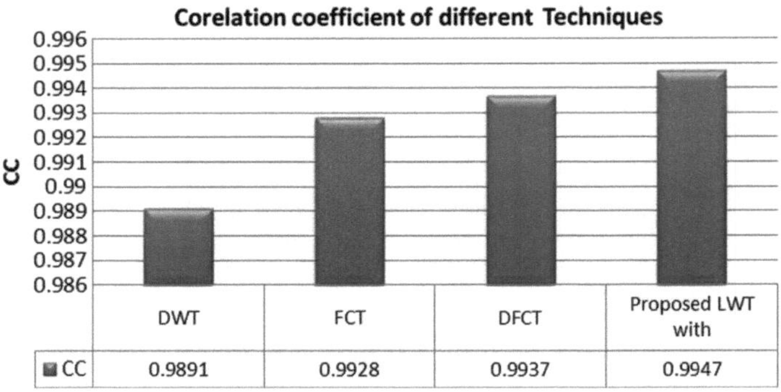

Fig. 8 Correlation coefficient comparison of different techniques with the proposed LWT technique

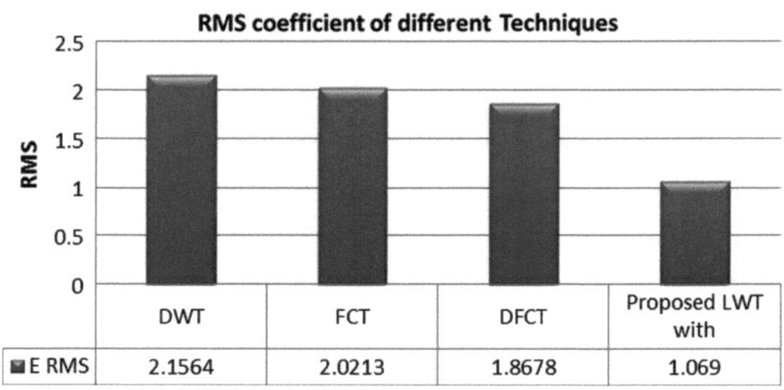

Fig. 9 RMS comparison of different techniques with the proposed LWT technique

7 Conclusion

In this proposed algorithm, lifting wavelet transform has been used to enhance the low-resolution images. Qualitative, quantitative, and visual performance analysis of the images had been done. The lifting wavelet transform has been applied to the input images by performing fifth level of decomposition and applying 9 combinations of Max–min–mean fusion rules. The proposed method for image fusion techniques is to provide good qualitative, quantitative, and visual results than the conventional as well as state-of-the-art methods.

In future, this proposed method may be combined with some other wavelet transform technique and by using some different wavelet functions. Also, the proposed method can be extended to be used in medical image data.

References

1. Petrovic, V.S., Xydis, C.S.: Gradient-based multiresolution image fusion. IEEE Trans. on Image Processing, Vol. 13, No. 2, 228–237 (2004)
2. Wang, Y., Lohmann, B.: Multisensor image fusion: concept, method and appli- cations. Tech. Rep., Institute of Automatic Technology, University of Bremen, Bremen, Germany, 2000
3. Swathi, A., Krishneswari, K., Jeevitha, N.: Multimodal Medical Image Fusion Based on Lifting Wavelet Transform and Neuro Fuzzy. African Journal of Basic and Applied Sciences, Vol. 7, No. 3, 176–180 (2015)
4. Joseph, J., Barhatte, A.: Medical Image Fusion Based on Wavelet Transform and Fast Curvelet Transform. International Journal of Engineering Development and Research, Vol. 2, No. 1, 284–288 (2014)
5. Manna, C., Rani, S., VijayaKumar, V.: Block based Image fusion technique using Lifting wavelet transform and Neural networks on Medical Images International Journal of Computer Science and Information Technology and Security, Vol. 2, No. 5, 980–986, (2012)
6. Sun, Y., Zhao C., Jiang, L.: A New Image Fusion Algorithm Based on Wavelet Transform and the Second Generation Curvelet Transform. Proceedings of IEEE International Conference on Image Analysis and Signal Processing, 438–441, (2010)
7. Sweldens, W.: The lifting scheme: A construction of second generation wavelets. Technical Report, University of South Carolina, (1996)
8. Sweldens, W.: The lifting scheme: A custom-design construction of biorthogonal wavelets. Technical Report, University of South Carolina, (1994)
9. Sweldens, W.: The lifting scheme: A construction of second generation wavelets. Technical Report, University of South Carolina, (1998)
10. Sweldens, W.: The Wavelets and the Lifting Scheme: A 5 Minute Tour. Technical Report, University of South Carolina, (1995)
11. Fattal, R.: A Brief Introduction To First And Second Generation Wavelets. Hebrew University of Jerusalem, Israel,(2009)
12. Kaur, P., Lalit, G.: Comparative Analysis of DCT, DWT and LWT for Image Compression. International Journal of Innovative Technology and Exploring Engineering, Vol. 1, No. 3, 90–94, (2012)
13. Mitianoudis, N., Stathaki, T.: Pixel-based and Region-based Image Fusion schemes using ICA bases,Information Fusion. Elsevier Journal of Image Fusion, Vol. 8, No. 2, 131–142, (2007).

Advanced Teaching Materials of Inverted Pendulum System by the PLC Sequence Method

Junichi Sugaya, Kuniaki Yajima and Yuta Iishiba

Abstract Manufacture for beginners of the teaching materials by student's proposals was done for fifth-year students' practical training classes, in Dept. of Electronic Control Engineering, from 2010 to 2013. So we propose this idea developing some basic teaching materials of sequence control for students of our college and international exchange students. The purpose of manufacturing the materials is to make students understand the foundations of sequence circuits, like logic circuit and self-hold circuit by using these teaching materials. Furthermore, as an advanced step-up course of fundamental teaching materials, we developed new teaching materials whose subject is inverted pendulum system to make students understand sequence control by the motion of pendulum visually. At present, we are testing the piecewise linear control for the inverted pendulum system, and we succeed in keeping it standing.

Keywords Fundamental teaching materials · PLC sequence method · Advanced teaching materials · Inverted pendulum system · Piecewise linear control

J. Sugaya (✉)
ICT Development Center, Hirose Campus, National Institute of Technology, Sendai College, Sendai, Miyagi, Japan
e-mail: sugaya@sendai-nct.ac.jp

K. Yajima
Department of Information Networks, Hirose Campus, National Institute of Technology, Sendai College, Sendai, Miyagi, Japan
e-mail: yajima@sendai-nct.ac.jp

Y. Iishiba
Graduate School Student of Informatics and Engineering, University of Electro-Communications, Chofu, Tokyo, Japan
e-mail: i1630006@edu.cc.uec.ac.jp

D.K. Mishra et al. (eds.), *Information and Communication Technology*, Advances in Intelligent Systems and Computing 625,
https://doi.org/10.1007/978-981-10-5508-9_14

1 Introduction

In modern technology society, we think that sequence control is a fundamental technology which is widely used and is an important one which engineers should study.

By the way, experiment and training courses for lower-class students in a National College of Technology are important because the courses raise their consciousness for "specialty," and they also improve their training motivation toward specialized experiment in upper classes [1], etc. In our college, "Creative Engineering" course [2] started in the first year student in the new Dept. of Information System Engineering in 2010. Sequence control in Creative Engineering is selected as one topic in the old Dept. of Electronic Control Engineering of our college.

In one of the fifth-year students' practical training of "Topics in Electronic Control" classes, students manufacture experiment materials of sequence control and also develop teaching materials for the class, based on the five donated FA learning kits. In the process, students' creativity is harnessed, and the teaching materials are further developed for first-year students of Creative Engineering class based on proposals of students. Sequence control engineering for beginners is tried by some National Colleges of Technology [3].

So first, we manufacture a new teaching material similar to the donated FA learning kit and report about it. This material is for beginners with which learners can make the easy sequence circuits and receive the basic training for sequence control.

On the other hand, we have been searching for the sequence control engineering for graduation research works and application themes for overseas trainees since 2010 [4]. So we think of producing the inverted pendulum system which is well known in the control field as an application theme. Because inverted pendulum system is unstable, it is possible to check the pendulum visually. Inverted pendulums utilized the apparatus in classes of automatic control in another college of technology [5]. And inverted pendulum model is applied to the human body to investigate the estimation method of the human postural control ability [6]. But the teaching materials of positioning control by sequence controller were only known on enterprise training of OMRON corp., and it was made for experts [7].

In this research, advanced teaching materials are made by using basic teaching materials which are applied sequence control engineering finally and are reported. It is expected to be used in classes of the advanced course or in classes for overseas trainees. It also seems effective for delivering lessons and for demonstrations.

2 Fundamental Experiment

Sequence control is taught in Creative Engineering class for first-year students, in Topics in Electronic Control for fifth-year students, and in lessons for foreign trainees, and these classes are experiment and training courses using the basic FA learning kit of OMRON Corp.

2.1 Basic Sequence Control Experiment

A package box of FA learning kit contains sensors, relays, switches, and so on, with which students can implement experiments and practical trainings with fundamental contents of sequence control. Students can assemble a self-hold circuit or other circuits only by setting and wiring each sequence element onto the conveyor equipment in the FA learning kit. They can also set a timer or a counter on it, and conduct experiments with photoelectronic sensors. We decided to perform this practical training in "Topics in Electronic Control" class for fifth-year students. Students will develop teaching materials within thirteen two-hour classes in the second semester.

2.2 Proposal-Type Teaching Materials in Practical Training

1. Basic experiment: Students implement basic experiments of "Creative Engineering" for first-year students in the first three classes.
2. Proposal of theme: Students propose creative applied themes based on basic knowledge of sequence control.
3. Teacher's advice: Teachers in charge scrutinize the contents of students' proposals.
4. Design of sequence program: Students put the new theme into exact writing and design the ladder diagram based on it.
5. Choice of parts: Students choose necessary parts on the Internet.
6. Manufacture: Students carry out actual manufacture on a circuit stationary plate.
7. Trial experiment: Students implement the trial experiment by the FA learning kit.
8. Presentation: Students present about the manufactured teaching materials and report it.

The cleaning robot that students manufactured in 2012 is given by these procedure 1–8, as an example, which is shown Fig. 1.

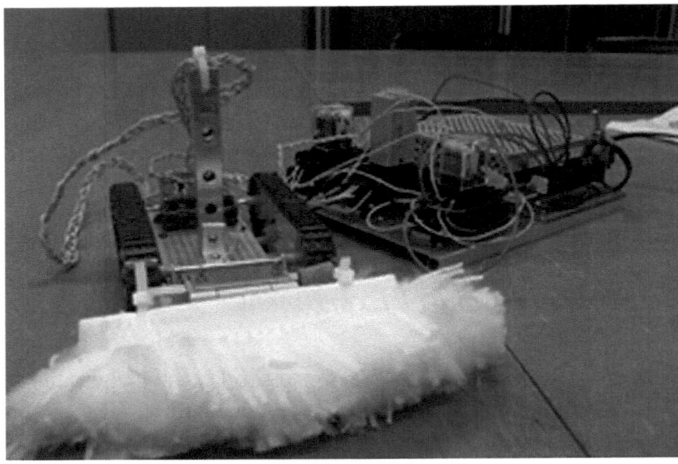

Fig. 1 Cleaning robot manufactured in 2010

However, the lesson is performed for one class unit in many cases, but we have only five sets of teaching materials, which are too few for the number of students. The students may be unable to study the contents correctly because the teaching materials are few. There is also another reason that the price of one FA learning kit of teaching materials is as expensive as about one million yen.

Therefore, we try to improve learning efficiency by making teaching materials by ourselves at low costs and in large numbers and having students use them.

3 Teaching Materials for Beginners

The electric power unit and the breaker which is used only at the time of power activation were installed in the topmost part so that an experimenter could not touch the fundamental parts. The switch and the display light were installed in the upper part, and each terminal box was grouped in the center so that an experimenter could wire easily. The DIN rail was installed in the lower part so that it would be easier to wire and to remove sockets.

3.1 Manufacture of First Experimental Model

The experimental model was manufactured. In this experimental model, parts were installed on a veneer board of depth 350 mm, width 450 mm, and height 5 mm.

Table 1 Result of questionnaire

Question		Where do you feel difficult to use?
Answer	1	Lack of electricity terminal
	2	Lack of switch and lamp
	3	Lack of the work space
	4	Feel heavily
	5	Lamp and switch is too small
	6	Wiring became complex

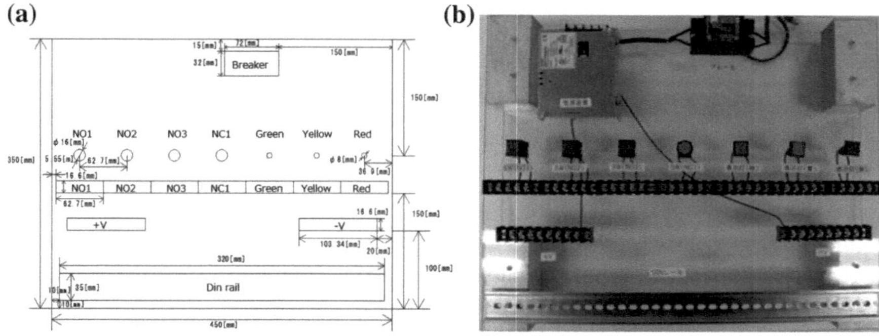

Fig. 2 **a** Design of teaching material, **b** top view of teaching material

Moreover, the questionnaire was carried out in order to add improvement. After, we actually had students make the push button circuit and a self-hold circuit in this questionnaire, and we investigated how easy it is to carry out the work by it.

As a result of having students answer to the question whether or not they have found any part hard to use, the five students who actually used these teaching materials answered like shown Table 1. We remade the improved teaching material based on these opinions that are answers of Table 1. New material is shown in next section.

3.2 *The Layout Manufacture of Improved Teaching Materials*

Here, the teaching materials which improved teaching materials for beginners are proposed and manufactured. In this model, parts were installed on an acrylic board of depth 350 mm, width 450 mm, and height 5 mm.

This model bases on the opinions that are answers of Table 1 in the questionnaire about the experimental model. The design of the actual equipment is shown in Fig. 2a. In order to avoid some accidents, such as wiring on the back being pulled and separated, it was tucked between two acrylic boards.

The 19 parts in 10 kinds were used for these teaching materials. The expense was 15,374 yen in total since we used the acrylic boards. Compared with basic FA learning kits, the cost was sharply held down for these teaching materials. The top view photograph of the completed teaching material is shown in Fig. 2b.

We provide the completed teaching material to student from the partner university and ask the feeling. As a result of having student answer to the question is no problem to use as teaching martial.

Also we made the new circuit manual to use as a text of the teaching materials instead of FA kit manual used as a text in class. It is correspondent to the improved sequence teaching materials and it isn't writing method of make the circuit.

4 Advanced Teaching Material of Inverted Pendulum System

An application system of inverted pendulum is manufactured by sequence control engineering, and an experiment on manufactures is implemented by using these basic teaching materials in this research. The layout drawing of inverted pendulum system composed in this research is shown in Fig. 3.

The objective of this inverted pendulum system is to keep pendulum standing upright. In order to achieve this objective, we divided the control into two parts: the one to swing up the pendulum [8] and the one to keep it standing upright. The control range of the former is within an angle area of ±15° based on pendulum standing, and the swing-up control part is within an angle area of excluding ±15°.

4.1 Experiment by Equipment

In this equipment, we will control by dividing the construction into two control parts: the standing upright control one and the swing-up control one. The control target of the main system is to swing up the pendulum in a suspension state and to

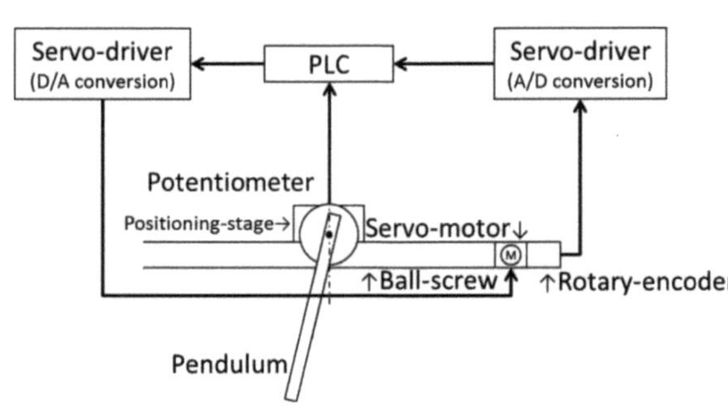

Fig. 3 Inverted pendulum system by PLC

keep it standing upright. A boundary in a standing upright control area (this area is used for the judgment of angle condition of the standing upright state) is compared with a change in standing upright keeping time when a control angle area is changed from −12° to −5° and from 5° to 12°.

The respective control areas are distinguished by the angle of the pendulum. Piecewise linear control is applied to PLC sequence control. The angle of the standing upright vertical state is set at 0°. For example, the standing upright control area is the range of ±10°, by all except for those angles it was made a control area for the pendulum is to be swung up.

We checked the longest standing upright time of the pendulum at angles of ±5°, ±7°, ±10°, and ±12°, respectively, with servomotor's revolution speed fixed at 2000 rpm. Table 2 shows the standing upright time of each control area. Under this condition, the longest standing time is 8.7 s at the angle of ±10°. Considering standing upright time, it can be said that the control result is good at the angle of ±10°.

4.2 Flowchart

Here, we explain it by using a flowchart.

A flowchart of swing-up control algorithm is shown in Fig. 4. The direction of rotation of the servomotor is decided from each range of the angle in this swing-up control. When it is in the range of the swing-up control (from −180° to −10°, and from 10° to 180°), we can distinguish whether the pendulum is up or down by comparing the current angle position (θ_N) with the previous angle position (θ_{N-1}).

4.3 Result

Figure 5 shows changes of the pendulum angle (degree) (the blue line) and servomotor revolutions (rpm) (the red line) which are indicated by *Y*-axis in a standing upright control area at ±10°. *X*-axis indicates time (s). Further, the angle of the pendulum is ±180° when it is hung down. The figure shows that the pendulum is swing up gradually and kept standing upright. The graph shows the change between 34 and 37 s so that a change in the servomotor might be easy to judge [9]. When the

Table 2 Comparison in standing upright time

Standing control range	Standing between times (s)	Standing time (s)
−5° to 5°	12.3 ~ 13.5	1.2
−7° to 7°	16.6 ~ 21.8	5.2
−10° to 10°	34.7 ~ 43.4	8.7
−12° to 12°	20.2 ~ 20.7	0.5

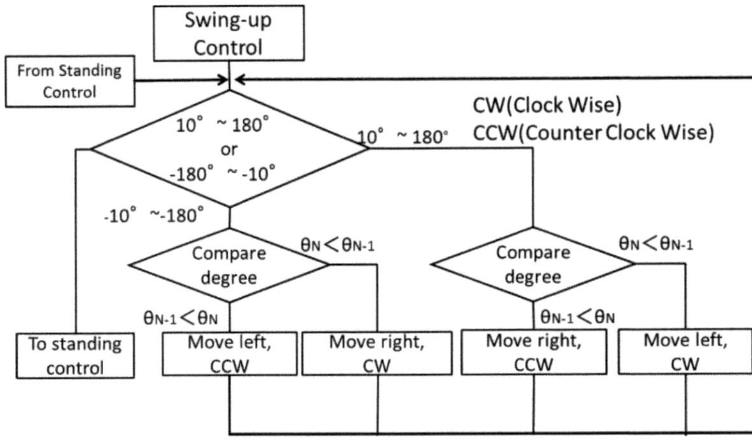

Fig. 4 Flowchart of swing-up control

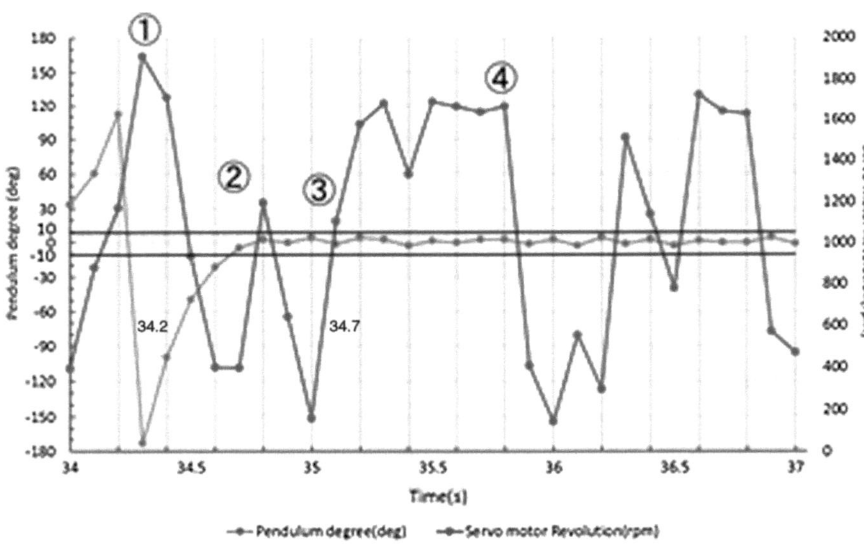

Fig. 5 Response of pendulum by sequence controller

pendulum is about to finish shaking, the servomotor revolutions go up about 1500 rpm and swing a pendulum up at around 34.2 s (Fig. 5 ①) and 34.7s (Fig. 5 ②).

5 Conclusion

We created the teaching materials for beginners such as a switch and a display light. As a next subject, we could introduce proposal-type teaching materials in practical training with using a motor, PLC programming, and so on. That is in order to manufacture the improved teaching materials which can make students study more deeply by the fifth students.

On the other hand, we manufactured an application system of inverted pendulum as developed course on the advanced teaching materials by sequence control engineering. And we did experiment by equipment of the inverted pendulum system. The control is divided into two parts, the swing-up and the standing upright control part. In standing upright control, we applied piecewise linear control to the inverted pendulum system with PLC sequence controller. As a result, the standing upright state is kept longest when control area is the range of $\pm 10°$.

As a future's problem, it is necessary to consider a detail simulation of an inverted pendulum system with swing-up control, PLC control methods of extension in standing upright time and last teaching materialization, etc.

Acknowledgements This work was supported by JSPS KAKENHI Grant Number 15K00938.

References

1. Junichi S., Kuniaki Y., Keimei K.: Questionnaire Surveys of Students in Department of Electrical Control Engineering. Research Report of SNCT, 36, 53–58 (2006)
2. Kimio S., Yasushi K., Yumi O., Yasuhisa N., Kouichi H., Keimei K., Toshiki A.: Activity for Series Educational Systems Improvement in Sendai Radio National College of Technology. Journal of Education in the College of Technology, 30, pp. 735–740 (2007)
3. Shoji E., Hidetaka K., Masatoshi T.: Lecture of the Sequence Control Engineering Utilizing Self-Developed Experimental Device for Each Student. Journal of Education the College of Technology, 28, 13–17 (2005)
4. Yuuta I., Junichi S., Kuniaki Y.: The Development of Fundamental Teaching Materials and the Inverted Pendulum System as Advanced Sequence Control. Proc. of IEEE, 7th ICITEE 2015 Conference, CS, 156–161 (2015)
5. Toshihiro H., Yuuichiroh M.: Educational Apparatus of Inverted Pendulum with Inertia Rotor. Journal of Education the College of Technology, 47, 415–420 (2013)
6. Hiroya G., Koichi S.: Control Characteristics Evaluation of Standing Posture during Swing Movement Using PID Control Law of One-link Inverted Pendulum Model. SICE, 49–12, 1113–1120 (2013)
7. Positioning control by controller in OMRON FA's training service, http://www.fa.omron.co.jp/seminar/area/Detail/207
8. Nobuaki F., Kazuo K., Ryuichi M.: Swinging-Up Motion and Upside-Down Standing of Pendulum with Inertia Rotor. Transactions of the JSME, Series C, 68-667, 810–816 (2002)
9 Junichi S., Yuuta I. and Kuniaki Y., "Advanced Teaching Materials for the Inverted Pendulum System by PLC Sequence Control", ICBIR2016 Conference Proceedings, AE, pp 41–46 (2016)

Intelligent Locker System

Robin Tommy, V. Vinesh Raja, Arun Jose and Hima Jose

Abstract The Smart locker is a security device that helps you to protect your locker. Mounted to a drawer in a table or to the door, the Smart locker provides the current status of the system to your smartphone. The Smart locker is capable of opening and closing in scheduled interval of time and storing the history of the locker activity in smartphone. The user receives alert notification when the locker is forcefully tried to open. The locker can be opened/closed via Internet. It has Bluetooth sense feature that opens the locker automatically when the user's phone is near the locker. The locker is enabled with motion detector and also sensors to understand burglary activity. The system is intelligent enough to identify the users. If any unauthorized access is detected, the system will throw indication/message to the admin.

Keywords Intelligent · Smart · Locker · IoT · Bluetooth · Android · Image processing · GSM

1 Introduction

The practice of locking the house manually with a mechanical lock and key is the protection that one can offer to the house. The control of anyone entering the house lies with only those persons having the physical key. But when the access needs to be granted to new persons immediately or if there comes the need to monitor your house remotely, the Smart locker serves the purpose. Smart locker is capable of

R. Tommy (✉) · V. Vinesh Raja · A. Jose · H. Jose
ILP Innovation Labs, Tata Consultancy Services, Thiruvananthapuram, India
e-mail: robin.tommy@tcs.com

A. Jose
e-mail: a.jos@tcs.com

H. Jose
e-mail: hima.jose@tcs.com

D.K. Mishra et al. (eds.), *Information and Communication Technology*,
Advances in Intelligent Systems and Computing 625,
https://doi.org/10.1007/978-981-10-5508-9_15

allowing any number of persons authorized with the digital key to access the house. Also it features remote monitoring of the status of the locker.

Currently, the bank locker systems are being robbed in multiple places and are due to less availability of intelligent locker systems. The current system developed is able to detect motion, temperature increase and track activity. The activity will be monitored for the user behavior with the locker system. If any spurious activity like breaking the locker or unauthorized access will be intimated and alarms systems will be sent off. The entire system is based on machine learning, IoT, and mobility solutions working on a collaborative network. The data is pushed to the cloud infrastructure, and then the classification algorithms are applied to create clusters (K-Means clustering). The clusters identified for each user and category.

2 Literature Survey

The literature survey for banking lockers is less. The Bluetooth-based locker systems [1, 2] as mentioned in 'Bluetooth Based Home Automation and Security System using ARM9' have been in the market for some time. The system enables opening/closing of doors using Bluetooth connectivity. Our system is intelligent enough to even identify the user and also spurious activity detection. Other studies in this field are based on biometric authentication which has been in the industry for a long time [3–6]. Our system performs well than the biometric system as the locker is not integrated with any such device and user phone biometric, pattern and password. Thus, the system is more secure and user-friendly. The gesture and other mechanisms of authentication [7] are discussed in the 'A Review Paper on Design of Highly Secured Automatic Teller Machine System by using Aadhaar card and Fingerprint' [8]. Our systems handle gesture as patterns, and also the system is intelligent enough to identify the users and produce analytics of the usage.

3 Locker Development

Smart locker was started to be built with network features such as Bluetooth and Internet to facilitate the primary features of Bluetooth unlocking and intruder alert notification. A prototype was built with an electromagnet attached to a drawer in a table. A webcam was attached to the locker, and person detection was implemented. An Android application was developed to provide the locker user with access to all the facilities of the locker.

3.1 Hardware

The system is integrated with temperature, PIR, camera, GPS sensors integrated to the Raspberry Pi device. The device is Bluetooth and Internet enabled (Wi-Fi).

Fig. 1 Smart locker

Any motion would be captured by the PIR sensor. The camera observes the face and detects is anyone who is in front of the locker. The locker is attached with a GPS unit to track if in case of theft. Figure 1 shows the physical locker device.

3.2 Software

The entire data from the sensors is uploaded to the cloud. The intelligent algorithm scans the data and makes decision regarding the current status. The statistics evaluation is carried out to understand the user behavior and usage of the locker. The android app for each user can be used to open the locker based on the fingerprint and password locks. The scheduling of the locker can also be done. Any unauthorized access can be monitored with analytics of usage.

4 System Functioning

The Smart locker consists of two systems—the physical locker and the Android application. The locker is fitted with accelerometer, relay, electromagnet, webcam, and a microcontroller. Accelerometer gives position feedback to the microcontroller.

If the locker is tried to be forcefully opened, then the change in the position of the locker is given as input signal to the microcontroller. Relay operates the electromagnet based on the signal it receives from the microcontroller. Webcam is used to capture the image of the intruder as soon as the unauthorized opening is detected. Microcontroller is attached with a Bluetooth and Wi-Fi module for communicating with the smartphone application. There should be a working Internet connection to the locker so that push notification can be sent to the user when an intruder is detected. The Android application is built with features such as Lock/Unlock button, Bluetooth access, Internet access, locker Status display, History Statistics, Schedule locker, and GCM Push Notification viewing. The home page of the Android application is given below (Fig. 2).

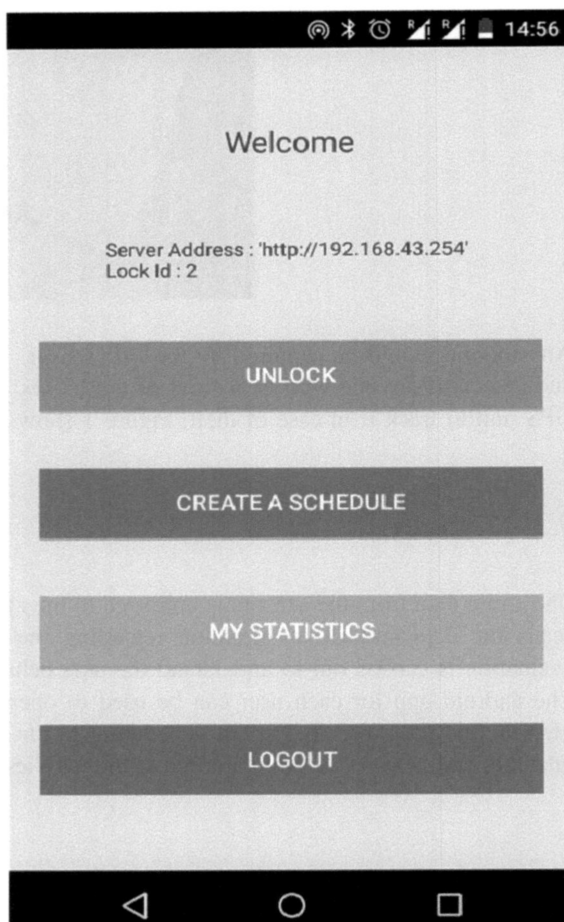

Fig. 2 Screenshot of Android application controlling the Smart locker

The Lock/Unlock button facilitates changing the locker status. Bluetooth access granted to the authorized phone automates the unlocking of locker in a close range. Internet access in the phone enables receiving push notifications and controlling the status of locker. Android application also features showing the statistics of locker status over a period of past 1 month.

5 Notable Aspects

5.1 Features of the Locker

A lock needs to be mechanically strong enough to withstand the force an intruder uses on the door. The project aims at providing a feedback system that enables the user to monitors the locker remotely. The Smart locker is capable of capturing the images of the intruder that can be used later for identifying the person. Immediate push notification to the user smartphone is an added feature that helps the user to take immediate action to secure the house. The locker features a scheduler that detects user lock patterns and suggests schedules to operate the locker automatically.

In the project, locker needs to act as a server and store the data regarding locker status in the memory. This data has to be transmitted over the network to the smartphone. A solution is provided with proper Internet and Bluetooth connection to the locker system. This has enabled functioning of the GCM push notification feature that is initiated by the locker when there is an intruder.

5.2 Digital Key

The domain of security holds a lot of potential in implementing novel ideas. With improved technology, the Smart locker can be aimed at providing more personalized features to the user. The user of the locker will be able to customize the application according to his/her needs and add or remove features. The user can monitor the locker usage statistics on the mobile app as shown in Fig. 3.

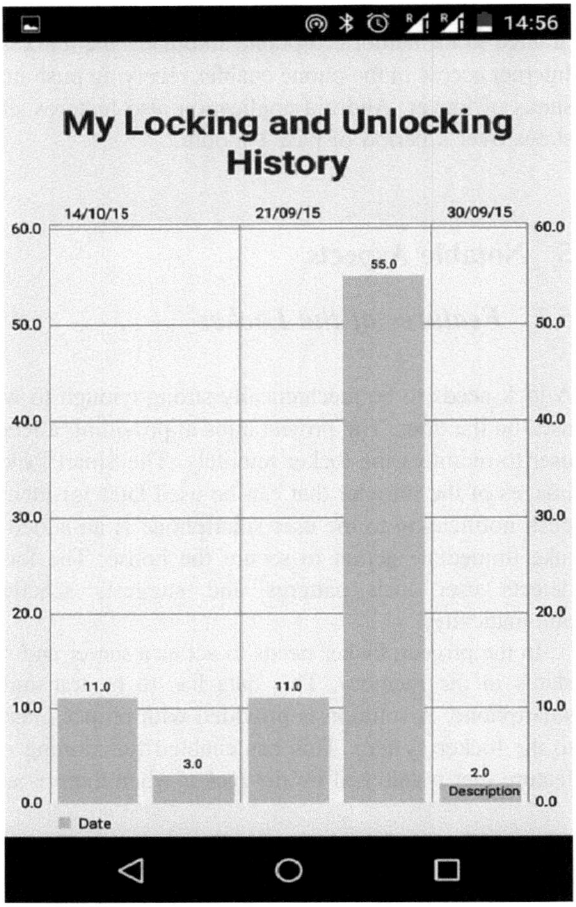

Fig. 3 Data logging and graph

6 Conclusion

Smart locker implementation will be successful in reaching out to house owners which will bring many people under the fold of increased security to their houses. The solution enables reduction in house theft and increased possibility of nabbing the culprit if a theft occurs. The incorporation of best practices such as the use of the digital key to open the door as authorized by the owner of the house in our solution will lead to enhanced functionality that the product delivers. This product is targeted toward improved security and reaching out to people's need. Any person with a smartphone and an idea to venture into new age technology for the house will benefit from the product.

References

1. D.H. Kang, K.H. Baek and K.Y. Kim "Bluetooth Security Technology", Weekly Technology Trends, National IT Industry Promotion Agency, vol. 1380, 2009, pp. 1–13.
2. D. Naresh, B. Chakradhar and S. Krishnaveni, "Bluetooth Based Home Automation and Security System using ARM9", vol. 4, issue 9, Internation Journal of Engineering Trends and Technology (IJETT), 2013, pp. 4052–4058.
3. D. Shekar and Goud and Ishaq Md and P.J. Saritha "A Secured Approach for Authentication System using Fingerprint and Iris" Global Journal of Advanced Engineering Technologies, Volume 1, Issue 3-2012.
4. Pramila D. Kamble and Dr. Bharti W. Gawali "Fingerprint Verification of ATM Security System by Using Biometric and Hybridization" International Journal of Scientific and Research Publications, Volume 2, Issue 11, November 2012.
5. Fei Zuo, Peter H.N. de "Real-time Embedded Face Recognition for Smart Home" IEEE, Volume: 51, Issue: 1.
6. Ravi J.K.B., Raja Venugopal K.R. "fingerprint recognition using minutia score matching". International Journal of Engineering Science and Technology Volume: 1(2), 2009, 35–42.
7. P. Viola, M. Jones. "Rapid object detection using a Boosted cascade of simple features". In: IEEE Conference on Computer Vision and Pattern Recognition, pp. 511–518, 2001.
8. Abhijeet S. Kale and Sunpreet Kaur Nanda "A Review Paper on Design of Highly Secured Automatic Teller Machine System by using Aadhaar card and Fingerprint" International Journal of Advance Research in Computer Science and Management Studies Research Paper. Volume 2, Issue 1, January 2014.

Precision Agriculture System Design Using Wireless Sensor Network

Arun M. Patokar and Vinaya V. Gohokar

Abstract This paper presents design of precision agriculture system infrastructure aiming at a multi-parameter monitoring system using wireless sensor network. Proposed infrastructure is based on low-power Intel's Galileo Gen-2 platform for monitoring, controlling and decision-making support using Internet of Things (IoT). Collection of different farm field parameters is to be done using sensor nodes deployed in the farmland. Each node is connected wirelessly to the base station for the collection of data using wireless transreciever hardware platform. Data is then fed to the personal computer and displayed on screen, e.g. temperature, humidity, sprinkler water flow and soil moisture. From the collected data, decision-making and controlling action can be taken by the use of Internet of Things.

Keywords Precision agriculture · Temperature · Humidity · Moisture · IoT

1 Introduction

E-agriculture is a recent topic of knowledge rising out of convergence of Information Technology and farming methodologies with different controlling techniques. It increases the agricultural value chain through the application of Internet and communication-related technologies. Use of IT will help farmers to have better productivity and access to information which increases the fertility rate. Use of such a wireless sensor network and Internet of Things technology improves the quality of production and efficiency and also reduces the environmental effects on the crop. Such a tools and technology in agriculture bring out the contribution to precision agriculture. The precision agriculture is the technique of applying the

A.M. Patokar (✉)
STC, School of Engineering & Research Technology, Shegaon, Maharashtra, India
e-mail: arunpatokar@gmail.com

V.V. Gohokar
MAEEER's Maharashtra Institute of Technology Pune, Pune, Maharashtra, India
e-mail: vinaya.gohokar@mit.edu.in

D.K. Mishra et al. (eds.), *Information and Communication Technology*,
Advances in Intelligent Systems and Computing 625,
https://doi.org/10.1007/978-981-10-5508-9_16

appropriate input amount (water, pesticides and fertilizer, etc.) in the right time at right location to enhance production and improve yield quality. WSN for precision agriculture is a collection of organized nodes in a cooperative network so as to collect the farm environment information. Each node in the network consists of processing capability, and whole system will comprised of microcontroller, CPUs or signal processing chip; it may contains antenna, power source, memory, RF transreciever, various sensors actuators; each node communicates wirelessly and often self-organizes after deployed in an ad hoc fashion.

Agriculture in India is the core for food security, sustainable development and poverty alleviation. It contributes approximately 14% GDP. Milestones in agro-development in India include white revolution, evergreen revolution, yellow revolution, blue revolution, green revolution, biotechnology revolution and the recent one which is information and communication revolution. Information Technology supports new methods for precision agriculture like computer-based farming machinery for fertilizers and pesticides. However, it is most important role is for communication. Internet has provided us an ideal opportunity for effective communication.

2 System Development

In this research, we proposed an efficient and robust wireless sensor network to monitor the temperature, humidity, soil moisture, rain/sprinkler water conditions of agriculture farm field. The proposed system is composed of Intel's Galileo Gen-2 platform. It has Quark processor, a 32-bit, speed 400 MHz's, 100 Mb Ethernet port, USB host port, micro SD slot and USB client port, 512 kb SRAM, 8 Mb flash, 256 Mb DDR3 and 12 V power-over-Ethernet capable.

The Intel IoT Developer Kit adds C, C++, Python and Node JavaScript support for developing connected sensor Internet-of-Things applications, Intel Galileo also supports C, Python, Node.js and Visual Programming supports from a remotely-connected browser. Temperature, moisture and humidity sensors are connected to data-collecting node. Data-collecting node will continuously send the node data to the base station and whenever there is an abnormality in sensed data the base station will send a message to farmer's cell phone through GSM. With the help of IoT support for Galileo, observation of data collected by sensor node at remote location is also possible. The proposed system development is comprised of (A) system hardware (B) system software.

2.1 Test Assembly

A test bed is created for the assembly and testing of different sensors for their working with real-time environment. Figure 1 shows the assembled unit of different

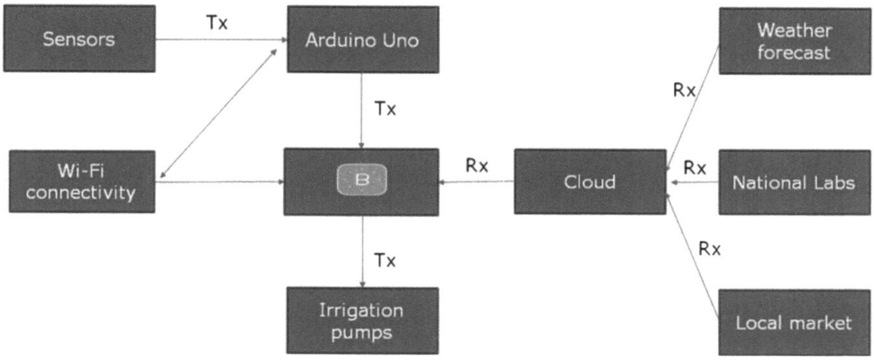

Fig. 1 Sensor test assembly

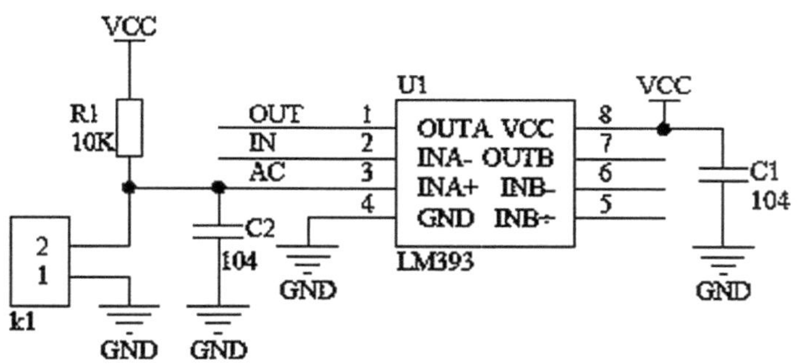

Fig. 2 Control module for rain sensor

sensors with Intel's Galileo Gen-2. Galileo supports for Ethernet, or one can attach a separate Wi-Fi card with it to send the data to base station. The different sensors used in the test assembly are as shown in the Fig. 1. Basic unit of the precision agriculture system is sensor node; it collects the field information to achieve perception, processing, collection, and wireless communication of field environment data.

2.2 *Sensor Nodes*

A. **Sprinkler/Rain Water Sensor**

The rain sensor is a tool for rain detection. It can be used for measuring rainfall intensity. The module comprises of two parts a rain-sensing board and the control circuit board with comparator that can adjust sensitivity though a potentiometer. The control circuit for rain sensor is as shown in Fig. 2, rain-sensing board having RF-04 Nickel material plate.

B. **Soil Moisture Sensor**

It gives us the information of plant's thirst or impersonation what plant roots are experiencing under the soil. They give tips when it is time to hold or irrigate watering. Use of this can avoid unnecessary watering, as it saves you money. Soil moisture module has adjustable potentiometer through which sensitivity can be adjusted.

C. **Plant Growth Monitoring Node**

Basic of monitoring growth is based on capturing plant images after a specified interval of time to monitor plant growth continuously. This is achieved with OV7670 CMOS camera image sensor module which is low-cost image sensor, very powerful and easy to interface with 8/16/32-bit controller/processor. OV7670 will take a snap after a specified interval and send the picture in real time, and also the images can be saved to the base station for further use, from such a system one can monitor the crop growth, crop destruction also.

By connecting all the irrigation motors to a smart circuit, we can simply access it from any remote place with the help of Internet connectivity. The various sensors installed in the farmland gives the real-time data about moisture, humidity and fertility of soil. With the help of this data, farmers would know which crops need to be watered, add fertilizers to and how much quantity of water is sufficient to crops as per the climate and their requirements. This result in saving large amount of water and achieving better yield; it also helps in conserving electricity.

3 Algorithms for Control Operation

The control modules are composed of a relay in connection of light, fans, sprinkler which are mounted in the control module. Switching of each module is done by the Galileo. Each relay is connected to the digital outputs of the Galileo Gen-2 which gives a low or high signal for opening and closing of the relay, respectively. The control modules are either switched OFF/ON according to the parameter range stated in the program.

The flow for control temperature is as shown in Fig. 3, which makes light ON/OFF depending on the threshold value set by the program.

When the detected value of temperature is less than lower threshold, the light is turned ON to increase the temperature of the farm field in greenhouse. When the sensed temperature value is more than its upper threshold, the fan is turned ON so as to decrease the temperature of the farm field of greenhouse. If the detected temperature is within its upper and lower threshold values, the temperature control modules are switched OFF.

Fig. 3 Program flow chart for temperature control

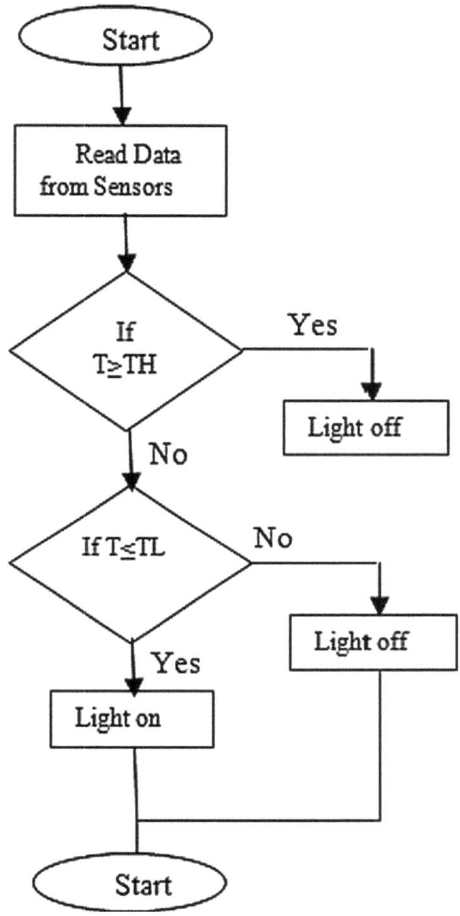

Sprinkler module control flow is as shown in Fig. 4 in this soil moisture is sensed by the sensor, depending on the soil moisture threshold value the sprinkler module works and making on and off sprinkler is to be achieved by the above programming flow.

Fig. 4 Program flow chart for sprinkler control

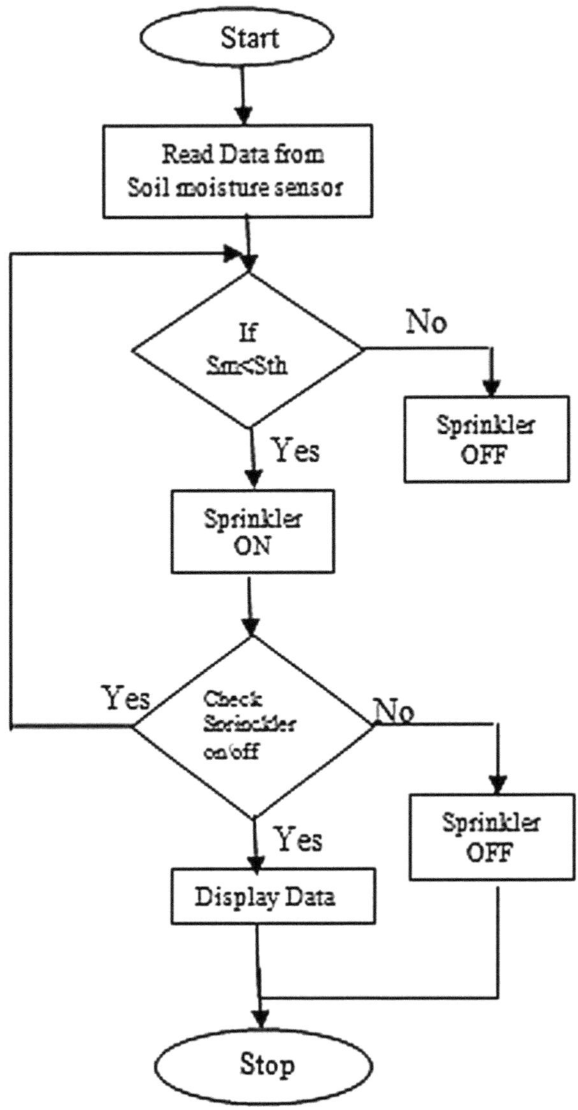

4 Outcomes

Our farmland can be divided into various parts like bajra, sugarcane, pulses, vegetables, etc. Each type of crop has different water and fertilizer requirement.

Each farm has its own pump which is connected to the main circuit and has a common source of water.

The various sensors placed in the farmland would continuously upload data to the cloud in real time (moisture, humidity and fertility of soil).

Water level sensor is fitted inside water reservoir which shows available water.

This system communicates with each other via Wi-Fi module.

5 Conclusion

The proposed system is designed for monitoring the parameter required for crop. For making precision agriculture system, the wireless sensor network infrastructure supports for acquiring and processing of information, recording the data of nodes. This is low-cost system; such a type of system improves the efficiency of resources used, which may improve the production. Adequate amount of water and fertilizers can be added, Qualitative and quantitative production of crops, Complete control is through a Smartphone which is highly affordable, flexible and user friendly, System is fast and responsive, If any error occurred farmers will be notified, Reduction in labour cost.

6 Future Scope

Data is sent to National Lab where scientists can read the nutrient values and soil-climatic condition, thus giving proper advice . Through smart app, farmers get connected to nearby markets which gives them idea about price and crop dealer which helps them to maintain profit. Farmer will continuously get update of sudden climatic changes which will help him to prevent his crop from damage.

Stronger the Agriculture sector, Stronger is the Nation!!

References

1. Paventhan, Sai Krishna Allu, Sameer Barve, V. Gayathri and N. Mohan Ram "soil property monitoring using 6lowpan-enabled wireless sensor networks" Proceedings of AIPA, India, 2012
2. Zhen Fang, Zhan Zhao, Jiangang Zhang, Du, Lidong "A new integrated temperature and humidity sensor" Nano/Micro Engineered and Molecular Systems (NEMS), IEEE International Conference pp. 788–91, 20–23 Feb. 2011
3. Qiang Wang, Terzis, A. Szalay A. "A novel soil measuring wireless sensor network" Instrumentation and Measurement Technology Conference (I2MTC), 2010 IEEE, pp. 412–415, 3–6 May 2010
4. Ying Zhang, "Design of the node system of wireless sensor network and its application in digital agriculture", IEEE, 2011

5. Sanbo Li "Application of the Internet of Things Technology in Precision Agriculture Irrigation Systems" Computer Science & Service System (CSSS), 2012 pp. 1009–1013, 11–13 Aug. 2012
6. Nandurkar S.R., Thool V.R., Thool R.C. "Design and development of precision agriculture system using wireless sensor network" Automation, Control, Energy and Systems (ACES), 2014 pp. 1–6, 2014
7. Lei Xiao, Lejiang guo, "The Realization of Precision Agriculture Monitoring System Based on Wireless Sensor Network", 2010 IEEE
8. N. Sakthipriya "An Effective Method for Crop Monitoring Using Wireless Sensor Network" Middle-East Journal of Scientific Research 20(9) pp. 1127–1132, 2014
9. Sprung, E., Tudor C., Meyer J., Tatro R. "Low power design of a wireless sensor node for indoor monitoring" Information Reuse and Integration (IRI), 2013 IEEE 14th International Conference, pp. 661–667, Aug. 2013
10. He Jianfeng, Qu Jinhui, Wang Yuan, Pan Hengya "The designing and porting of temperature & humidity sensor node driver based on ARM-Linux" Electronics, Computer and Applications, IEEE Workshop pp. 127–130, May 2014
11. F.L. Lewis "Wireless Sensor Networks" Smart Environments: Technologies, Protocols, and Applications ed. D.J. Cook and S.K. Das, John Wiley, New York, 2004
12. I.F. Akyildiz, W. Su, Y. Sankarasubramaniam, E. Cayirci "Wireless sensor networks: a survey" Computer Networks 38 (2002) pp. 393–422, 2002 Published by Elsevier Science B.V. 2002
13. Pinaki Mondal, Manisha Basu "Adoption of precision agriculture technologies in India and in some developing countries: Scope, present status and strategies" Progress in Natural Science 19 (2009) pp. 659–666, 2009
14. Mohamed A. Tawhid, and Hasan Mahmood "Recent Trends in Wireless Sensor Networks with Applications" Hindawi Publishing Corporation International Journal of Distributed Sensor Networks Volume 2014, Article ID 912574, 4 pages, 2014
15. Subrata Kr. Mandal and Atanu Maity "Precision Farming for Small Agricultural Farm: Indian Scenario" American Journal of Experimental Agriculture 3(1): 200–217, 2013
16. D.D. Chaudhary, S.P.Nayse, L.M. Waghmare, Application of wireless sensor network for greenhouse parameter control in precision agriculture. IJWMN, 2011
17. Kshitij Shinghal, Dr. Arti Noor, Dr. Neelam Srivastava, Dr. Raghuvir Singh "Wireless sensor networks in agriculture: for potato farming" International Journal of Engineering Science and Technology Vol. 2(8), 2010, pp. 3955–3963, 2010
18. Zhao Liqiang, Yin Shouyi, Liu Leibo, Zhang Zhen, Wei Shaojun "A Crop Monitoring System Based on Wireless Sensor Network" © 2011 Published by Elsevier Ltd, Procedia Environmental Sciences 11 (2011) pp. 558–565, 2011
19. Pedro Andrade-Sanchez and John T. Heun" Things to Know About Applying Precision Agriculture Technologies in Arizona" The University of Arizona Cooperative Extension, article AZ1535, 2010
20. Sheikh Ferdoush, Xinrong Li "Wireless Sensor Network System Design using Raspberry Pi and Arduino for Environmental Monitoring Applications" The 9th International Conference on Future Networks and Communications (FNC-2014, Procedia Computer Science 34 (2014) pp. 103–110, 2014
21. A. Gaddam, M. Al-Hrooby, W.F. Esmael, "Designing a Wireless Sensors Network for Monitoring and Predicting Droughts" 8th international conference on sensing technology livepoor, UK., pp. 210–215 Sept. 2–4, 2014
22. A. Kaloxylos et al. "Farm management systems and the Future Internet era" Computers and Electronics in Agriculture 89 (2012) pp. 130–144, 2012
23. Suryakant A. Sawant, J. Adinarayana and Surya S. Durbha "krishisense: a semantically aware web enabled wireless sensor network system for precision agriculture applications" IGARSS, pp. 4090–4093, IEEE 2014

24. Scott Fazackerley Ramon Lawrence "Reducing Turfgrass Water Consumption using Sensor Nodes and an Adaptive Irrigation Controller" IEEE Sensors Applications Symposium. February 2010
25. Sherine M. Abd El-kader, Basma M. Mohammad El-Basioni, Precision farming solution in Egypt using the wireless sensor network technology" Egyptian Informatics Journal 2013, 14, pp. 221–233, 2013
26. Jun Jiao, Huimin Ma, Yan Qiao, Yulin Du, Wen Kong and Zhongcheng Wu, Tongling Puji "Design of Farm Environmental Monitoring System Based on the Internet of Things" Advance Journal of Food Science and Technology 6(3) © Maxwell Scientific Organization: pp. 368–373, 2014
27. Juan A. López, Fulgencio Soto, Andrés Iborra, Pedro Sánchez, and Juan Suardíaz "Design and Implementation of a Wireless Sensor Network for Precision Horticulture" SENSAPPEAL 2009, LNICST 29, pp. 27–42, 2010
28. Gunjan Pandey, Ratnesh Kumar, and Robert J. Weber "A low RF-band impedance spectroscopy based sensor for in-situ, wireless soil sensing" Sensors Journal, IEEE, Volume:14, Issue: 6, pp. 1995–2005, April 2014
29. Pasquale Daponte, Luca De Vito, Francesco Picariello, Sergio Rapuano, Ioan Tudosa "Design and prototyping of wireless sensor nodes for non-contact road-level climate monitoring" 20th IMEKO TC4 International Symposium, Italy, pp. 528–533, September 15–17, 2014
30. N.G. Shah, U.B. Desai, I. Das, S.N. Merchant and S.S. Yadav "In-field wireless sensor network (WSN) for estimating evapotranspiration and leaf wetness" international agricultural engineering journal 2009, 18 (3–4)-51, 2009
31. Atis Elsts, Rihards Balass, Janis Judvaitis, Leo Selavo " SAD: wireless sensor network system for microclimate monitoring in precision agriculture" International Conference on Applied Information and Communication Technologies (AICT2012), Jelgava, Latvia 26–27. April, 2012
32. Valery Butenko, Anatoly Nazarenko, Viliam Sarian, Nikolay Sushchenko and Aleksandr Lutokhin. "Applications of Wireless Sensor Networks in Next Generation Networks" Technical Paper of telecommunication standardization sector of ITU, ITU-T series y.2000: next generation networks, Feb. 2014
33. A. Alonso-Arroyo, A. Camps, A. Aguasca, G. Forte, A. Monerris, C. Rüdiger, J. P. Walker, H. Park D. Pascual, and R. Onrubia "Improving the Accuracy of Soil Moisture Retrievals Using the Phase Difference of the Dual-Polarization GNSS-R Interference Patterns" (IEEE) geoscience and remote sensing letters, vol. 11, no. 12, December 2014
34. Arun M. Patokar, Dr. Vinaya V. Gohokar "Development of precision agriculture system based on internet of things: A Review" International Conference on Telecommunication Technology & Management (ICTTM 2015) IIT Delhi during 11th–12th April, 2015

Optimal Tree Search by a Swarm of Mobile Robots

Maitry Sinha and Srabani Mukhopadhyaya

Abstract Swarm robotic research in discrete domain assumes that robots in the swarm are randomly deployed over any graph and the robots can move only through the edges of the graph. In *target searching*, all the robots in the swarm are required to gather at the specially designated node, termed as *target node*. Moreover, during target search if the graph is guaranteed to be explored completely, the scope of the solution increases. An algorithm for tree searching by swarm of asynchronous robots of limited visibility has been proposed in this paper. An $O(1)$ memory is assumed to be attached to each node of the tree. The target node is initially visible to at least one robot in the swarm. However, if it is executed on synchronous system, the algorithm takes $O(n)$ computational cycles to gather all the robots at the target node after exploration of the tree completely, where n is the number of nodes in the graph.

Keywords Swarm robots · Target searching · Weak perception · Whiteboard

1 Introduction

Swarm robots are a system of small, identical, anonymous, oblivious, simple mobile robots which function in a group. The idea of swarm robots emerges from the behavior of small social insects like bees, ants. Most promising applications of swarm robots are in disaster rescue missions, in mining, in agricultural foraging tasks, etc. The jobs that are carried out by a robot swarm are very basic in nature, like gathering, flocking, converging. It is expected that these basic jobs form the foundation of a complex task.

M. Sinha (✉)
Calcutta Institute of Engineering and Management, Kolkata, India
e-mail: sinhamaitry@gmail.com

S. Mukhopadhyaya
Birla Institute of Technology, Mesra, Kolkata Extension Centre, Kolkata, India
e-mail: smukhopadhyaya@bitmesra.ac.in

D.K. Mishra et al. (eds.), *Information and Communication Technology*,
Advances in Intelligent Systems and Computing 625,
https://doi.org/10.1007/978-981-10-5508-9_17

For theoretical research, these robots are usually assumed to be deployed over a two-dimensional plane where they can move freely over the plane. However, very recently researchers are considering swarm robots in discrete domain also. In discrete domain, the working place is modeled as a graph. The robots are deployed over the nodes of a graph and they can move only along the edges of the graph.

In discrete domain, the researchers have mainly addressed the gathering problem. In gathering problem, robots in a swarm are required to gather at a single location which is determined by the robots unanimously during the process. The gathering problem has been addressed extensively in the continuous domain [1, 2]. The solutions to this problem under different models and characteristics of the robots already exist in the literature. However, there are some practical models, under which the solutions of gathering are yet to be found. On the other hand, in discrete domain it is observed that researchers have addressed gathering problem for some selected types of graphs like ring, grid, tree, regular bipartite graphs [3–7]. Moreover, it has also been shown that for these specific topologies also gathering is not possible always, in particular, for all kinds of initial distributions.

When gathering problem is considered under unlimited visibility model, one obvious solution is to identify a point (node) of invariance in the graph and make all robots gather at that point. Since the robots can see the whole graph, they can easily identify such a node, if exists. For example, an odd $\times$ odd grid has a unique center [3, 5]. A tree can have at most two centers. In case of bicentric trees, if the centers are distinguishable then gathering is possible. In case of general graphs, the option of finding point of invariance does not arise at all. This problem is partly resolved when initial distribution of the robots is also taken into consideration along with the topology of the graph. For example, when the number of robots is odd or initial distribution is asymmetric, gathering is possible for a ring, provided robots have multiplicity detection capability [3, 4]. The term "multiplicity" is used to describe the presence of more than one robot at a particular location. From the above discussion, it is clear that in case where initial distribution of robots is symmetric over a symmetric structure like, ring, grid gathering of even number of robots is really difficult even when the robots have unlimited visibility. In these studies on ring, grid, tree, etc., the robots are usually assumed to have unlimited visibility.

In the area of swarm robotics, robots are expected to have limited capacity for cheap mass production. In that scenario, a robot having a very strong visibility capacity (unlimited visibility) is not a very practical assumption. In discrete domain, when robots can view only up to a fixed distance, gathering is addressed only for regular bipartite graphs and trees. In [8], the authors have established a very clear direction of research in solving gathering problem under limited visibility model. In this paper, it has been proved that on regular bipartite graphs, gathering is possible only when the initial configuration is a star of size at least 3, for both local and global multiplicity detection capacity of the robots. An obvious extension of some of the adversaries discussed in [8] can prove that gathering is impossible when regularity criterion is removed and hence for a general graph. A tree is a bipartite graph but not regular. Hence, gathering is not possible on tree under such minimal criterion when robots are oblivious and have limited visibility capacity. Alternative ways of addressing this problem are either to make the robots more powerful or to slightly relax the problem environment. In this paper, we have assumed that the robots are required to gather at a specific node which is distinguishable from others. Let us call this problem as *target searching problem*, where the robots search for the target node and finally they all gather at that node. Moreover, we enhance the scope of the problem as while searching the target node, the robots collaboratively explore the complete graph. In this paper, we mention this problem as *graph searching*. In our solution, we assume that robots are oblivious but an $O(1)$ memory is attached to each node of the graph.

Of late in 2015, Bhaumik and Gan Chaudhuri [9] proposed a solution for a problem similar to target searching on trees. The assumptions made in this solution raise some serious issues. The authors assumed the existence of memory but did not explicitly mention where the memories are attached to, nodes or links. Constant amount of memory in each node is not sufficient for their solution. Secondly, the solution involves some movement of the robots where a robot "turns 180 degrees" and takes a move along an edge of the graph. The weakness of this concept lies on the implication that the nodes in the graph are distributed on a physical plane. Software agents moving in a network would not have such a notion. In other words, this assumption ruins the abstraction provided by the notion of a graph. Finally, the solution proposed in [9] does not guarantee exploration of the complete tree.

In this paper, we propose an algorithm for graph searching problem on a connected acyclic graph, a tree, with $O(1)$ memory attached to each node of the tree.

In Sect. 2, we give a formal definition of the problem along with the basic assumptions made regarding problem environment and the characteristics of the robots. The proposed algorithm is discussed in Sect. 3 along with its correctness. An estimation of time is also presented here. Finally, we conclude in Sect. 4.

2 Target Searching Problem on Tree

Given a connected acyclic graph, $G(V, E)$, i.e., a tree with n number of nodes, a node in the graph is specially marked as the target node. Target node can be located anywhere in the graph. m number of asynchronous robots, with limited visibility, are randomly deployed over the graph so that robots are placed on the nodes only. The robots are required to gather at the predefined target node within finite amount of time after exploring (collectively) the tree completely.

2.1 Basic Models and Assumptions

The solvability of a problem in the area of swarm robotics greatly depends on the basic models assumed regarding problem environment and robot capacity.

The environment where the robots are deployed is as follows:

Graph: The robots are deployed on a connected acyclic graph, a tree, whose nodes and links are not labeled. A node in the tree (located anywhere) is specially marked as target node.
Initial distribution of robots: Initially, m robots are randomly deployed over the n nodes of the tree with a restriction that at least one robot must be deployed among the nodes adjacent to the target node, including the target node.
Node memory: Each node of the tree has $O(1)$ amount of memory, called *white-board* (WB). These whiteboards can be accessed for writing only by the robots which are present at the local nodes. If more than one robot tries to write on these boards, only one is allowed to write. However, these WBs are visible from all of its adjacent nodes. Initially, all the whiteboards are blank, unmarked. In target node, the whiteboard is marked specially.

Characteristics of the robots in the swarm:
Identical and anonymous robots are assumed to have the following characteristics:

Visibility: Robots can observe only its adjacent nodes. Robots which are residing at any of its adjacent nodes are visible to it. The robots can also read the whiteboards placed at those nodes which are visible to it.
Node memory: The robots are oblivious; they do not retain past memory.

Computational model:
Robots are assumed to follow CORDA model [2]. Robots carry out a sequence of computational cycles throughout the execution of the algorithm. Each cycle consists of three phases, observe, compute, and move. These phases are not overlapped. In compute phase, a robot computes its next destination after processing the information gathered at observe phase and then moves to that destination in move phase.

The robots are assumed to be asynchronous. The robots can go to sleep state anytime, only with the assumption that they cannot be in sleep state for infinite amount of time.

A node, which is not occupied by any robot, is mentioned as an empty node or an unoccupied node. An occupied node may have more than one robot. A node of which the whiteboard is not marked is mentioned as unmarked node or otherwise marked.

3 The Algorithm

A robot which is at any adjacent node of the target node initiates the algorithm by marking its local whiteboard (WB) as 1 (distance of the node from target). A robot does not start its action unless it finds one of its adjacent nodes is marked.

Whenever a robot starts its action, its first responsibility is to explore the subtree rooted at any of its unmarked, unoccupied neighbors. In course of exploring, if a robot finds another robot present in its adjacent node, it passes on the responsibility of exploring the subtree rooted at that node to that robot. When all the adjacent nodes are marked or occupied, then only a robot starts approaching toward the target node by following a path along the smaller WB marks. After reaching the target node, if a robot finds another unexplored direction, it explores that subtree and so on. A formal description is given below.

Algorithm *Tree_Search*
In the observe phase, a robot takes a snapshot of its surroundings. Depending on the status of the adjacent nodes where the robot is currently residing, the robot *R* decides its next move according to the following cases.

Case 1: ***The robot R is at target node***.

Subcase 1.1: *All visible whiteboards are either marked or all adjacent nodes are occupied by some robots or both.*

- Robot *R* stays at the current node and stops executing the algorithm.

Subcase 1.2: *At least one visible unoccupied neighbor with unmarked whiteboard.*

- Robot *R* sets one of these nodes as its next destination.

Case 2: ***The target node is visible to the robot R, but R is not at the target node***

- Issue a write request to ensure that its local whiteboard be marked as 1. Now two subcases may arise.

Subcase 2.1: *All visible whiteboards (except the one at the target node) are marked.*

- Set the target node as its next destination.

Subcase 2.2: *At least one visible whiteboard is not marked.*

2.2.1 *All nodes with unmarked whiteboards are occupied.*

- Set the target node as its next destination.

2.2.2 *At least one unoccupied neighbor with unmarked whiteboard.*

- Set one such unmarked empty node as its next destination.

Case 3: ***The target node is not visible to R***

Subcase 3.1: *All visible whiteboards (at occupied or unoccupied neighbors) are unmarked.*

- *R* stays idle at the current node.

Subcase 3.2: *At least one visible whiteboard is marked.*

- Among all the visible marked whiteboards, let k be the least value marked on a whiteboard. R compares the local whiteboard value with k. If the local whiteboard is unmarked or contains a value greater than k, then R issues a write request to ensure that its local whiteboard be marked as $k + 1$. Now two cases may arise.

3.2.1 *At least one visible unoccupied neighbor with unmarked whiteboard.*

- Set the next destination to an unoccupied visible node with empty whiteboard.

3.2.2 *All the visible neighbors with empty whiteboard are occupied.*

- Set the next destination to a node with whiteboard value k.

3.1 Discussion and Correctness

Before formally proving the correctness of the algorithm, let us state the following observation regarding movement of the robots in the algorithm.

Observation: Each robot takes two types of movements which may be termed as *backward* movement and *exploring* movement.

Backward movements are taken to approach the destination, whereas exploring movements are for exploring the unexplored portion of the graph reachable from that node. In backward movement, a robot moves in the direction of decreasing order of the whiteboard marks. In exploring movement, a robot moves in the direction from marked whiteboard to an unmarked one. Thus, in exploring movement, a robot moves in the direction of increasing order of the whiteboard marks.

Theorem 1 *If a robot R, initially placed at a node v, starts its action, then the subtree T_v rooted at v is completely explored by R (with the help of other robots placed at that subtree T_v) and R, along with other robots placed in T_v, comes back to v in subsequent steps. Moreover, in T_v, if a node is at a distance r from v, then its WB is marked as $k + r$, if k is the mark at v.*

Proof Since we consider a general tree, not a rooted tree, without loss of generality, let us consider the target node as the root node. We prove the result by the method of induction on the height of the node v.

Base cases: The result is trivially true when v is at a height 0. Now, we consider the node v at height 1. All the children of v say $u_1, u_2, \ldots, u_r$ are leaf nodes. Some of them may contain robots. Since these children are connected with the root node by exactly one path passing through v, none of the robots present at u_is start any action unless R starts its action at v. If R starts its action, the following steps are executed:

R marks its local whiteboard as k (say) and then moves to an unoccupied neighbor (if any). Mark that whiteboard as $k + 1$ and go back to its original position

v taking a backward movement. R continues to do so until WBs of all unoccupied children are marked appropriately. Subsequently, when the robots (if any) at u_is observe the WB mark at v as k, they mark their local WBs as $k + 1$ and move to the node v taking a backward movement. There they act in the same way as R. Thus, here also the subtree rooted at v is explored completely, and in subsequent steps, all robots initially placed at that subtree reach v.

Induction hypothesis: Let us assume that the result is true when R is initially placed at a node u of height at most h.

To prove: The same is true for a robot R which is initially placed at a node v of height $(h + 1)$.

According to the algorithm, R starts execution as soon as it finds one of its adjacent nodes (in particular its parent) has a mark on its local whiteboard.

If R starts its action, it first marks its local WB appropriately as k (say). Let u be a child of v, which is at a height h. Depending on the status of u, the following actions are taken place.

Case a: The node u is occupied by a robot R^1.

R ignores this child. However, by looking at the mark of WB at v, R^1 starts its action by marking the local WB as $k + 1$. By induction hypothesis, R^1 completely explores the subtree T_u, rooted at u, jointly with other robots placed in T_u and come back to u in subsequent steps with all other robots in T_u. If a node in T_u is at a distance r from u (i.e., at the distance $r + 1$ from v), it is marked as $(k + 1) + r = k + (r + 1)$.

Case b: The node u is unoccupied and not marked.

R moves to u and marks the local WB as $k + 1$. Since R is now at height h and it is already in action, by induction hypothesis we can say that R completely explores the subtree T_u jointly with other robots placed in T_u and comes back to u with all other robots in T_u in subsequent steps after proper marking of the whiteboards.

Case c: The node u is unoccupied but marked.

R ignores this child. Since it is marked, it has already been explored by some robot in T_v.

At v, R continues to take action as long as there is an unoccupied and unmarked child. Other than R, if any other robot R^1 arrives at v and finds an unmarked and unoccupied child, it takes the same action as R. Once a robot reaches any child of v, it can see the node v with a smaller WB mark. Hence, it finally moves to v.

Now, we conclude that all the subtrees T_u rooted at u, u being a child of v, are explored completely by R (with the help of other robots initially placed somewhere in T_v). Thus, the subtree T_v is completely explored by the robots initially placed in T_v, and subsequently, all these robots (including R) come back to v.

Hence, by the principle of induction we can say that if a robot starts its action at a particular node v, it completely explores the subtree rooted at v and comes back to v after marking all the nodes in the subtree by their distance from v. Moreover, all the robots initially placed in the subtree rooted at v finally move to v.

Theorem 2 *The algorithm "Tree_Search" gathers all the robots at the target node after exploring the tree completely.*

Proof According to the algorithm, if a robot is initially placed at the target node or at a node adjacent to the target node, it initiates the algorithm. Since we have assumed that there is at least one robot adjacent to the target node, that particular robot initiates the execution.

If the initiator is at the target node initially, the result is true by direct application of the Theorem 1. Otherwise, the initiator after complete exploration of the subtree rooted at its initial position again comes back to its original position, adjacent to the target node as follows from Theorem 1. According to the algorithm, the initiator then moves to the target node. Since the initiator is already in the process of execution of the algorithm and it is at the root, the whole tree is explored, and finally, all robots gather at the target node.

3.2 Completion Time

For an asynchronous system, estimation of time is not possible. However, assuming robots to be synchronous, estimation on the number of computational cycles can be obtained for the same algorithm.

Once a robot initiates its execution, in each successive computational cycle it makes exactly one move until it terminates. An initiator (a robot which is initially adjacent to or on the target node) starts its action in the very first computational cycle. Thus, even if an initiator requires to explore the whole tree and then go back to the target node, it takes $\theta(n)$ cycles since a node is traversed by exploring movement of a robot exactly once and in the backward movement, a robot takes a path of length $O(n)$ to reach the target from any position of the tree. Now, we are only to establish that any robot (irrespective of its initial position) requires $O(n)$ cycles to get initiated. This is obvious since in the worst case an initiator can reach (if requires at all) a robot, wherever it is initially residing, in $\theta(n)$ cycles. Since gathering at the target node cannot be done in time lesser than $\theta(n)$ (in the worst case, a node may need to traverse an $\theta(n)$ distance), we claim our solution to be cost optimal.

4 Conclusion

In this paper, we present an algorithm for tree searching problem. After searching the tree completely, the robots gather at a specially designated node. This problem differs from the conventional gathering problem in the sense that the target node is marked, whereas, in the gathering problem, the gathering point is determined in run time. However, real-life situations may demand gathering of the robots on some predefined location rather than any location. These types of requirements and impossibility results of gathering problem in discrete domain drive us to think the problem from a different perspective. The proposed tree searching algorithm gathers all the robots in a tree to a predefined target node after the complete exploration of the tree. The assumptions made for solving the problem are also minimal. The required time has also been estimated to gather all the robots after collectively exploring the tree.

References

1. Kamei S., Lamani, A., Ooshita, F. and Tixeuil, S.: Asynchronous Mobile Robot Gathering from Symmetric Configurations without Global Multiplicity Detection. In: Proc. 37th International Symposium on Mathematical Foundations of Computer Science (MFCS) vol. 7464, pp. 542–553. Springer-Verlag (2012).
2. Floccihni, P., Prencipe, G., Santoro, N. and Widmayer, P.: Gathering of Asynchronous Robots with Limited Visibility. Theoretical Comp. Sc., 337(1–3), pp. 147-168, (2005).
3. D'Angelo, G., Di Stefano, G. and Navarra, A.: Gathering Asynchronous and Oblivious Robots on Basic Graph Topologies under the Look –Compute –Move Model. HAL (2012).
4. Klasing, R., Markou, E. and Pelc, A.: Gathering Asynchronous Oblivious Mobile Robots in a Ring. Theoretical Comp. Sc., 390(1), pp. 27–39 (2008).
5. D'Angelo, G., Di Stefano, G. and Navarra, A.: Gathering of Robots on Anonymous Grids without Multiplicity Detection. In: Proc.19th International Colloquium on Structural Information and Communication Complexity, LNCS, vol. 7355, pp. 327–338 (2012).
6. Fraigniaud, P., Pelc A.: Deterministic Rendezvous in Trees with Little Memory. In: Taubenfeld (ed.): DISC 2008, LNCS vol. 5218, pp. 242–256, (2008) Springer- Verlag Berlin Heidelberg (2008).
7. Kowalski, D. R. and Pelc, A.: Polynomial Deterministic Rendezvous in Arbitrary Graphs. In: Proc. 15th Annual Symposium on Algorithms and Computation, ISSAC, LNCS, vol. 3341, pp. 644–656, (2004).
8. Guilbault, S. and Pelc, A.: Gathering Asynchronous Oblivious Agents with Local Vision in Regular Bipartite Graphs. In Struct. Inf. and Com. Complexity, pp. 162–173, (2011).
9. Bhaumik, S. and Gan Chaudhuri, S.: Gathering of Asynchronous Mobile Robots in a Tree. In Proc. IEEE 2nd International Conference, Applications and Innovations in Mobile Computing (AIMoC) (2015).

Achieving Guaranteed Service with Fault-Tolerant Resources in Grid

Sukalyan Goswami and Ajanta Das

Abstract Composed of loosely coupled virtual resources, grid, being highly distinguished from traditional high-performance computing, is extensively used in computation-intensive problem solving in the arenas of science and technology. Maintaining performance or balancing load of each resource in grid is always more challenging with high chances of resource failure. The objective of this paper is to improve the efficiency of the Nearest Deadline First Scheduled (NDFS) algorithm considering resource failure a sudden occurrence in grid. The algorithm introduces periodical runtime backup to another available resource for retaining Quality-of-Service as approved in service quality agreement. This paper presents multiple job execution cases through implementation of benchmark codes executed in local grid test bed using Globus Toolkit middleware, with an emphasis on resource failure phenomenon of grid. These experimental results establish the requirements of the proposed algorithm to ensure the job deadline misses get reduced even if unexpected resource failures happen.

Keywords Grid computing · Job allocation · Runtime backup · Quality-of-service · Resource failure

1 Introduction

With the advancement of technology, grid computing supports computations across the multiple administrative domains. Computation-intensive problems are easily solved in computational grid [1] environment with successful coordination of

S. Goswami
Institute of Engineering & Management, Kolkata, India
e-mail: sukalyan.goswami@gmail.com

A. Das (✉)
Department of Computer Science & Engineering, Birla Institute of Technology, Mesra, Kolkata Campus, Kolkata 700107, India
e-mail: ajantadas@bitmesra.ac.in

D.K. Mishra et al. (eds.), *Information and Communication Technology*, Advances in Intelligent Systems and Computing 625,
https://doi.org/10.1007/978-981-10-5508-9_18

sharing resources among autonomous groups. Still today, resource scheduling and reliability is challenging in computational grid. Because of the heterogeneous and dynamic nature of the grid, the scheduling of resources and load balancing across the grid are two very difficult problems to handle, hence considered as two major research areas.

The objective of this paper is to improve the efficiency of the Nearest Deadline First Scheduled (NDFS) algorithm [2] considering resource failure, a sudden occurrence in grid. The algorithm provides backup facility to compete with the failure of resources. It introduces periodical runtime backup to another available resource for retaining Quality-of-Service as approved in service quality agreement (SQA). SQA is a bipartite agreement between the clients and broker in the grid. In this proposed fault-tolerant NDFS approach, if failure happens, the job need not be resubmitted. The job can resume its execution from the last backup point into some other active resource. This approach is advantageous, because the job submission is done by the clients only once. The same SQA, which is signed at the time of submission, is considered for resumption of the job execution till its last backup point. The job continued its execution to the newly allocated resource and successfully completes the execution. Finally, the results are sent back to client after verifying the SQA.

This research work also puts emphasis on execution of benchmark codes in real grid environment. The benchmark codes of Matrix Multiplication [3] are considered as single and multiple jobs. The grid environment is set up using Globus Toolkit middleware. This result proves that proposed fault-tolerant NDFS algorithm is a novel solution in case of resource failure.

Organization of the paper is as follows. The related works are presented in Sect. 2. Section 3 presents the proposed fault-tolerant NDFS algorithm. Experimental results are depicted in Sect. 4. Section 5 concludes the paper.

2 Related Works

The relevant research works are studied thoroughly and presented here. Equal workload distribution achieving is the primary challenge in computational grid. To cater to the above problems in computational grid, a load balancing algorithm, NDFS, is proposed in [2, 4, 5]. The simulations were run in the GridSim [6], and the results were compared with other scheduling algorithms in [4]. The simulation results depict that NDFS algorithm performs better than many available algorithms.

Consideration of resource failure is another promising area while meeting SQA is necessary. Modified *ELISA* algorithm proposed by Ruchir, Bharadwaj, and Manoj [7] considers job migration cost but fails to achieve better utilization of resources in the grid. Balasangameshwara and Raju [8] used passive replication scheme in their backup approach designed for grid, with multiple same job replica increasing the overhead during resource failures. Resource allocation heuristics proposed by Shestak et al. [9] did not guarantee task allocation to fastest machine with fastest execution time. This approach, therefore, has scope for improvement in reducing the application execution time.

Hence, resource failure is another major area of concern in computational grid. Few backup approaches, to implement fault tolerance, have been studied, and one method was proposed for grid in [10]. In [2], our previous research considers resource failure, but the resubmission approach incurs overhead. The novelty of this current research work contributes to the phenomenon of resource failure in the grid and finding out a solution without compromising on the performance of the grid.

3 Proposed Fault-Tolerant NDFS Algorithm

In NDFS algorithm, the resources are categorized as *underutilized*, *less loaded*, and *over loaded*. Initial version of the proposed NDFS algorithm is responsible for only single job execution [5], and it advises for resubmission of jobs if failure happens in grid. The execution process of this proposed algorithm is based on CBR model where the resources are considered to be heterogeneous. Currently, this research deals with multiple jobs and identifies that instead of resubmitting the job it could be better option to take partial backup during partial execution, f (where $0 < f < 1$), of job simultaneously to support fault tolerance of the algorithm.

The jobs submitted can have differing deadline and processing power requirements. The responsibility of smooth operation of the grid is bestowed upon the broker, which works as the middleware.

In this proposed fault-tolerant workload scheduling algorithm, NDFS, the clients are given highest priority, as parameters related to the submitted jobs play an important role for proper workload balancing. The broker then searches for the resources which are having matching or higher capacity of these parameters. NDFS algorithm is implemented in five major phases which are described in [11] followed by an important addition of "competence enhancement" of the NDFS algorithm.

Phase I: *Submission of Job Parameters (by clients).*
Phase II: *Ranking of Resources (by broker).*
Phase III: *Signing of SQA (between client and broker).*
Phase IV: *Job Allocation and Execution (by broker and subsequently resources).*
Phase V: *Compliance of SQA (between broker and client).*

These above-mentioned phases explain the process starting from submission of single job to allocation to resource and execution of the same job in computational grid environment. However, in grid, heterogeneous types of multiple jobs are getting submitted simultaneously and almost continuously. Therefore, the job queue is always updated and remaining jobs in the job queue will be handled accordingly.

Competence Enhancement:

Competence of the proposed NDFS algorithm is enhanced with its fault-tolerant approach in the phase of job execution, i.e., *Phase IV*. Moreover, if resource fails, in which the job is currently allocated and being executed according to the above-mentioned phases, meeting of SQA is challenging. In this circumstance, during execution of the job, the partial backup approach is initiated with initialization fraction f as 0.3. This modified version of the algorithm is named as "*fault-tolerant NDFS*." The total system state of the resource after 30% completion of execution, along with the copy of the job, is sent to the second-ranked resource for ensuring availability of backup. The same phenomena of sending the whole system state to backup the intermediate result are carried out up to 60 and 90% completion of the job execution at first resource. The following steps depict the process of fault-tolerant NDFS algorithm.

Step 1 (a): Successful allocation of job to the 1st ranked resource.
(b): Broker passed on the detailed address of 2nd ranked resource to the allotted resource.
Step 2 (a): Initialize $f = 0.3$
(b): Repeat until completion of job
{
Execution of job till f
Pause execution
Increment f by 0.3, $0 < f < 1$
Take backup
Send backup to 2nd ranked resource
Resume execution
}
Step 3: Send execution results to broker.
Step 4: Broker verifies SQA and sends result to client.

In this proposed fault-tolerant version of NDFS algorithm, if allocated resource fails, then job can always be resumed just from the point of last taken backup in the second-ranked resource. Hence, it saves the resubmission time and execution time till the phase from which it resumes.

The next section establishes the result of the algorithm with execution of heterogeneous multiple jobs in real grid environment.

4 Experimental Results

A grid test bed is set up, for the purpose of this research work, consisting of three clients, two resources and a grid broker. All client nodes are having dual-core processor, 2 GB RAM, and 160 GB HDD. Broker and resources have quad-core processor, 4 GB RAM, and 500 GB HDD. Globus Toolkit 5.2 is used to set up the real grid environment. Java 6 has been used as the programming language for implementation of this research work because of its extensive and robust support in the networking environment. The System Information Gatherer and Reporter (SIGAR) [12] API is used to retrieve system parameters.

Experimental results are represented for various cases of the proposed fault-tolerant version of NDFS with the execution of single job.

Fault-Tolerant Execution of Single Job

In this section, the benchmark code of Matrix Multiplication [3] is considered as single job and executed in the above-mentioned grid test bed according to the fault-tolerant NDFS algorithm. This benchmark code named as "MatMul.java" is executed for four different cases.

- *Case I*: Resources are active; execution of job is successful "according to without fault-tolerant version of NDFS algorithm" and SQA is met.
- *Case II*: Resource fails; execution of job is initiated according to "without fault-tolerant version of NDFS algorithm" and SQA is not met.
- *Case III*: Resources are active; execution of job is successful according to "fault-tolerant version of NDFS algorithm" and SQA is met.
- *Case IV*: Resource fails; execution of job is initiated according to "fault-tolerant version of NDFS algorithm" and SQA is met.

Case I:

It represents successful execution without fault tolerance; hence, SQA for "MatMul.java" is met successfully after allocating to Resource 2 as represented in Fig. 1.

Case II:

It shows that submitted job is executed according to NDFS without fault tolerance. But, resource fails and job needs to be resubmitted, reallocated, and then it is executed successfully. So, the total time taken for completion of execution exceeds

Fig. 1 Execution results of single job execution

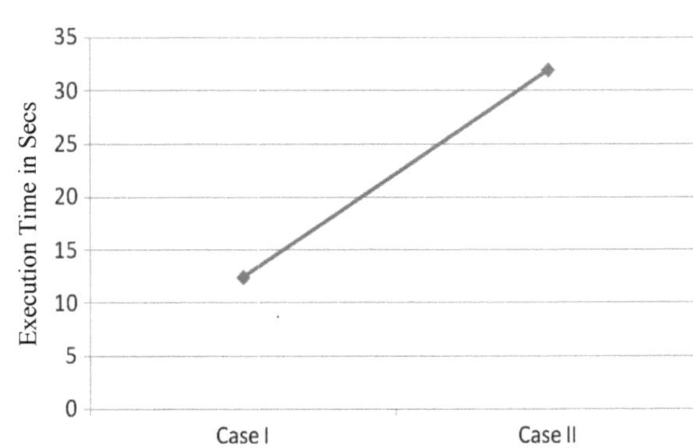

the specified deadline of the job. Figure 1 depicts this situation for non-fault-tolerant approach of NDFS. SQA is not met in this case.

To cater to this problem, fault-tolerant NDFS algorithm is introduced. Case III and Case IV discuss the fault-tolerant approach for single job execution.

Case III:

The Matrix Multiplication job, namely MatMul.java, is submitted by Client 2 and is allocated to highest ranked resource, namely Resource 2.

It presents successful execution of the job according to fault-tolerant NDFS algorithm.

- Naming convention used in Table 1: Execution Completion Time stamp (ECT), Backup Completion Time stamp (BCT), and Execution Completion Final Time stamp (ECFT).
- The MatMul.java job is executed up to 30% completion in 3.71 s, and then, the job is paused and backup is taken in second-ranked resource, Resource 1, in the next 0.61 s. The cumulative time stamp values are presented in Table 1.
- Execution of job is resumed and continued till 60% completion. Again backup is taken at Resource 1 after pausing execution. Time stamp values for execution and backup are presented in Table 1.
- Similarly, backup is taken after completion of 90% job execution and time stamp values are added in the same table.
- Rest of the 10% execution is completed by Resource 2, and the results are sent to Client 2 through broker. Although total execution time increases compared to non-fault-tolerant approach of NDFS, still SQA is met.

Case IV:

It presents successful execution of the job according to fault-tolerant NDFS algorithm even after resource failure happens. This scenario is presented as a bar diagram in Fig. 2.

- Execution continues at Resource 2 till 30% and consumes 3.72 s. Then, backup is taken in second-ranked resource, Resource 1, at backup time 0.6 s.
- Execution continues for 13% more by consuming 1.62 s. But, Resource 2 fails after 43% execution.
- Hence, the job resumes its execution from the last backup point (30%), in second-ranked resource, Resource 1.
- Remaining (70%) execution time of the job is 10.37 s, and the execution is completed by Resource 1.

Table 1 Time stamp values for MatMul.java through fault-tolerant NDFS algorithm

Fractional coefficient ($f = 0.3$)						
30% (s)		60% (s)		90% (s)		Remaining 10% (s)
ECT	BCT	ECT	BCT	ECT	BCT	ECFT
3.71	4.32	8.16	9.02	12.78	13.81	15.06

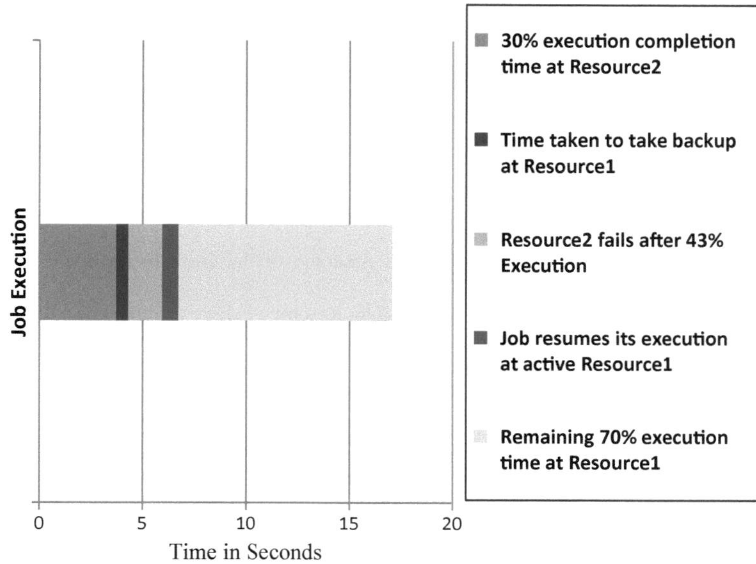

Fig. 2 Resource failure scenario of fault-tolerant NDFS algorithm

5 Discussion

The results ensure that SQA is met for the Case I, III, and IV. However, SQA could not be met for Case II. Moreover, for Case IV, resource failure happens after 43% job completion. Since backup was taken at 30%, resumption of job execution happens from that point, i.e., just after 30%. Hence, only 13% of job execution happens twice with extra cost of 2.04 s (including execution migration time). Hence, fault-tolerant version of NDFS actually enhances its performance.

6 Conclusion

The two most difficult problems encountered in computational grid environment are workload balancing among the participating resources and handling resource failure situation in order to meet Quality-of-Service. Both these problems are solved by the proposed approach depicted in this research. Fault-tolerant approach is introduced in NDFS, which enhances the efficiency of the algorithm. Globus Toolkit is used to set up the grid test bed. The execution results of the computation-intensive benchmark code of Matrix Multiplication are presented in this work. The results ensure that the SQA could be met by this approach even if unexpected resource failures happen. Moreover, these experimental results demonstrate the enhanced competency of the proposed algorithm to ensure the job deadline misses get reduced even if unexpected resource failures happen.

References

1. Foster, I., Kesselman, C., Tuccke, S.: The Anatomy of the Grid. International Journal of Supercomputer Applications (2001).
2. Goswami, S., Das (nee De Sarkar), A.: Handling Resource Failure towards Load Balancing in Computational Grid Environment. In: Proceedings of the Fourth International Conference on Emerging Applications of Information Technology (EAIT 2014), pages 133–138 (2014).
3. Matrix multiplication and FFT benchmark codes. http://introcs.cs.princeton.edu.
4. Goswami, S., Das, A.: Deadline Stringency Based Job Scheduling in Computational Grid Environment. In: Proceedings of the International Conference on Computing for Sustainable Global Development, 9th IndiaCom - 2015, pages 531–536, (2015).
5. Goswami, S., Das, A.: Resource Prioritization Technique in Computational Grid Environment. In: Proceedings of the Second International Conference on Computer and Communication Technologies - IC3T 2015, Advances in Intelligent Systems and Computing 380, pages 765–772, (2015).
6. Buyya, R., Murshed, M.: GridSim: a toolkit for the modeling and simulation of distributed management and scheduling for Grid computing. The Journal of Concurrency and Computation: Practice and Experience vol. 14 issue 13–15, (2002).
7. Ruchir, S., Bharadwaj, V., Manoj, M.: On the design of adaptive and de-centralized load balancing algorithms with load estimation for computational grid environments. IEEE Transactions on Parallel and Distributed Systems, vol. 18, no. 12 (2007).
8. Balasangameshwara, J., Raju, N.: Performance-Driven Load Balancing with Primary-Backup Approach for Computational Grids with Low Communication Cost and Replication Cost. IEEE Transactions on Computers, Digital Object Identifier 10.1109/TC.2012.44, (2012).
9. Shestak, V., Smith, J., Siegel, J. H., Maciejewski, A. A.: Stochastic robustness metric and its use for static resource allocations. Journal of Parallel and Distributed Computing, vol. 68, no. 8, pp. 1157–1173, (2008).
10. Abawajy, J. H.: Fault-Tolerant Scheduling Policy for Grid Computing Systems. In: Proceedings of the International Parallel and Distributed Processing Symposium (IPDPS) (2004).
11. Goswami, S., Das, A.: Optimisation of Workload Scheduling in Computational Grid. In: Proceedings of the Fifth International Conference on Frontiers in Intelligent Computing; Theory and Applications (FICTA 2016) (Springer AISC Series Publication Details Awaited).
12. SIGAR. https://github.com/hyperic/sigar.

A Smart Air Pollution Analytics Framework

Anindita Desarkar and Ajanta Das

Abstract Air pollution which is the worst environmental health risk across the world takes millions of lives every year both in developing and developed countries. These huge premature deaths happen due to long-term exposure to air pollutants as most of the cities do not meet the acceptable pollution level suggested by World Health Organization (WHO). So there is an urgent need to reduce the air pollution level across the globe. This paper proposes a state-of-the-art approach and proposes a layered air pollution reduction framework. The methodology of the proposed framework also suggests the action plans to reduce air pollution level with an innovative Rule Base and mining appropriate data from the huge dataset which is basically a data warehouse. It also discusses the expected outcome of the proposed framework beneficial to the citizens.

Keywords Predictive analysis · Air pollution · Action plan · Data mining · Data warehouse · Knowledge discovery

1 Introduction

Air pollution is basically contaminated air includes NO_2, SPM, photochemical oxidant, sulfur dioxide, carbon monoxide, and fine particulate matter are vulnerable to public health [1]. The major sources of air pollution include traffic sector, industrial domain, power plants, and fossil fuel burning. People who live in the polluted areas have increased risk of various heart and acute respiratory diseases, lung cancer, and other chronic problems which cause huge premature deaths, especially in developing countries [2, 3].

A. Desarkar · A. Das (✉)
Department of Computer Science and Engineering, Birla Institute of Technology, Mesra, Kolkata Campus, Kolkata 700107, India
e-mail: ajantadas@bitmesra.ac.in

A. Desarkar
e-mail: aninditadesarkar@gmail.com

D.K. Mishra et al. (eds.), *Information and Communication Technology*, Advances in Intelligent Systems and Computing 625,
https://doi.org/10.1007/978-981-10-5508-9_19

The Central Pollution Control Board of India has taken up an initiative named NAQMP which stands for National Air Quality Monitoring Program. NAQMP takes into consideration 341 stations across 126 cities/towns and 4 union territories in India including Kolkata [4, 5]. In all of these locations, the amount of sulfur dioxide (SO_2), oxides of nitrogen (NO_2, etc.), suspended particulate matter (SPM), and respirable suspended particulate matter (RSPM/PM10) are regularly monitored. The monitoring is done continuously for 24 h in a frequency of twice a week, to have around 104 observations on a yearly basis. This monitoring is being carried out in association with SPCBs, National Environmental Engineering Research Institute (NEERI), Nagpur, etc. [6, 7].

The US Consulate General has set up an air pollution monitoring system in Park Street, Kolkata [8]. The data obtained from this monitoring system is displayed in raw concentrations only but this data is yet to be converted to an air quality index (AQI) [6, 9].

The major objective of this paper is to monitor air pollution level in Metropolitan cities and proposes a layered air pollution reducing framework. Novelty of this proposed air pollution monitoring framework is that it will collect various air pollution parameters from sample air and then it analyzes the level with respect to the AQI. This paper gives the emphasis on analytics and knowledge discovery part by which it will be able to forecast the forthcoming pollution level. Finally it suggests some action plan based on proposed Rule Base to reduce the pollution level if the prediction is higher than the threshold. The organization of rest of the paper as follows. Section 2 explains air pollution monitoring approach and methodology. Expected outcome is listed in Sect. 3, and Sect. 4 concludes the paper.

2 Air Pollution Monitoring Approach

According to Centre for Science and Environment (CSE), the level of air pollution in Kolkata far exceeds the permissible limit. Therefore the CSE recommends that Kolkata should take up emergency air pollution control measures as adopted in Delhi [9].

2.1 Proposed Air Pollution Framework

This section proposes a data analytics based layered framework toward monitoring and reducing air pollution in the metropolitan city. This proposed framework is presented in Fig. 1 consists of four layers and the layered-wise functionalities are described in the following.

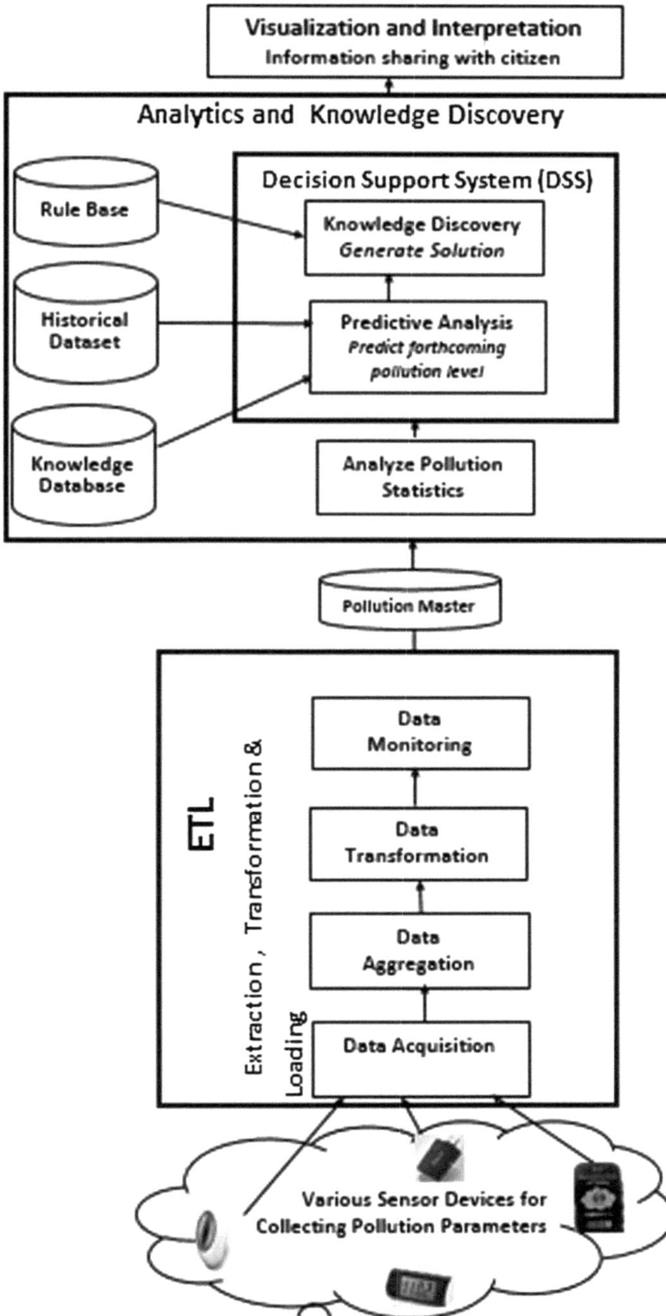

Fig. 1 Proposed air pollution monitoring framework

2.1.1 Source of Data

It includes various types of sensor devices

- Sensors, cameras, and other devices should be placed in various parts of the city for sample air collection.
- Various parameters for measuring air pollution like SO_2, NO_2, RSPM, and SPM will be calculated from the sample dataset.

2.1.2 Extraction, Transformation, and Loading (ETL)

- Sample data generated from various air samples would be aggregated for further analysis.
- The aggregated data should be transformed in proper format based on the monitoring requirement.
- After successful monitoring, the result dataset will be loaded into the pollution master database.

2.1.3 Analytics and Knowledge Discovery

- Specific analysis will be carried out from the pollution master database and statistical result can be interpreted.
- Predictive Analysis or *Analytics* will be performed on the output dataset to forecast the pollution level of the forthcoming days. Historical dataset and knowledge database both will be treated as two main sources of input generation for better prediction. Historical dataset would contain the pollution history and corresponding incidents occurred, as example, *how many accidents happened in a particular pollution level due to low visibility, how the pollution level affected the city life*, etc. Knowledge database will store information regarding the forthcoming events which may affect the pollution level of the city, like *information about the general holiday list*—assuming that city will encounter less traffic in the holidays which leads to lesser pollution level.
- The next part is *knowledge discovery* which would be performed based on the prediction generated in the previous level and with the help of *Rule Base*. Rule Base is basically a set of predefined rules along with their application criteria based on the pollution level. For example, Rule 1 will be applied if pollution parameters exceed a certain threshold value. Sample Rule Base is discussed in later section.

2.1.4 Visualization and Interpretation

- The last layer will represent the result after taking proper actions of the government authorities and city Management.
- The result must reflect the highly polluted region and laws or rules to be followed for those specific regions.
- It also suggests the safer region or less polluted region so that citizens can avail those routes to reach their destinations.

2.2 *Framework Methodology*

Major source of emission of carbon monoxide is transportation system. In order to control the air pollution, it needs to be monitored and vehicles responsible for polluting should be identified. Air quality index (AQI) plays major role in measuring or monitoring air pollution. It is a number provided by the government agencies to the citizen for their proper awareness. This number must act as a threshold value and should be displayed to the public who are traveling the region.

In India, the six categories of AQI are: *Good*, *Satisfactory*, *Moderately Polluted*, *Poor*, *Very Poor*, and *Severe,* depending upon the air quality. Associate health impacts (are numbered accordingly, like Impact 1, etc.) related to these AQI is presented in Table 1.

Suggested steps or plan of action to monitor air quality as well as pollution reduction are as follows.

Table 1 Categories of AQI and its associated health impacts [2, 10]

AQI	Associated health impacts
Good (0–50)	Impact 1: Safe
Satisfactory (51–100)	Impact 2: Slight uneasiness in breathing for sensitive people
Moderately polluted (101–200)	Impact 3: Uneasiness to people with heart disease, children and older adults
Poor (201–300)	Impact 4: Uneasiness to all kinds of people having heart disease
Very poor (301–400)	Impact 4 Impact 5: Respiratory illness could be found to the people on prolonged exposure. More vulnerable in people with lung and heart diseases
Severe (401–500)	Impact 5 Impact 6: May cause respiratory impact even on healthy people. The health impacts ma\be experienced even during light physical activity

2.2.1 Collection of Pollution Parameters from Sample Air

Different Air Pollution measurement parameters will be calculated from the samples collected through various sensors placed across the city. Collection of pollution is the main input to the framework. These ambient concentrations of the pollutants are used to measure the air quality index. Air pollutants can be SO_2, NO_2, RSPM, and SPM. Indices for single or individual pollutants should be collected and then summed up for generating the total air quality index. This process is iterative to standardize the value for the specific region at specific instant of time. The calculated value then needs to be stored in an "air pollution data file" (presented in Table 2) according to the location and time, respectively.

2.2.2 Analysis of Pollution Statistics

The air sample should be analyzed to find whether the present condition is vulnerable for city life. The mapping is developed between AQI and pollution percent violation based on the AQI impact on city life. Then a report can be prepared and minimal attributes for this analysis report are displayed in Table 3.

2.2.3 Predictive Analysis of the Forthcoming Pollution Level

This decision support system will predict the pollution level for forthcoming days by analyzing the pollution statistics collected from air sample along with historical dataset and knowledge database. As Example:

- Historical dataset (as presented in Table 4) includes various pollution history and their corresponding phenomenon, like 5 accidents occurred in a day due to low visibility if pollution percent violation is greater than 20. So it captures the event of creating smog in the air due to air pollution which effects visibility and hampers the safety of city life.

Table 2 Air pollution data file

Location	Date_Time	SO_2 ()	NO_2 ()	RSPM ()	SPM ()	AQI

Table 3 Sample statistics analysis report

Location	AQI	Weekly average	Standard deviation	Valid monitoring days	Percent violation

Table 4 Sample structure of historical dataset

Location	Pollution parameters	Period	Pollution percent violation	Overall pollution percent violation

Table 5 Sample structure of knowledge database

Location	Date	Major incident	Holiday

Table 6 Pollution metrics related rule set

Pollution percent violation	Selection of rule
<20%	1. Suggestive measure 1 2. Suggestive measure 2
≥ 20 and $\leq 40\%$	1. Rule 3 and (rule 4 or rule 5) 2. Suggestive measure 1 3. Suggestive measure 2
>40 and $\leq 60\%$	1. Rule 2
>60%	1. Rule 1

Table 7 Sample Rule Base

S. No.	Rule name
Rule 1	No private vehicle should pass touching the region. Emergency situation would be excused
Rule 2	Private vehicles carrying four people can only pass through. Emergency situation would be excused
Rule 3	Only senior citizens would be allowed to pass through. Emergency situation would be excused
Rule 4	Only odd numbered private vehicles should pass the place. Emergency situation would be excused
Rule 5	Only even numbered private vehicles should pass the place. Emergency situation would be excused
Suggestive measure 1	Shared service cab providers would get extra benefit in tax
Suggestive measure 2	Solar energy enabled vehicles will receive special discount on paying road tax

- Knowledge database (as presented in Table 5) includes various events like the incoming holidays, other events like any large people gathering events which affect traffic situation.

2.2.4 Designing Sample Rule Base

The sample table, Table 6, describes rule name and their application criteria based on the parameter "pollution percent violation" value. Table 7 describes few sample rules which can be applied depending on the pollution forecast for the incoming days. New rules can be added or removed depending on the requirement. It is suggested that adequate laws and corresponding measures should be applied so that rules can be successfully implemented across the city.

3 Expected Outcome

Expected benefits from the proposed air pollution framework are presented in the following:

(a) Citizens will lead life in a better healthy environment with clean air.
(b) Risk of diseases will be reduced with reduction of SO_2 and NO_2 in air. It will ensure relief from unnecessary smog, acid rain, and lung diseases.
(c) The amount of carbon emissions and also the fatalities caused due to air pollution will be reduced.
(d) Number of accidents will be reduced with increasing the visibility in road by reducing the pollution level accordingly.
(e) India will become a cleaner planet by its Swachh Bharat Abhiyan (Clean India Mission) campaign by 2019.

4 Conclusion

Air pollution monitoring study reveals that concentration of air pollutants is increasing in the air due to various reasons where growing number of vehicles dominate the other sources. Several initiatives have been taken to tackle the issue effectively and efficiently by the government authorities as well as various NGOs which includes both policy reformation and technological innovation [11]. This paper proposes a smart framework for metropolitan cities toward reducing air pollution. It presents a novel Rule Base for better healthy environment. Finally it also lists the expected outcome for the study.

References

1. "Air Quality Assessment" developed by Central Pollution Control Board, Retrieved from http://www.cpcb.nic.in/15-44.pdf on 17-06-2016.
2. Jana Spiroska, Md. Asif Rahman and Saptarshi Pal, "Air Pollution in Kolkata: An Analysis of Current Status and Interrelation between Different Factors", SEEU Review. Volume 8, Issue 1, Pages 182–214, ISSN (Print) 1409–7001, DOI: 10.2478/v10306-012-0012-7, February 2013
3. Brauer, M., Lencar, C., Tamburic, L., Koehoorn, M., Demers, P., & Karr, C. (2015). A cohort study of traffic-related air pollution impacts on birth outcomes (Doctoral dissertation, University of British Columbia).
4. "Citizens Report – Air Quality and Mobility in Kolkata" developed by Center for Science and Environment, New Delhi 2011 Retrieved from http://www.cseindia.org/userfiles/Kolkata%20Report.pdf on 17-06-2016.
5. A report on "National Air Quality Index (AQI)" launched by the Environment Minister under initiative 'Swachh Bharat' on 17th Oct-2014 Retrieved from http://pib.nic.in/newsite/PrintRelease.aspx?relid=110654 on 17–06-2016

6. Mansi Shah, "Waiting for Health Care: A Survey of a Public Hospital in Kolkata". Accessed from http://ccs.in/internship_papers/2008/Waiting-for-Healthcare-A-survey-of-a-public-hospital-in-Kolkata-Mansi.pdf on 22nd February 2016.
7. Pethuru Raj and Anupama C. Raman, (2015) "Intelligent Cities: Enabling Tools and Technology", CRC Press. (c) 2015
8. "Air Pollution level in Kolkata is much higher than permissible limit: CSE" Retrieved from http://articles.economictimes.indiatimes.com/2016-01-31/news/70222631_1_air-quality-anumita-roychowdhury-pm-2-5-levels on 17-06-2016
9. SuklaBhaduri, "Vehicular Growth and Air Quality at Major Traffic Intersection Points in Kolkata City, An Efficient Intervention Strategies", The SIJ Transactions on Advances in Space Research & Earth Exploration (ASREE), Vol. 1, No. 1, September-October 2013
10. "Air Quality Data" Retrieved from http://kolkata.usconsulate.gov/airqualitydata.html on 17-06-2016.
11. Schnelle Jr, K. B., Dunn, R. F., & Ternes, M. E. (2015). Air pollution control technology handbook. CRC press.

An Investigation of the Classifiers to Detect Android Malicious Apps

Ashu Sharma and Sanjay Kumar Sahay

Abstract Android devices are growing exponentially and are connected through the Internet accessing billion of online Websites. The popularity of these devices encourages malware developer to penetrate the market with malicious apps to annoy and disrupt the victim. Although for the detection of malicious apps different approaches are discussed. However, proposed approaches are not sufficed to detect the advanced malware to limit/prevent the damages. In this, very few approaches are based on opcode occurrence to classify the malicious apps. Therefore, this paper investigates the five classifiers using opcode occurrence as the prominent features for the detection of malicious apps. For the analysis, we use WEKA tool and found that FT detection accuracy ($\sim$79.27%) is best among the investigated classifiers. However, true positives rate, i.e. malware detection rate is highest ($\sim$99.91%) by RF and fluctuate least with the different number of prominent features compared to other studied classifiers. The analysis shows that overall accuracy is majorly affected by the false positives of the classifier.

Keywords Android security · Malware detection · Machine learning · Static analysis

1 Introduction

Android is one of the most popular operating systems for smart devices and is connected through the Internet accessing billions of online Websites. The exponential increase in android apps is basically due to the open source, third-party

A. Sharma · S.K. Sahay (✉)
Department of Computer Science and Information System, Birla Institute of Technology and Science, K.K. Birla Goa Campus, NH-17B, By Pass Road, Zuarinagar 403726, Goa, India
e-mail: ssahay@goa.bits-pilani.ac.in

A. Sharma
e-mail: p2012011@goa.bits-pilani.ac.in

D.K. Mishra et al. (eds.), *Information and Communication Technology*, Advances in Intelligent Systems and Computing 625,
https://doi.org/10.1007/978-981-10-5508-9_20

distribution, free rich SDK and the very much suited Java language. In this growing android apps market, it is very hard to know which apps are spam or malware content. As per Statista [1] $\sim 2 \times 10^6$ android apps are available at Google play store. Also, there are many third-party android apps available for the users [2], which may be malicious. Hence, potential of the malicious apps or malware entering these systems is now at never seen before levels.

Due to ease of use, these devices hold sensitive information such as personal data, browsing history, shopping history, financial details [3], i.e. users are even more frequent to use the Internet, as a consequence, these devices are vulnerable to cyber threats/attacks. In this, Quick Heal Threat Research Labs in the third quarter of 2015 reported that they have received samples of files at the rate of $\sim 4.2 \times 10^5$ samples per day for the Android and Windows platforms, and the Data security experts expect a rapid increase in number of new malware samples in 2016 compared to previous years [4].

The traditional approach, i.e. signature-based techniques, to detect the advanced malicious android apps is no longer effective, as it uses code obfuscation techniques. However, a number of methods have been proposed on static and dynamic analyses for analysing and detecting Android malware prior to their installation [5–9]. It appears that so far proposed approaches are not suffice to detect the advanced malware to limit/prevent the damages [10]. Therefore, we investigated the five classifiers (FT, Random forest, J48, LMT and NBT) and present a novel approach to combat malware threat/attack by analysing the opcode occurrence in the apps. The remaining paper is organised as follows. In the next section, we discuss the related work. Section 3 describes our approach to detect the malicious apps based on static analysis. The results of our approach are discussed in Sect. 4. Finally, Sect. 5 contains the conclusion and direction for the future work.

2 Related Work

Static and dynamic analyses are the two main approaches applied for detection of android malware [10]. In static analysis, without executing the apps codes are analysed to find a malicious pattern by extracting the features such as permissions, APIs used, control flow, data flow, broadcast receivers, intents, hardware components, etc., whereas in the dynamic analysis the apps are examined in run-time environment by monitoring the dynamic behaviour (network connections, system calls, resources usage, etc.) of the apps and the system response. However, in both the approaches selected classifiers are trained with a known data set to differentiate the benign and malicious apps. In this Seo et al. by analysing the permissions, dangerous APIs and keywords associated with malicious behaviours detected potential malicious scripts in Android apps [11]. A lightweight framework was discussed by Arp et al. which uses AndroidManifest.xml file and disassembled code to generate a joint vector space [12]. Wu et al. approach detects the malware by analysing AndroidManifest.xml and tracing the systems calls [13]. Sanz et al. analysed five classifiers with machine learning (DT, KNN, BN, RF and SVM) for automatic malware detection by analysing different sets of Android market permissions, ratings and a number of ratings. They found that among five classifiers BN performs the best while RF second and DT worst [14]. Vidas et al. developed a tool which automatically analyses the apps to find the least permissions/privileges that are required to run the apps [15]. In this, Fuchs et al. method analyses the data flow across the android apps components [16]. Daniel et al. did a broad static analysis by embedding the features in a joint vector space, such that the typical patterns of malware can be automatically identified [12]. In the DREBIN project, a study has been done with 123,453 benign and 5,560 malware apps. Based on a set of characteristics derived from binary and metadata, Gonzalez et al. proposed a method named as DroidKin, which can detect the similarity among the apps under various levels of obfuscation (code reordering, register reassignment, etc. [10, 17]) [18]. SVM-based malware detection scheme given by Gugian et al. integrates both risky permission combinations and vulnerable API calls and used them as features for the classification [19]. Saracino et al. [20] proposed a novel host-based malware detection system called MADAM which simultaneously analyses and correlates the features at four levels (kernel, application, user and package) to detect and stop the malicious behaviours. Quentin et al. uses opcode sequences to detect the malicious apps; however, the approach will not detect completely different malware [21]. Later on using N-opcode, BooJoong et al. classified the malware and reported F-measure 98% [22].

3 Our Approach

A novel approach to classify the unknown android malware is shown in Fig. 1, which involves finding the promising features (Algorithm 1), classifiers training and its detection.

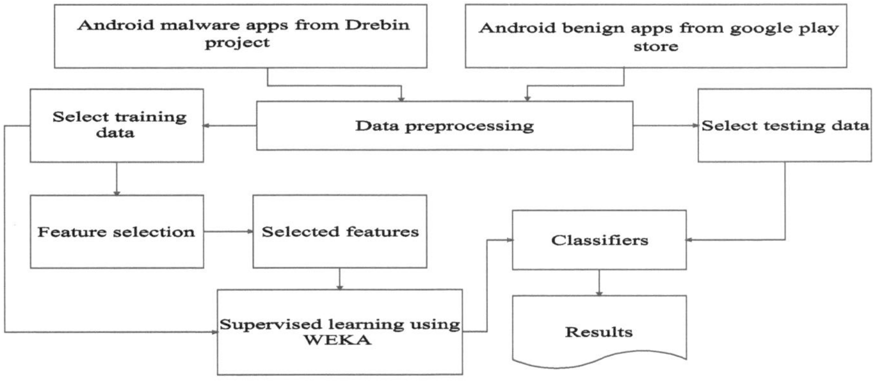

Fig. 1 Flow chart of the proposed approach for detection of android malicious apps

3.1 Data Pre-processing and Feature Selection

For the classification of unknown android malware apps, we downloaded 5531 android malware from DREBIN [12] and 2691 benign apps from Google play store. The benign apps are cross-verified from virustotal.com [23].

To understand the logic of android malware apps, we use freely available *apktool* [24] to decompress the android *.apk* files. After decompressing, we kept *.smali* files and discarded other created files/folders. The *.smali* files contain only one class information and are equivalent to *.class* file. To find the prominent features for classification of android malware and benign, we extracted the opcodes (list of the android opcodes is available at http://pallergabor.uw.hu/androidblog/dalvik_opcodes.html) of the apps from the obtained *.smali* files. We analysed the opcode occurrence of all the android apps and found that the occurrence of many opcodes in malware and benign apps differs in large. The normalized opcode occurrence of both the apps is shown in Fig. 2. The mapping of the opcodes with hexadecimal representation has been kept same as given by the android developers [25]. The prominent opcodes (features), which suppose to distinguish the malicious and benign android apps, are obtained as described in the Algorithm 1. For the classification, we have used Waikato Environment for Knowledge Analysis (WEKA) tool, a collection of visualisation tools and algorithms for data analysis and predictive modelling, together with graphical user interfaces for easy access to this functionality [26], in which many inbuilt classifiers are available. On the basis of studies done by Sharma and Sahay [27, 28], we have selected the best classifier (Random forest [29], Logistic model trees (LMT) [30], Naive Bayes tree (NBT) [31], J48 [32] and Functional Tree (FT) [33]) for in-depth analysis by using *K*-fold cross-validation technique.

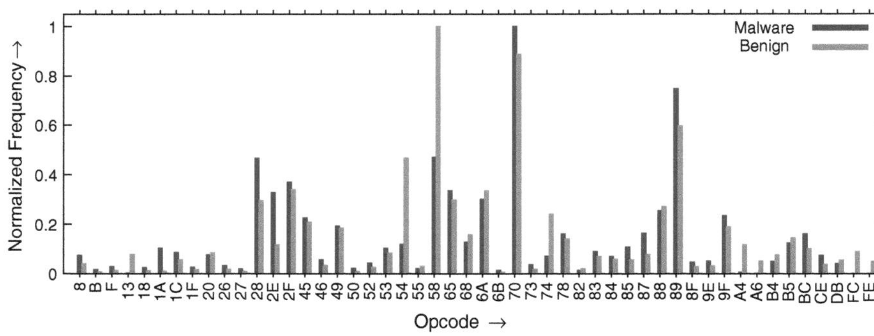

Fig. 2 Dominant opcodes of malicious and benign android apps

Algorithm 1 Feature Selection

INPUT: Pre-processed data
$\mathbf{N_B}$: Number of benign android apps, $\mathbf{N_M}$: Number of malware android apps,
n: Total number of prominent features required.
OUTPUT: List of prominent features

BEGIN
for all benign apps **do**

Compute sum of the frequencies $\mathbf{f_i}$ of each opcode **Op** and normalize it.

$$F_B(\mathrm{Op}_j) = \left(\sum f_i(\mathrm{Op}_j)\right)/N_B$$

end for
for all malware data **do**

Compute sum of the frequencies $\mathbf{f_i}$ of each opcode **Op** and normalize it.

$$F_M(\mathrm{Op}_j) = \left(\sum f_i(\mathrm{Op}_j)\right)/N_M$$

end for
for all opcode $\mathbf{Op_j}$ **do**

Find the difference of the normalized frequencies for each opcode $\mathbf{D(Op_j)}$.

$$D(\mathrm{Op}_j) = \left|F_B(\mathrm{Op}_j) - F_M(\mathrm{Op}_j)\right|$$

end for
return n number of prominent opcodes as features with high **D**(**Op**).

4 Result Analysis

The five selected classifiers are analysed by applying supervised machine learning technique with K-fold cross-validation for $k = 10$. For the analysis, we first obtained the top 200 promising features (Algorithm 1). The accuracy of the classifiers is obtained by varying the promising features and is measured by the equation

$$\text{Accuracy} = \frac{\text{TP} + \text{TN}}{\text{TP} + \text{FN} + \text{TN} + \text{FP}} \times 100 \qquad (1)$$

where

TP True positive, the number of malware apps correctly classified.
FN False negative, the number of malware apps incorrectly classified.
TN True negative, the number of benign apps correctly classified.
FP False positives, the number of benign apps incorrectly classified.

The performance of the classifier has been studied by taking 20% of available data (not used for training) with 20–200 best features, incrementing 20 features at each step, and the result obtained are shown in Fig. 3. From the analysis, we find that the best accuracy obtained by FT, random forest, J48, LMT and NBT is approximately 79.27, 74.95, 71.73, 70.51 and 68.67% (Fig. 4). Among these classifiers the least fluctuation in the accuracy by varying the features is observed in random forest. Figure 5 shows the TPR (malware detection rate) of all five classifiers with a different number of features. We found that the RF gives maximum TPR with least fluctuation compared to other classifiers.

Figure 6 shows the TNR (benign detection rate) for all five classifiers with a different number of features. Here with some exception, we observed that FT detected the benign better than the other classifiers with a different number of features. Figure 7 shows the false negatives of all selected classifier, in which

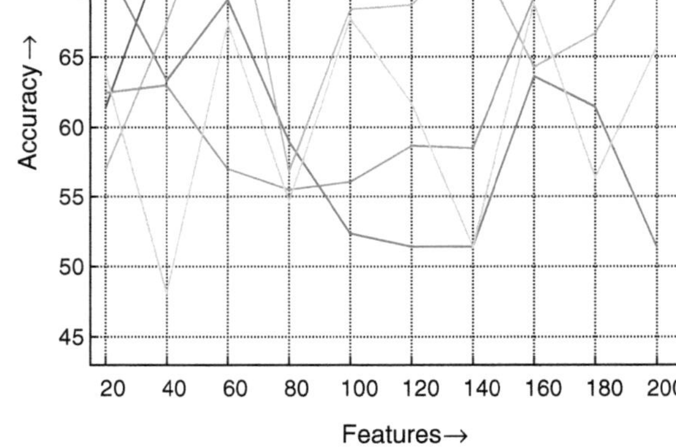

Fig. 3 Detection accuracy obtained by the selected five classifiers with different number of prominent features

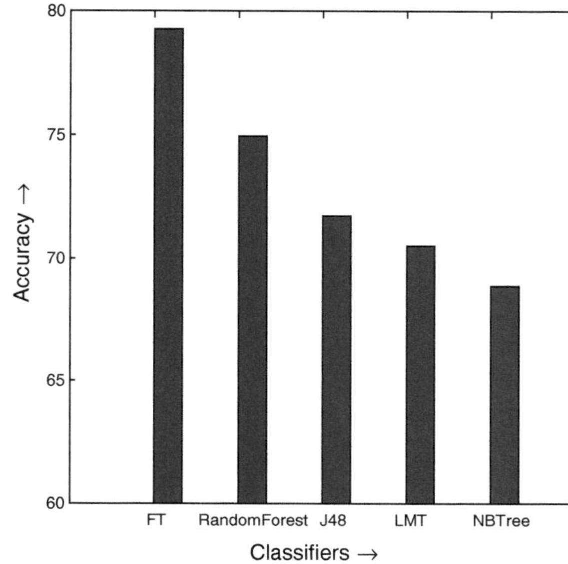

Fig. 4 Best accuracy obtained by the selected five classifiers

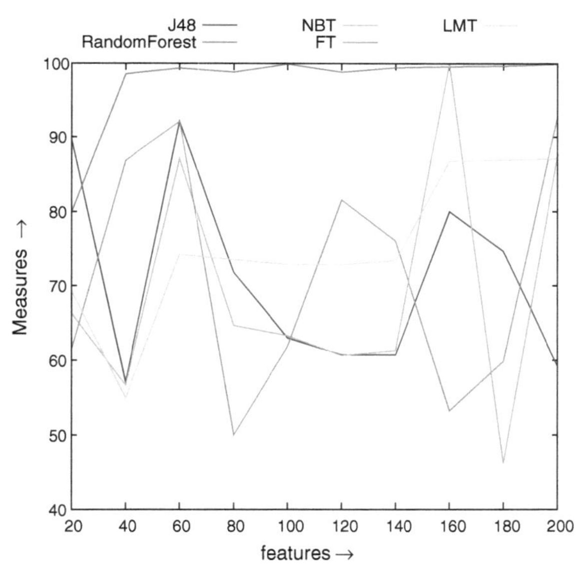

Fig. 5 True positives obtained by selected five classifiers with different number of prominent features

compared to other classifiers the RF is good and also fluctuation is least with the number of features. Figure 8 shows the false positives of the analysed classifiers, and here we observed that all the five classifiers do not give a good result; hence, they very much affect the final accuracy. However, the false negative of RF is not as par but the fluctuation with the number of features is least compared to other classifiers.

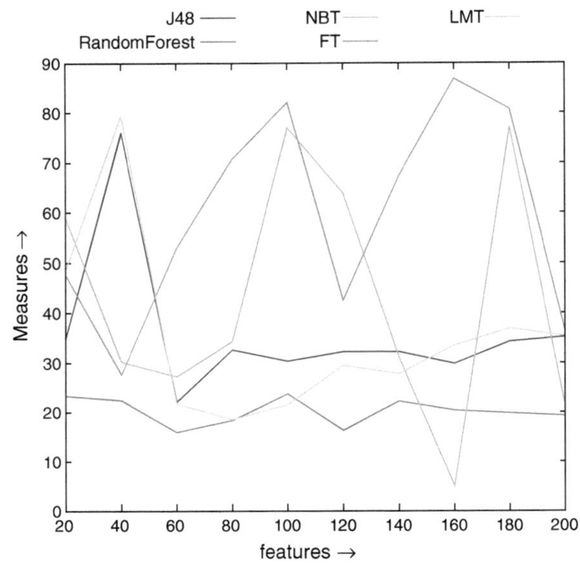

Fig. 6 True negatives obtained by selected five classifiers with different number of prominent features

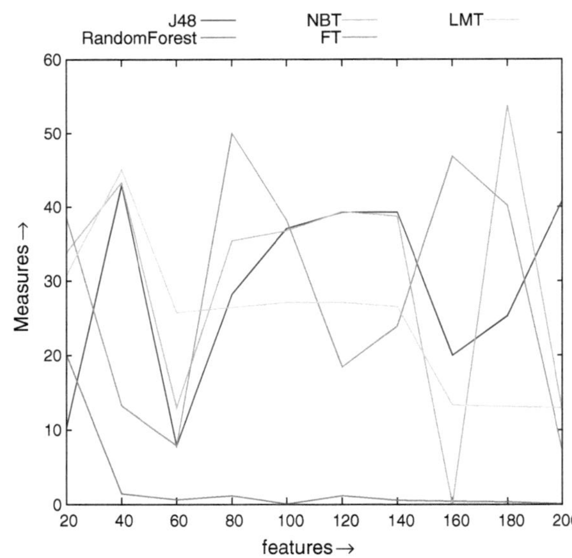

Fig. 7 False negatives obtained by selected five classifiers with different number of prominent features

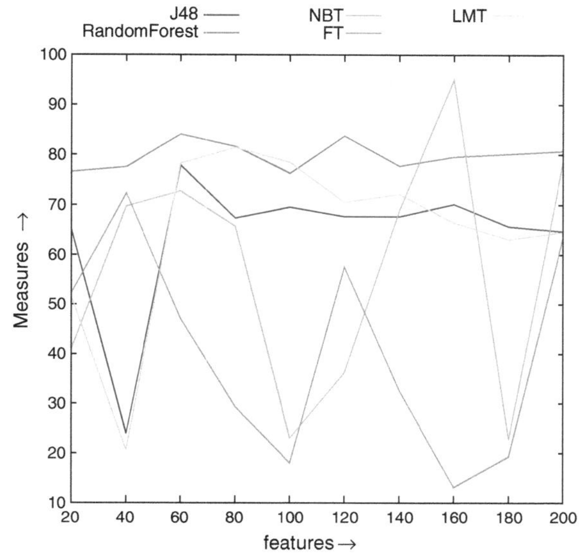

Fig. 8 False positives obtained by selected five classifiers with different number of prominent features

5 Conclusion

The threat/attack from the malicious apps in android devices is now never seen at before levels, as millions of android apps are available officially (Google play store) and unofficially. Some of these available apps may be malicious; hence, these devices are very much vulnerable to cyber threat/attack. The consequence will be devastating if in time counter-measures are not developed. Therefore, in this paper, we investigated five classifiers FT, Random forest, J48, LMT and NBT for the detection of malicious apps. We found that among the studied classifiers, FT is the best classifier and detect the malware with $\sim$79.27% accuracy. However, true positives, i.e. malware detection rate, is highest $\sim$79.27% by RF and fluctuates least with the different number of prominent features compared to other studied classifiers, which is better than BooJoong et al., *F*-measure (98%) [22]. The analysis shows that overall accuracy is majorly affected by the false positives of the classifier. Hence, in future more detailed study is required to decrease the false-positive and false-negative ratio for overall good accuracy and in this direction work is in progress, showing impressive results.

Acknowledgements Mr. Ashu Sharma is thankful to BITS, Pilani, K.K. Birla Goa Campus, for the support to carry out this work through Ph.D. scholarship No. Ph603226/Jul. 2012/01.

References

1. Statista: Number of available applications in the google play store from December 2009 to February 2016 (August 2016), https://developer.android.com/guide/topics/security/permissions.html
2. 9apps: Free android apps download (August 2016), http://www.9apps.com/
3. Threat report 3rd quarter, 2015 (2015), http://www.quickheal.co.in/resources/threat-reports
4. Data, G.: Mobile malware report. Tech. rep., G DATA (2015)
5. Enck, W., Gilbert, P., Han, S., Tendulkar, V., Chun, B.G., Cox, L.P., Jung, J., McDaniel, P., Sheth, A.N.: Taintdroid: an information-flow tracking system for realtime privacy monitoring on smartphones. ACM Transactions on Computer Systems (TOCS) 32(2), 5 (2014)
6. Felt, A.P., Chin, E., Hanna, S., Song, D., Wagner, D.: Android permissions demystified. In: Proceedings of the 18th ACM conference on Computer and communications security. pp. 627–638. ACM (2011)
7. Grace, M., Zhou, Y., Zhang, Q., Zou, S., Jiang, X.: Riskranker: scalable and accurate zero-day android malware detection. In: Proceedings of the 10th international conference on Mobile systems, applications, and services. pp. 281–294. ACM (2012)
8. Reina, A., Fattori, A., Cavallaro, L.: A system call-centric analysis and stimulation technique to automatically reconstruct android malware behaviors. EuroSec, April (2013)
9. Yan, L.K., Yin, H.: Droidscope: seamlessly reconstructing the os and dalvik semantic views for dynamic android malware analysis. In: Presented as part of the 21st USENIX Security Symposium (USENIX Security 12). pp. 569–584 (2012)
10. Sharma, A., Sahay, S.K.: Evolution and detection of polymorphic and metamorphic malwares: a survey. International Journal of Computer Applications 90(2), 7–11 (March 2014)
11. Seo, S.H., Gupta, A., Sallam, A.M., Bertino, E., Yim, K.: Detecting mobile malware threats to homeland security through static analysis. Journal of Network and Computer Applications 38, 43–53 (2014)
12. Arp, D., Spreitzenbarth, M., Hubner, M., Gascon, H., Rieck, K.: Drebin: Effective and explainable detection of android malware in your pocket. In: NDSS (2014)
13. Wu, D.J., Mao, C.H., Wei, T.E., Lee, H.M., Wu, K.P.: Droidmat: Android malware detection through manifest and api calls tracing. In: Information Security (Asia JCIS), 2012 Seventh Asia Joint Conference on. pp. 62–69. IEEE (2012)
14. Sanz, B., Santos, I., Laorden, C., Ugarte-Pedrero, X., Bringas, P.G.: On the automatic categorisation of android applications. In: 2012 IEEE Consumer communications and networking conference (CCNC). pp. 149–153. IEEE (2012)
15. Vidas, T., Christin, N., Cranor, L.: Curbing android permission creep. In: Proceedings of the Web. vol. 2, pp. 91–96 (2011)
16. Fuchs, A.P., Chaudhuri, A., Foster, J.S.: Scandroid: Automated security certification of android. Tech. rep., University of Maryland Department of Computer Science (2009)
17. Sharma, A., Sahay, S.K., Kumar, A.: Improving the detection accuracy of unknown malware by partitioning the executables in groups. In: Advanced Computing and Communication Technologies, pp. 421–431. Springer (2016)
18. Gonzalez, H., Stakhanova, N., Ghorbani, A.A.: Droidkin: Lightweight detection of android apps similarity. In: International Conference on Security and Privacy in Communication Systems. pp. 436–453. Springer (2014)
19. Schölkopf, B., Platt, J.C., Shawe-Taylor, J., Smola, A.J., Williamson, R.C.: Estimating the support of a high-dimensional distribution. Neural computation 13(7), 1443–1471 (2001)
20. Saracino, A., Sgandurra, D., Dini, G., Martinelli, F.: Madam: Effective and efficient behavior-based android malware detection and prevention (2016)
21. Jerome, Q., Allix, K., State, R., Engel, T.: Using opcode-sequences to detect malicious android applications. In: 2014 IEEE International Conference on Communications (ICC). pp. 914–919. IEEE (2014)

22. Kang, B., Yerima, S.Y., McLaughlin, K., Sezer, S.: N-opcode analysis for android malware classification and categorization. In: Cyber Security And Protection Of Digital Services (Cyber Security), 2016 International Conference On. pp. 1–7. IEEE (2016)
23. Virustotal - free online virus, malware and url scanner (June 2016), https://www.virustotal.com/
24. Winsniewski, R.: Android–apktool: A tool for reverse engineering android apk files (2012)
25. Paller, G.: Dalvik opcodes, http://pallergabor.uw.hu/androidblog/dalvik_opcodes.html
26. Holmes, G., Donkin, A., Witten, I.H.: Weka: A machine learning workbench. In: Intelligent Information Systems, 1994. Proceedings of the 1994 Second Australian and New Zealand Conference on. pp. 357–361. IEEE (1994)
27. Sahay, S.K., Sharma, A.: Grouping the executables to detect malwares with high accuracy. Procedia Computer Science 78, 667–674 (2016)
28. Sharma, A., Sahay, S.K.: An effective approach for classification of advanced malware with high accuracy. International Journal of Security and Its Applications 10(4), 249–266 (2016)
29. Rodriguez, J.J., Kuncheva, L.I., Alonso, C.J.: Rotation forest: A new classifier ensemble method. IEEE transactions on pattern analysis and machine intelligence 28(10), 1619–1630 (2006)
30. Landwehr, N., Hall, M., Frank, E.: Logistic model trees. Machine Learning 59(1–2), 161–205 (2005)
31. Kohavi, R.: Scaling up the accuracy of naive-bayes classifiers: A decision-tree hybrid. In: KDD. vol. 96, pp. 202–207. Citeseer (1996)
32. Bhargava, N., Sharma, G., Bhargava, R., Mathuria, M.: Decision tree analysis on j48 algorithm for data mining. Proceedings of International Journal of Advanced Research in Computer Science and Software Engineering 3(6) (2013)
33. Gama, J.: Functional trees. Machine Learning 55(3), 219–250 (2004)

A Case-Based Reasoning Framework for Prediction of Stroke

Pattanapong Chantamit-o-pas and Madhu Goyal

Abstract Case-based reasoning (CBR) has been a popular method in health care sector from the last two decades. It is used for analysis, prediction, diagnosis and recommending treatment for patients. This research purposes a conceptual CBR framework for stroke disease prediction that uses previous case-based knowledge. The outcomes of this approach not only assist in stroke disease decision-making, but also will be very useful for prevention and early treatment of patients.

Keywords Case-based reasoning · Stroke disease · Decision-making · Prediction

1 Introduction

Stroke is the second or third most common cause of death in most countries [1, 2]. The patients who survived usually have poor quality of life because of serious illness and long-term disability and become burden to their families and health care system. There is a strong demand for the management focused on prevention and early treatment of diseases by analysing different factors. Several health conditions and lifestyle factors have been identified as risk factors for stroke. These factors have three groups as follows: the risk factors cannot be change, the risk factors can be changed (treated or controlled) and other risk factors are less well-documented. The risk factor cannot be changed is focused on demographic data such as age, heredity (family history), race, sex (gender), and prior stroke, transient ischaemic attack (TIA) or heart attack. Some patients have had some behaviour and/or other disease before stroke attack. Furthermore, they are trying to control health and

P. Chantamit-o-pas (✉) · M. Goyal
Faculty of Engineering and Information Technology,
University of Technology Sydney, Broadway, PO BOX 123,
Sydney, NSW 2007, Australia
e-mail: Pattanapong.Chantamit-o-pas@student.uts.edu.au

M. Goyal
e-mail: Madhu.Goyal-2@uts.edu.au

D.K. Mishra et al. (eds.), *Information and Communication Technology*,
Advances in Intelligent Systems and Computing 625,
https://doi.org/10.1007/978-981-10-5508-9_21

behaviour (as personality behaviour and eating behaviour) and change the quality of life that can prevent from stroke disease, for example hypertension, many kinds of heart diseases such as myocardial infarction (MI), diabetes mellitus (DM), valvular heart diseases and atrial fibrillation, asymptomatic carotid artery disease, blood lipids and smoking. The other risk factors that less well-documented are geographic location, socio-economic factor, alcohol abuse and drug abuse [3]. Recognition of these risk factors is important to reduce the incidence of stroke, which has been increasing [4].

Case-Based Reasoning (CBR) is a methodology for solving problem that uses a previous data or memorized problem situations called cases. The processes of CBR system proceed in four main steps such as *retrieve, reuse, revise and retain* (Fig. 1) [5]. The new case starts at the top of stage, where an input is entered into the system. The previous case is compared to the new case and start *retrieve* step. In practical, CBR system is a comparison between all the cases in the system and a new case, and then the result will list the ranking of similar cases.

In this research, we propose a conceptual case-based reasoning framework to predict from patient risk factors and to recognize case that probably develops stroke or preparing patients to handle diseases burden outcome. This framework is comparing stroke patients in database and predicting patients who have risk factors

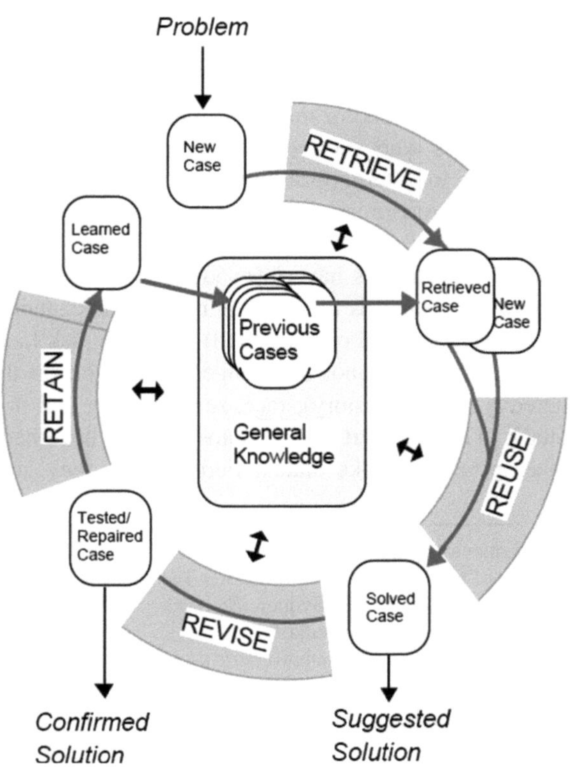

Fig. 1 The CBR cycle implemented by Aamodt and Plaza [5]

which are related to stroke disease such as smoking, high blood pressure and so on. It would not only support medical professionals for stroke disease decision-making, but also provide suggestion and warnings to patients before they visit a hospital or go for costly medical check-ups.

The rest of this paper is organized as follows. Section 2 is the related work which reviews case-based reasoning (CBR) in health care sector as well as in other domains. Section 3 proposes the conceptual CBR framework for stroke disease. The conclusion and future work are presented in Sect. 4 of this paper.

2 Related Work

This section reviews the research done on case-based reasoning in various domains and also case-based reasoning in health care sector.

2.1 Case-Based Reasoning in Health Care

The Case-Based Reasoning Systems have many application areas in health care sector which have provided solutions for diagnosis and treatment of diseases based on past experiences. For example, the mixture of experts for case-based reasoning (MOE4CBR) [6] is an application for high-dimensional biological domains to prediction to disease. The data sets are used in ovarian mass spectrometry, leukaemia and lung microarray data sets. The biomedical domains are complex, but also a system is unsuitable method for this research. They used data mining and logistic regression methods applied in a system and also improved the classification performance. A case is defined by logistic regression approach that supports to filter the important feature in CBR. Similar cases are also grouped by data mining technique. Thus, the system also supports for the "dimensionality" problem in this domain. For complex medical diagnosis, if patients have a complex disease, more medical domains have to be used for this. For example, the premenstrual syndrome (PMS) is related among gynaecology and psychiatry and also needs complex algorithm for diagnosis. The CBR-based expert system used the k-nearest neighbour (k-NN) algorithm to search *k* similar case that focusing on the Euclidean distance measure [7]. A CBR in treatment and management of diabetes is also represented in an application. It solved the problem by using patient health record such as demographic data, laboratory result and physical examination. Those are compared with previous case by using k-NN algorithm [8]. For a complex data, a CBR has applied by using machine learning and data mining techniques that based on gene expression profiles. This method used k-NN with weighted-feature based technique to retrieve and compare among previous cases and new cases. The herein-proposed methodology is used on several data sets in this framework. The results show how many percentage of gene expression profile of new patients are similar to previous cases and help to predict the risk of disease [9].

Moreover, Sharaf-el-deen et al. [10] introduced the automated adaptation process, which applies the adaptation rules for solving the new case. To evaluate the approach, the researchers develop the prototype for diagnosing breast cancer and thyroid diseases. They proposed a hybrid-based medical diagnosis approach in order to enhance the performance of the CBR retrieval system. The main idea of the proposed approach is to combine both case-based reasoning and rule-based reasoning. In addition, Ahmed et al. [11] apply various data processing and feature extraction techniques by considering time and frequency domains for disease prediction. Given input data, the CBR system discovers the relevant cases and then creates an alarm based on the output. To evaluate the proposed system, the researchers compared it with the classification results from experts.

Furthermore, Amin et al. [12] proposed clinical decision support systems for disease prediction and diagnosis. These approaches are able to extract hidden pattern and relationships among medical data. This leads the proposed approaches to be efficient for designing the decision support systems.

2.2 *Case-Based Reasoning in Other Domains*

The CBR had been applied in other sectors such as information technology, educational technology, bankruptcy prediction modelling and so on. Jonassen and Hernandez-Serrano [13] stated that problem solving on organization is complex that can solve the program from previous case or similar case. Normally, organization had been applied the lessons learned from their old stories to the new problems that has significance to decision-making and to justify the use of previous case as instruction support. In addition, Bryant [14] proposed a CBR to bankruptcy prediction modelling. He stated that financial company has risk to many factors as stakeholders, customers, investors, managers and employees. This model used various factors from financial statement in 500 firms from 1990 to 1994 in non-bankrupt and 14 firms between 1990 and 1994 in bankruptcy case and also used clustering and decision tree techniques for analyse and prediction to bankruptcy in organization.

Moreover, a CBR is integrated with a fuzzy decision tree (FDT) and genetic algorithm (GA), called “The hybrid classification model” [15]. The approach aimed to develop a decision-making system for solving classification problems among various databases. The case-based approach is used for clustering data into small cases, and then the genetic algorithms are applied for enhancing the fuzzy decision tree.

3 Case-Based Reasoning Framework for Stroke Patients

The overview of framework proposed for prediction method is shown in Fig. 2. The framework consists of 6 processes for prediction of stroke patient: clustering, retrieval, reusing, prediction, retain and review, and store, respectively. This framework has two processes: clustering and prediction (which are shown in dot line) which differ from the original framework.

Flow chart of the proposed framework is presented in Fig. 3. The detail of processes is described below as follows:

(a) Clustering process—this process aims to cluster stroke patient records, based on age, gender and race of patients. Those clusters are important factors to predict stroke disease. *K*-mean clustering technique is applied for finding groups to partition *n* observations into *k* clusters. The basic algorithm is given by Eq. (1) [16]:

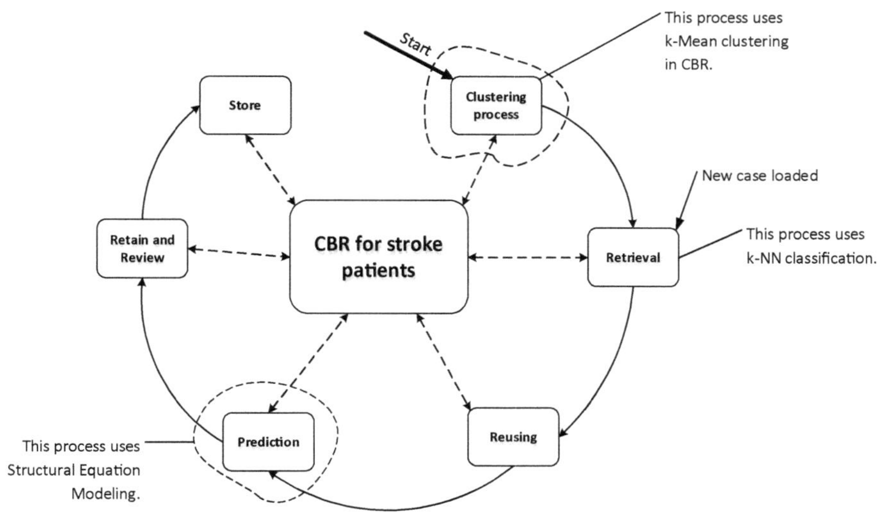

Fig. 2 An overview of the case-based reasoning framework for prediction of stroke disease

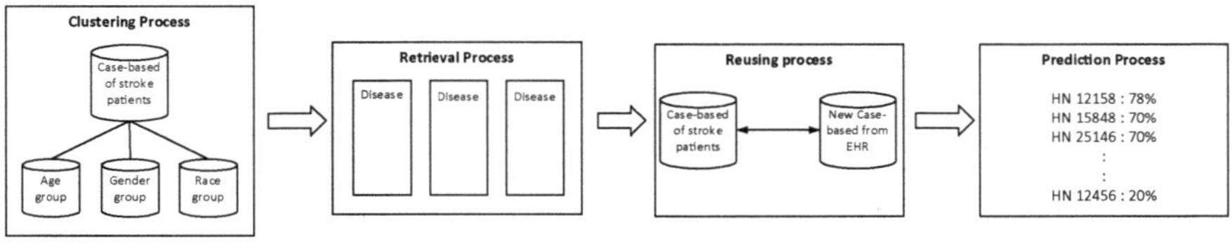

Fig. 3 Flow chart of the case-based reasoning for prediction of stroke patients

$$j = \sum_{i=1}^{k} \sum_{x \in S_i} \|x_i - c_i\|^2 \quad (1)$$

We assume that $(x_1, x_2, x_3, \ldots, x_n)$ is a collection of observations, where x_i is the ith dimensional real vector. The observations are partitioned into k groups; $s = \{s_1, s_2, s_3, \ldots, s_k\}$, and c_j is mean of s_j.

(b) Retrieval process—this process is retrieval in which an electronic health care records (EHR) are compared with information stored in "*knowledge containers*" [17]. A CBR system for stroke patients includes a cased-based knowledge, two medical knowledge databases (medical vocabulary knowledge and medical or clinical knowledge) and EHR (Fig. 4). The medical vocabulary knowledge contains stroke vocabularies and related diseases. Knowledge from experts is represented in clinical knowledge for a hospital with an acute stroke unit. Given output from the previous process, this process uses k-nearest neighbour (k-NN) approach based on medical and vocabulary knowledge to classify each patient group into risk factors.

To calculate the distance between p and q in k-NN algorithm, Eq. (2) is applied [18].

$$\text{dist} = \sqrt{\sum\nolimits_{i=1}^{n} (p_i - q_i)^2} \quad (2)$$

where $p = (p_1, p_2, p_3, \ldots, p_n)$, $q = (q_1, q_2, q_3, \ldots, q_n)$ and n is the number of dimensions.

(c) Reusing process—this process aims to match cases that are relevant to the given risk factors from the previous process. As we mentioned above, cases are collected from the real cases in the hospital and stored in the knowledge container. In this paper, we use those cases for stroke prediction in the next process.

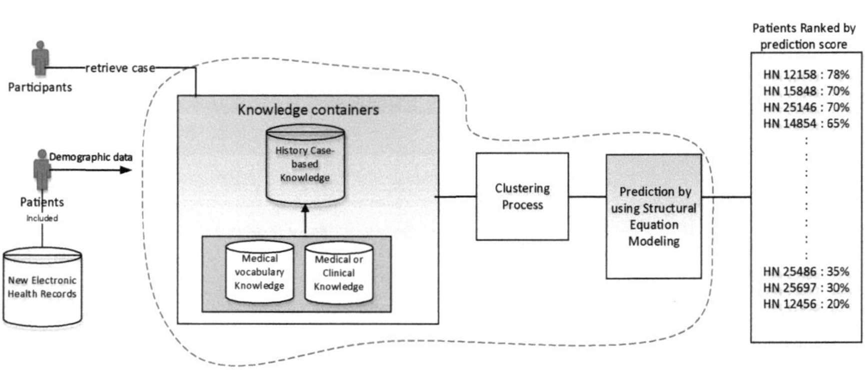

Fig. 4 The detail of the case-based reasoning for prediction of stroke patients

(d) Prediction process—data mining and statistical methods are well known for dealing with medical data analysis and prediction. To properly select tools and develop prediction models, general and incomplex guidelines are necessary and required [19]. This process uses structural equation modelling (SEM) because the data type has shown to multiple groups in current patient records and risk factors if they got it such as diabetes data set, heart disease data set, behaviour data set and so on, which is a data set that depends on stroke and other diseases. The SEM supports multiple values to prediction and can be described "The basic statistic of SEM is the covariance, which is defined for two continuous observed variables X and Y, where r_{XY} is the Pearson correlation and SD_X and SD_Y are their standard deviations. A covariance represents the strength of the association between X and Y and their variabilities, although with a single number. Because the covariance is an unstandardized statistic, its value has no upper or lower bound [20]". The formula is given by (3):

$$\mathrm{COV}_{xy} = r_{xy}\mathrm{SD}_x\mathrm{SD}_y \tag{3}$$

In term of stroke disease, there exist various risk factors that are useful for effectively predicting disease. The analysis process identifies variables. The age values are independence variables (called "primary variables") and risk factors are dependence variables, then the algorithm analyses and predicts between previous cases and current patient records. It processes case-by-case with other disease groups that relate to risk factors. After that, the output presents in terms of stroke risk estimation. For data sets of stroke that used in three main groups such as the risk factor cannot be changed, the risk factor can be changed, treated or controlled, and other risk factors are less well-documented. The first group is demographic data such as age, race, gender and prior stroke. The second group consists of behaviour and historical disease from EHR such as hypertension, heart disease, atrial fibrillation, peripheral artery disease, carotid, diabetes mellitus, obesity, high blood cholesterol, sickle cell disease, first stroke, alcohol abuse, poor diet, physical inactivity, drug abuse and smoking. The last group includes hometown of patient, socio-economic factor, alcohol abuse and drug abuse. These are loaded from current patient records. For stroke patients, incidence of stroke is required and loaded from historical records.

(e) Retain and Review process—After that, the output will be verified and sent to participants or nurses. The result shows the percentage of stroke for individual patient.

(f) Store process—The prediction results of patients who have risk factors in stroke disease will be stored in CBR system for reuse in the future. This information can help in decision-making for participants in order to make a suggestion and warnings to patients as care plan, lifestyle, quality of life, behaviour and so on. Finally, the outputs are updated in historical case-based knowledge.

4 Conclusion and Future Work

A Case-Based Reasoning has been applied for diagnosis diseases such as diabetes, leukaemia and lung, premenstrual syndrome, breast cancer and thyroid. In this paper, we have purposed the CBR framework for stroke disease. There are two processes which differ from the original case-based framework (clustering process and prediction process). The result of CBR framework is quite significant decision-making for patients. Specially, it can give suggestions and warnings to patient in spite of the fact that stroke does not have warning signs. Consequently, the proposed framework is beneficial for stoke disease management. In future, we will compare our framework with other prediction techniques and implement an e-stroke application.

References

1. Langhorne, P., Bernhardt, J., Kwakkel, G.: Stroke rehabilitation. The Lancet 377, 1693–1702 (2011)
2. Khosla, A., Cao, Y., Lin, C.C.-Y., Chiu, H.-K., Hu, J., Lee, H.: An integrated machine learning approach to stroke prediction. Proceedings of the 16th ACM SIGKDD international conference on Knowledge discovery and data mining, pp. 183–192. ACM, Washington, DC, USA (2010)
3. The American Heart Association, http://www.strokeassociation.org/STROKEORG/AboutStroke/UnderstandingRisk/Understanding-Stroke-Risk_UCM_308539_SubHomePage.jsp
4. Gorelick, P.B., Sacco, R.L., Smith, D.B., Alberts, M., Mustone-Alexander, L., Rader, D., Ross, J.L., Raps, E., Ozer, M.N., Brass, L.M.: Prevention of a first stroke: a review of guidelines and a multidisciplinary consensus statement from the National Stroke Association. Jama 281, 1112–1120 (1999)
5. Aamodt, A., Plaza, E.: Case-based reasoning: Foundational issues, methodological variations, and system approaches. AI communications 7, 39–59 (1994)
6. Arshadi, N., Jurisica, I.: Data mining for case-based reasoning in high-dimensional biological domains. IEEE Transactions on Knowledge and Data Engineering 17, 1127–1137 (2005)
7. Chattopadhyay, S., Banerjee, S., Rabhi, F.A., Acharya, U.R.: A Case-Based Reasoning system for complex medical diagnosis. Expert Systems 30, 12–20 (2013)
8. Kiragu, M.K., Waiganjo, P.W.: Case based Reasoning for Treatment and Management of Diabetes. Diabetes 145, (2016)
9. Anaissi, A., Goyal, M., Catchpoole, D.R., Braytee, A., Kennedy, P.J.: Case-Based Retrieval Framework for Gene Expression Data. Cancer Informatics 14, 21–31 (2015)
10. Sharaf-el-deen, D.A., Moawad, I.F., Khalifa, M.E.: A New Hybrid Case-Based Reasoning Approach for Medical Diagnosis Systems. J Med Syst 38, 1–9 (2014)
11. Ahmed, M.U., Banaee, H., Loutfi, A.: Health monitoring for elderly: An application using case-based reasoning and cluster analysis. ISRN Artificial Intelligence 2013, (2013)
12. Amin, S.U., Agarwal, K., Beg, R.: Genetic neural network based data mining in prediction of heart disease using risk factors. In: Information & Communication Technologies (ICT), 2013 IEEE Conference on, pp. 1227–1231. (Year)
13. Jonassen, D.H., Hernandez-Serrano, J.: Case-based reasoning and instructional design: Using stories to support problem solving. Educational Technology Research and Development 50, 65–77 (2002)

14. Bryant, S.M.: A case-based reasoning approach to bankruptcy prediction modeling. Intelligent Systems in Accounting, Finance & Management 6, 195–214 (1997)
15. Chang, P.-C., Fan, C.-Y., Dzan, W.-Y.: A CBR-based fuzzy decision tree approach for database classification. Expert Systems with Applications 37, 214–225 (2010)
16. MacQueen, J.: Some methods for classification and analysis of multivariate observations. In: Proceedings of the Fifth Berkeley Symposium on Mathematical Statistics and Probability, pp. 281–297. University of California Press, (1967)
17. Richter, M.M., Weber, R.: Case-Based Reasoning: A Textbook. Springer Science & Business Media (2013)
18. Han, J., Pei, J., Kamber, M.: Data mining: concepts and techniques. Elsevier (2011)
19. Bellazzi, R., Zupan, B.: Predictive data mining in clinical medicine: Current issues and guidelines. International Journal of Medical Informatics 77, 81–97 (2008)
20. Kline, R.B.: Principles and practice of structural equation modeling (2015)

Cognitive Radio: A Network Structure Perspective

Tapan Kumar, Vansha Kher and Pooja Jain

Abstract The ideal utilization of radio spectra is a major issue of concern in the field of wireless communication. Increasing demand for wireless radio services has led to the issue of frequency scarcity. Therefore, in order to accommodate more and more users, cognitive radio technology came into existence. The adaptive nature of cognitive radio helps them enhance the spectral efficiency, thereby utilizing the available spectra without causing any interference for the licensed users. The primary task of cognitive radio lies in the spectrum sensing and identification of holes. But the presence of a single CR and multiple secondary users in the network can lead to delay and collision. Therefore, the algorithm named "multiple CRs single-hop (MCSH) secondary user cognitive radio network architecture" has been formulated and proposed in which multiple CRs can coordinate with each other via single hop as well as with unlicensed users in order to diminish the delay, jitter, and packet loss.

Keywords Cognitive radio · Multiple CRs · Secondary user · Sensing · Multi-hop

1 Introduction

The electromagnetic spectrum is a wholesome, natural resource meant for data transmission and reception of data, and the exploitation of it by a large number of transmitters and receivers is strictly licensed by the government [1]. The federal communications commission (FCC) is the central agency that is solely responsible for the maintenance, control, and regulation of interstate telecommunication,

T. Kumar (✉) · V. Kher · P. Jain
Indian Institute of Information Technology, Kota, India
e-mail: tapankumarjain@gmail.com

V. Kher
e-mail: vansha26may@gmail.com

P. Jain
e-mail: poojaalld@yahoo.com

D.K. Mishra et al. (eds.), *Information and Communication Technology*,
Advances in Intelligent Systems and Computing 625,
https://doi.org/10.1007/978-981-10-5508-9_22

licensing as well as management, and of electromagnetic radio spectrum within the USA, and it also audits time-to-time interstation interference in all radio frequency bands [2]. All conventional wireless connection services, substantially based on fixed spectrum allocation methodology, are much constrained by the factors such as wastage of static spectrum allocation, restricted and limited wireless functionality, leading to inefficient utilization of radio spectrum. Therefore, spectrum efficiency can be improved by manifesting the concept of frequency reuse where the secondary users (SUs) are being permitted to ingress the spectrum when the spectrum is temporarily being not utilized by the primary users (PUs). The basic idea in order to manage RF resources in such a way is that SU can be permitted to access the licensed frequencies following the condition that they can guarantee minimum interference perceived by the PU allocated in the RF spectrum.

The basic aim is to consider the architecture of cognitive radio (CR) in which all PU and SU will send their data to the CR either in the licensed or in the unlicensed mode (when spectrum holes are present). Several cognitive users are existing in a distributed fashion and coordinate with SU for data communication in a single hop in order to increase the throughput, maximize the spectrum efficiency, and to decrease the delay.

With the advances in software and technology, CR can smartly sense and adapt to the changing environment by modifying its transmitting parameters, such as modulation, frequency, frame format [3]. In the early days of communication, there were fixed radios in existence in which the transmitter parameters were static and were fixed deliberately that set up by their operators. The new era of communication inculcates the concept of software-defined radio (SDR) [4–8].

Software Defined Radio (SDR) is a radio in which the transmitter's operating parameters are the frequency range and type of modulation. The maximum radiated or output power can be altered by initializing a change in software without performing any hardware changes. It is used to reduce hardware requirements since it provides user an inexpensive and reliable solution. But it will not take into consideration the area of spectrum availability. CR is basically a recent version of SDR in which all the transmitter parameters modify and update like SDR, but it will also adapt its parameters as per the spectrum availability. The primary network is totally unaware and unknown regarding the capabilities of the cognitive network behavior and does not necessitate any specific functionalities to coexist with it. When a PU arrives in the spectrum, the secondary users ought to vacate the spectrum and should immediately react by altering their parameters such as frequency rate, baud rate, power, capacity, channel used, codebook so that PU quality of service (QoS) might not degrade.

The proposed technique is to design a CR network architecture named as "Multiple-CR Single-Hop (MSCH) Secondary User CR Network Architecture" in which multiple CRs will act as a heterogeneous node that can perform the diverse functions of spectrum sharing, allocation, management, mobility, decision making at the same time. The unlicensed secondary users will behave as homogeneous nodes that can transmit their data on the licensed bands via multiple CRs in a single-hop fashion [9].

2 Related Work

2.1 Cognitive Radio Network

Fixed spectrum allocation policy is employed in wireless networks. Spectrum can remain underutilized in some area or for some period of time, whereas some frequencies will be highly utilized. Therefore, some underutilized wireless spectrum should be exploited for maximizing the spectrum usage. CR acts as secondary-tier networks in order to access the spectrum. While the licensed users or PUs are not using the spectrum, CR user completes the spectrum in order to maximize the spectrum utilization throughput [5]. Thus, CR can be defined as a radio that can change its transmission parameters based on the active environment in which it operates. The CR determines the portion of the spectrum, which remains available and thus detects the availability of licensed users, and selects the best available channel. CR also coordinates access to this channel with others. The ultimate objective of CR lies in the fact that it needs to obtain the best available spectrum due to its property of reconfigurability and cognitive capability. The most challenging situations for CR is to share the licensed spectrum keeping the condition that it will not interfere with the transmission of other licensed users since most of the spectrum is legally shared between several PUs. The CR enables the process of usage of temporarily unused spectrum gaps called as spectrum hole or white space. In the case of again using the spectrum by the licensed user, the CR moves in another spectrum hole or remains in the same band. By altering its modulation technique, power level is transmitted in order to mitigate the chances of interference (Fig. 1).

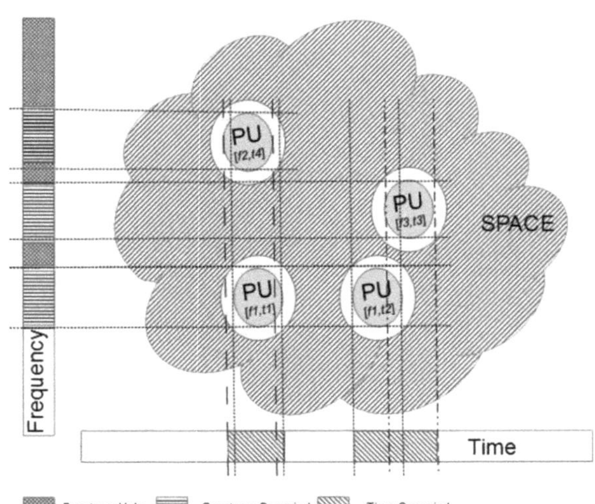

Fig. 1 Spectrum hole concept

2.2 Cognitive Network Architecture

The reference CR architecture includes different types of networks such as the primary network, an infrastructure-based secondary network, and an ad hoc-based secondary network. These CR-based networks are operated under the mixed spectrum environment that consists of both licensed and non-licensed frequencies. As quoted by multiple authors in the literature for cognitive networks, multiple secondary networks can communicate with each other in a multi-hop manner or across the base station, leading to collision of data and a large amount of delay between different SUs during the data transmission. Therefore, in this proposed technique, we are relying on the fundamentals of single-hop technique between different CR users and SUs in order to maximize throughput and reduce the delay and collisions using MS-SH [9] network architecture.

There are three different access types which are shown in Fig. 2:

1. Secondary network access: SUs have the ability to access their own secondary base station both on licensed and on unlicensed spectrum bands.
2. Secondary ad hoc access: SUs are free to communicate with all other secondary users through ad hoc connection on licensed as well as unlicensed spectrum bands.
3. Primary network access: The SUs are capable of accessing the primary base station through the licensed band [10].

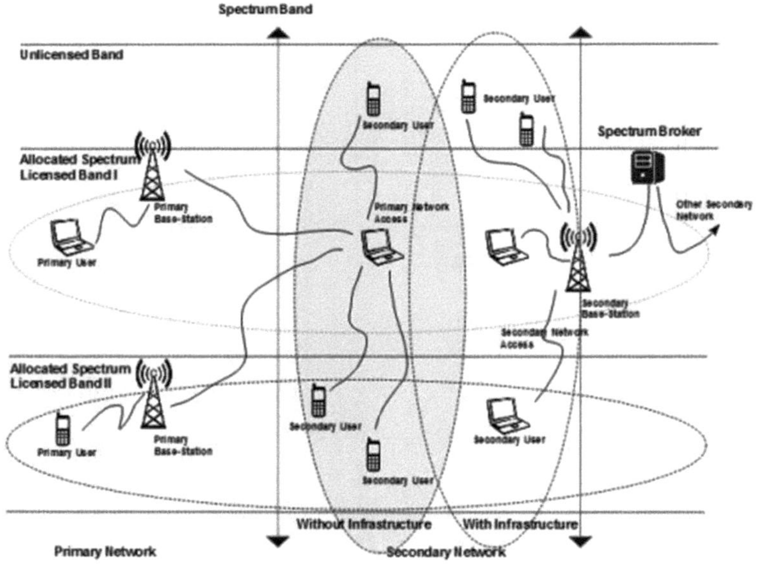

Fig. 2 Cognitive radio architecture [10]

There are two basic groups in CR, namely primary user network and secondary user network. With the help of cognitive radio, the secondary user interacts with the primary user for spectrum allocation.

The primary network: It has an exclusive right over a certain spectrum band, like for cellular networks and TV broadcast networks, since it is a licensed user. The basic components of primary networks are: PU, which is also called as licensed user that is having all rights to operate in a licensed band, remains unaffected by the activities of CR; primary base station, which is also called as licensed base station, is a fixed infrastructure network component having spectrum license.

CR network: The CR network does not possess license to operate in a licensed band, and its spectrum access is allowed according to the opportunistic environmental conditions. The components of CR network are as follows:

- CR user: It is basically an unlicensed user that is possessing no license over the spectrum. CR can use the spectrum only when PU is not present, and CR has to vacate the spectrum/channel whenever the PU will be detected.
- CR base station: It is an unlicensed base station meaning a fixed infrastructure component with CR capabilities that provide a single-hop connection to CR users.
- Spectrum broker: It is a central network entity that is capable of managing and controlling the spectrum resource sharing among the CR base stations.
- The SU comes in picture along with the cognitive user only when the PU is not present in the spectrum.

2.3 *Cognitive Radio Cycle*

The important areas of CR cycle are mainly categorized into four following steps as shown in Fig. 3:

Step 1: Spectrum sensing: It is used to sense the spectrum holes when only for the unused portion of the spectrum, CR allocation can be done. Therefore, continuous monitoring of the available spectrum bands is important, and hence, the spectrum holes can be detected. Spectrum sensing is basically performed on the physical layer and is closely related to spectrum allocation problems. Three main potential approaches are recognized such as beacon signals, database registry, and spectrum sensing [11] in order to identify the spectrum opportunities. The database registry technique inculcates the method of global positioning system (GPS) that is mounted on secondary devices to locate its respective location and for accessing the database of primary network for locating the channels that are unused and vacant at that time. Two spectrum sensing methods are widely used in the CR architecture:

- Non-cooperative/transmitter detection and
- Cooperative detection.

Fig. 3 Cognitive radio cycle

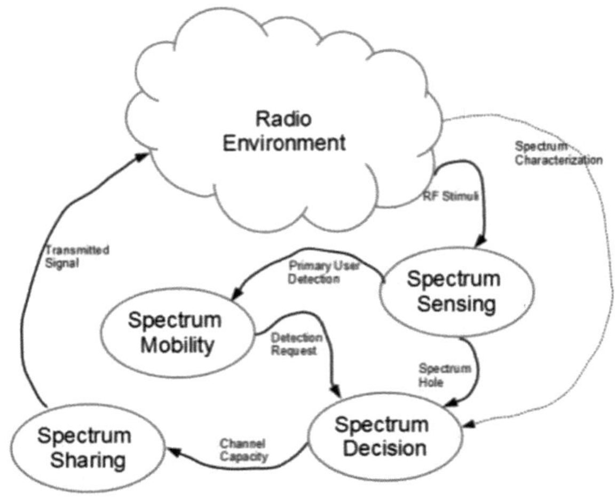

Step 2: Spectrum decision: According to the QoS requirements of different bands, the CR user identifies the most suitable band after the process of identification of available spectrums in the network. The statistical behavior of the PUs and the radio environment decides the characteristics of the spectrum band. For dynamic spectrum characteristics, it is important to have a priori information about the PU activity and this entire process is done in the link layer and the network layer.

Step 3: Spectrum sharing: Since multiple CRs are coordinating with each other interconnected with different SUs in order to avoid collisions in the overlapped portions of the spectrum, the technique of spectrum sharing provides the capability to have resource allocation in order to mitigate interference caused on the primary network. Therefore, physical layer and MAC protocols are being applied that can easily facilitate the sensing control to distribute the sensing task among the coordinating nodes necessary for transmission.

Step 4: Spectrum mobility: In case of detection of PU in the network, the CR should vacate the licensed spectrum and should continue its transmission in another unutilized bands, thereby connecting to other CRs lying in the vicinity of that particular SU. Therefore, the spectrum mobility techniques utilize the scheme of spectrum handoff in order to detect the failure in any link and in order to decrease the packet drops. Also, the connection management scheme is added in order to sustain the performance of upper layer protocols.

2.4 CR Network Capability

The capabilities of a CR network include reconfigurability, operating frequency, modulation, transmission power, and communication technology. Reconfigurability can be defined as the ability of adjusting some operating parameters for the purpose of transmission without any mandatory modification. For the CR user to adapt easily to the dynamic radio environment, reconfigurability is an important feature in CR networks [12, 13].

3 Cognitive Radio Architecture Framework

In order to decrease the end-to-end delay and to fulfill the connection requirements, single hopping is preferred in CR networks among corresponding SUs. In the case of arrival of PUs in the RF spectrum, spectrum handoff occurs leading to dynamic spectrum allocation in which SU relocates to another CR through spectrum broker, thereby maximizing the spectral efficiency. It would not be possible in "single CR single hop" and "single CR multiple hop" networks as SU has to vacate the spectrum as the PU comes into the picture. Hence, "multiple CRs single-hop (MCSH) secondary user cognitive radio network architecture" is the most optimal CR architecture reducing the delay, jitter, and packet loss.

The CR ensures three basic fundamental cases to incur Multiple CRs Single-Hop Architecture:

Case 1: If one of the SUs is connected to multiple secondary BS/primary network access, then the unlicensed user will relinquish its control over one CR so that it can allocate the unused spectrum to any other unlicensed user in the network.

Case 2: The restructuring of network should be done like energy balancing, bandwidth allocation in order to decrease delay and to increase the network coverage.

Case 3: To calculate the optimum number of CRs so that no packet loss and delay can be incurred in the network.

3.1 Simulation Step

Following are the simulation steps and results shown in Table 1:

1. All the CRs (secondary base stations/primary network access) behave as heterogeneous and cover defined geographical area and provide the connection to all the secondary users. Simulation was performed for 10–15 CRs.

Table 1 Simulation results

No. of secondary BS/primary N/W access	Max. SU connected to single Sec. BS	After restructuring the Max. SU connected to single Sec. BS
10	378	186
11	359	152
12	364	137
13	353	118
14	352	106
15	361	110

2. SUs are randomly deployed and are stationary. For simulation, the total 1000 SUs are used.
3. The optimal number of CRs is also found out using connection restructuring of SU. Maximum 250 SU are supported by each CRs.
4. Initially, one SU is connected to multiple CRs, but after the connection, restructuring SU is connected to the single CR and enhances the spectral efficiency. Results are shown in Table 1.

4 Conclusion

On the basis of above three cases, we conclude that multiple CRs single-hop secondary user cognitive radio network architecture has been found out to be the optimum structure as per the design parameters of CRs such as bandwidth, collision avoidance, connectivity, optimum number of CRs required as compared to Single CR single-hop and multiple CR multiple-hop CR network architecture design.

Acknowledgements The authors are sincerely thankful to the funding agency of Indian Institute of Information Technology Kota for their financial support.

References

1. FCC, ET Docket No 03-222 Notice of proposed rule making and order, December 2003.
2. S. Haykin, Cognitive radio: brain-empowered wireless communications, IEEE Journal on Selected Areas in Communications 23 (2) (2005) 201–220.
3. J. Mitola III, Cognitive radio: an integrated agent architecture for software defined radio, Ph.D Thesis, KTH Royal Institute of Technology, 2000.
4. Joseph Mitola III "Cognitive radio: an integrated agent architecture for software defined radio," Ph.D. Thesis, Royal Institute of Technology (KTH) Sweden, May 2000.

5. F. K. Jondral, "Software-defined radio: basics and evolution to cognitive radio," EURASIP Journal on Wireless Communications and Networking, vol. 5, no. 3, pp. 275–283, 2005.
6. U. Ramacher, "Software-defined radio prospects for multi-standard mobile phones," IEEE J. of Computer, vol. 40, no. 10, pp. 62–69, 2007.
7. R. Bagheri, A. Mirzaei, M.- E. Heidari, S. Chehrazi, M. Lee, M. Mikhemar, W. K. Tang, and A. A. Abidi, "Software-defined radio receiver: dream to reality," IEEE Communications Magazine, vol. 44, no. 8, pp. 111–118, 2006.
8. H. Arslan and S. Yarkan, "Cognitive Radio, Software Defined Radio, and Adaptive Wireless Systems," Springer: Netherlands, 2007.
9. Jain, Tapan Kumar, Davinder Singh Saini, and Sunil Vidya Bhooshan. "Lifetime optimization of a multiple sink wireless sensor network through energy balancing." Journal of Sensors 2015 (2015).
10. Akyildiz, Ian F., Won-Yeol Lee, Mehmet C. Vuran, and Shantidev Mohanty. "NeXt generation/dynamic spectrum access/cognitive radio wireless networks: a survey." Computer networks 50, no. 13 (2006): 2127–2159.
11. E. Hoossain, D. Niyato, and Z. Han, "Dynamic Spectrum Access and Management in Cognitive Radio Networks," Cambridge University Press: New York, 2009.
12. Pandit, Shweta, and G. Singh. "An overview of spectrum sharing techniques in cognitive radio communication system." Wireless Networks (2015): 1–22.
13. Akyildiz, Ian F., Won-Yeol Lee, and Kaushik R. Chowdhury. "CRAHNs: Cognitive radio ad hoc networks." AD hoc networks 7, no. 5 (2009): 810–836.

Comparative Study of N-Tier and Cloud-Based Web Application Using Automated Load Testing Tool

Manisha Jailia, Manisha Agarwal and Ashok Kumar

Abstract These days people cannot think a single day without online world. In online world, Web applications play very important role in all sectors, let it be for searching, shopping, and education too. Too many Web sites are launched daily. But what matters a lot for a user is all about performance. In this paper, we have discussed the comparison between N-tier-based Web application and cloud-based Web application, so that while designing of Web site one can efficiently select Web architecture. At last with the help of loadcomplete tool, performance is to be evaluated on various metrics.

Keywords N-tier · Cloud · Web · Loadcomplete · Architecture

1 Introduction

The Web applications are defined as the applications that use Web browser to fulfill the requirements of users and are written in browser compatible programming languages (such as HTML, JavaScript, and CSS). Most Web applications work on the principle of Client–Server Architecture. In this architecture, the client and the server communicate over the network. The client computer uses an interface to send request to the server and the server responds which results in display of results on client's computer. In this paper, we have discussed other architecture, i.e., Cloud Architecture. We have also examined the performance of both the architectures on the basis of their response by testing tool.

M. Jailia (✉) · M. Agarwal · A. Kumar
Banasthali Vidyapith, Rajasthan, India
e-mail: manishajailia@yahoo.co.in

M. Agarwal
e-mail: mani1811@yahoo.com

A. Kumar
e-mail: dynamic6091@gmail.com

D.K. Mishra et al. (eds.), *Information and Communication Technology*,
Advances in Intelligent Systems and Computing 625,
https://doi.org/10.1007/978-981-10-5508-9_23

1.1 Client–Server Architecture

Client–Server Architecture is distributing computing architecture that segregates jobs or workloads between servers and clients. If we go its most elementary configuration, Client–Server Architecture involves a program unit called client which makes a request to another program unit called server which accomplishes it [1]. Figure 1 shows the communication between client and server.

The clients commence the communication by making requests to the server. The client uses the personal computers to make their calls. These computers have network software applications installed in them. Servers are the machines who have files and databases stored in them. They have features like highly powered central processors, large disk drives, and more memory capacity in comparison with the clients [2]. After receiving service requests from client, the server resolves the requests and then tries to accomplish the requests. A client can send requests to several servers, and a server can handle many clients. The client and server hold a command and control association between them. The server can never initiate any request, and the client can never fulfill any requests. There are many examples of Client–Server Architecture, but few of them are e-mail, Web browser, Internet, database, etc.

1.1.1 N-Tier Architecture

N-tier Architecture is the architecture which separates the applications into many tiers. Tiers are defined as computers which are physically deployed like one-tier, two-tier, and three-tier. This is shown in Figs. 2 and 3.

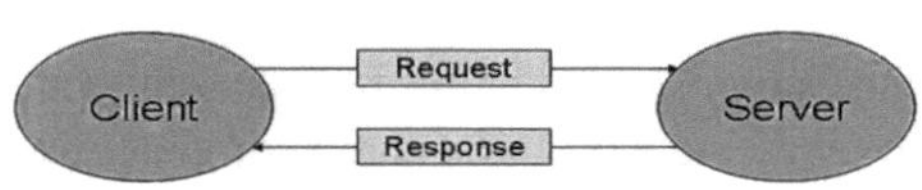

Fig. 1 Client–server architecture

Fig. 2 Two-tier architecture

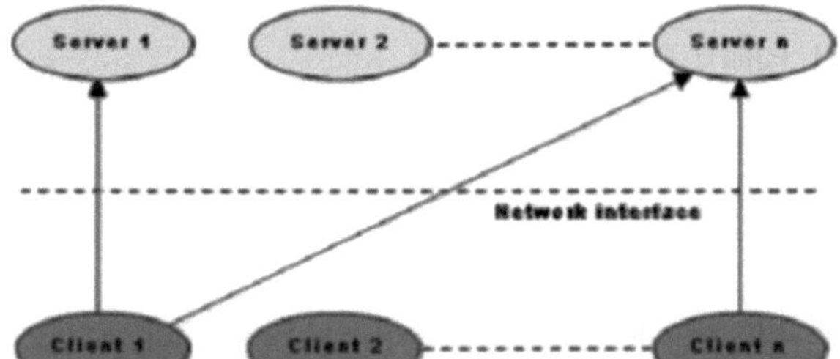

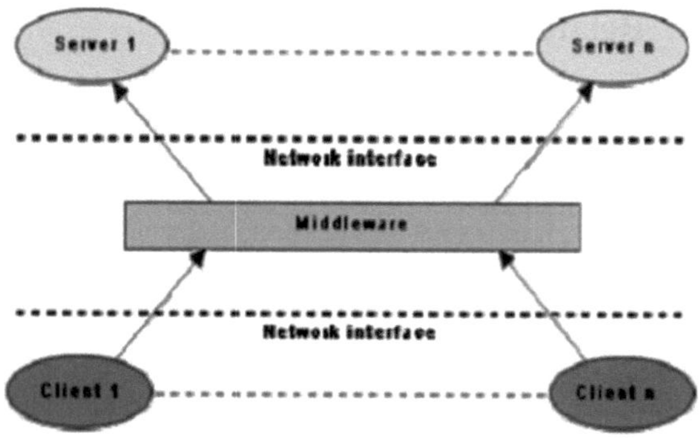

Fig. 3 Three-tier architecture

1.2 *Cloud Architecture*

The front-end platform, back-end platform, network, and delivery model merge to form Cloud Architecture. The front-end platform involves fat client, thin client, and mobile devices. On the other hand, the back-end platform involves servers and database storage. The network involves Internet, cloud network, and intranet. The delivery model involves SaaS, PaaS, and IaaS (Fig. 4).

Software as a service (SaaS)—In this cloud-based delivery model, the cloud service providers install and maintain software on the cloud and provide the accessibility to the users via Internet. Examples are Salesforce, LinkedIn, Oracle, SAP, etc.

Platform as a service (PaaS)—In this cloud-based delivery model, the cloud service providers provide users the application platform equipped with the database. Examples are Amazon Web services, OpenStack, ThinkGrid, etc.

Infrastructure as a service (IaaS)—In this cloud-based delivery model, the cloud service providers provide the whole system but in virtual form. Examples are HP, Verizon, Amazon Web Services, etc.

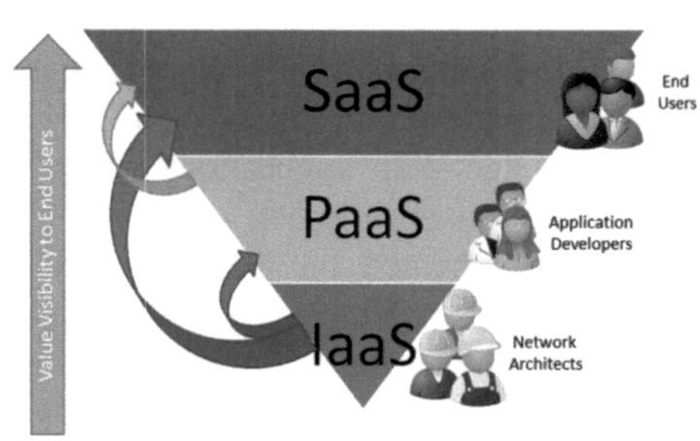

Fig. 4 Cloud deployment model

2 Related Work

As we are moving forward, the Web technologies are also moving at pace. With the advancement in technologies, we are getting new ideas and works in this domain. Many people have worked on these new technologies and architectures. Among them, the Cloud Architecture is in great demand.

It is widely accepted approach for developing optimized and better performing Web applications. It is also now in demand for developing mobile-based application because of its potential to provide different views for different devices without altering any other component of the application. The National Institute of Standards and Technology (NIST) which defines cloud computing as "model for enabling convenient, on-demand network access to a shared pool of configurable computing resources (e.g., networks, servers, storage, applications, and services) that can be rapidly provisioned and released with minimal management effort or service provider interaction" [3].

Cloud computing architecture is an arising trend which is changing the way of using everything in computer domain. Many big tech giants have already adopted this trend. Some of them are Google, IBM, Microsoft, and Amazon. It helps faster deployment of Web applications, and applications are more efficient in usage. The N-tier architecture provides us the freedom to develop flexible and reusable Web applications due to its property of finer modularity [4]. These applications are fault tolerable, efficient, secure, etc.

There are other Web architectures also present like Model View Controller [5]. The business logic issues in N-tier are evaluated on the basis of performance testing for the validation of Web applications also in MVC architecture using ASP.NET technology [6]. Performance of Web crawler on different architectures is also discussed by various authors [7].

3 Implementation Strategy

The implementation of Web Applications, using ASP.NET technology, is accomplished with the aid of Visual Studio 2012 inbuilt SQL Server LocalDB. So we are comparing the Web applications which are developed on different architectures using different technologies.

We have created the Web applications with the INSERT, UPDATE, and DELETE functionalities. The Web applications based on N-tier and cloud are created to retrieve data from database and display it on user interface. We have created two Web applications so far. The two Web applications are created on the architectures, N-tier and cloud using ASP.NET technology.

Interfaces of Web Application's on different Web architecture.

3.1 N-tier Architecture

See Figs. 5, 6 and 7.

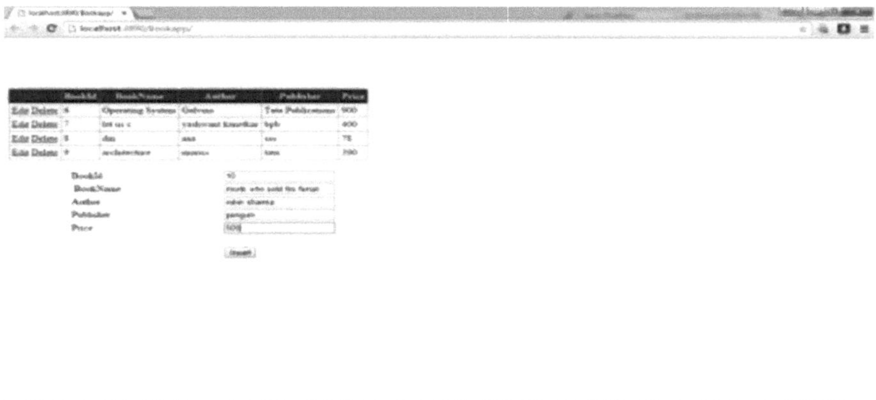

Fig. 5 INSERT operation

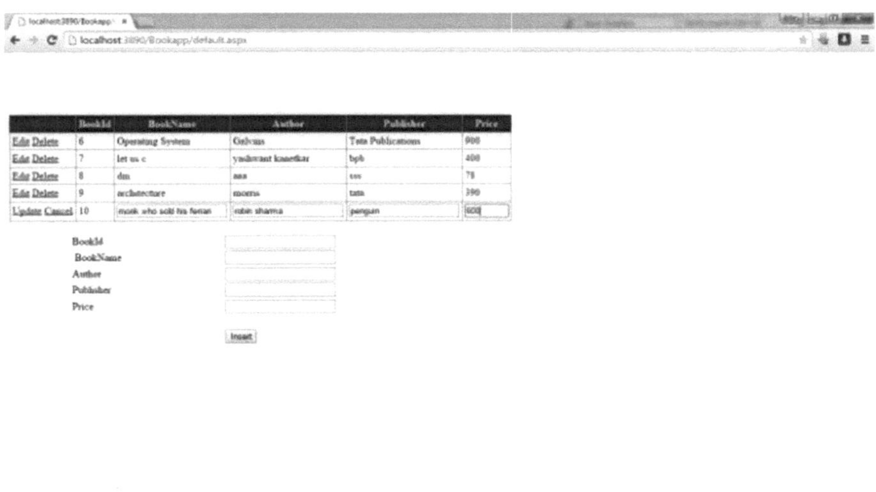

Fig. 6 UPDATE operation

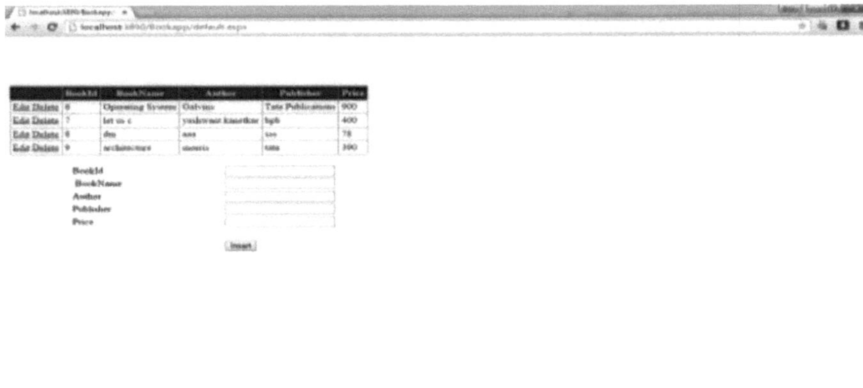

Fig. 7 DELETE operation

3.2 Cloud Architecture

See Figs. 8, 9 and 10.

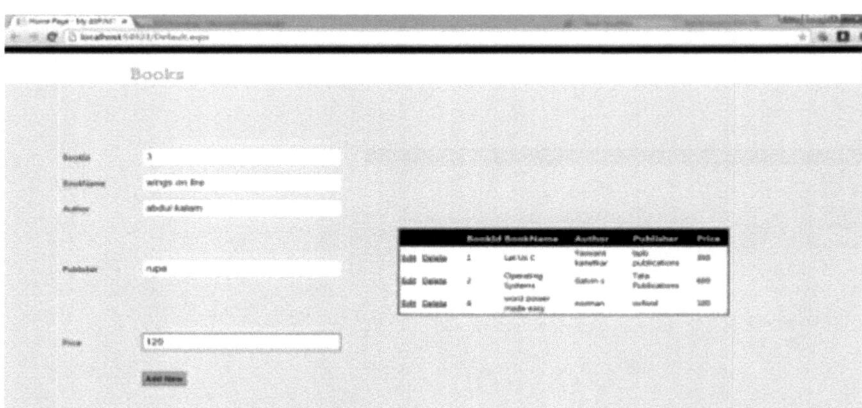

Fig. 8 INSERT operation

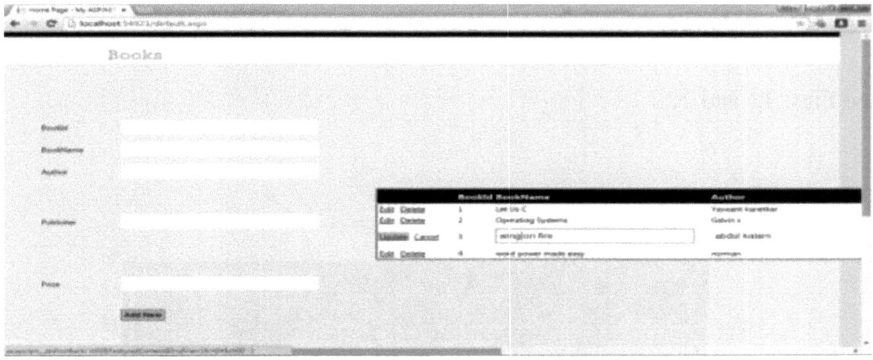

Fig. 9 UPDATE operation

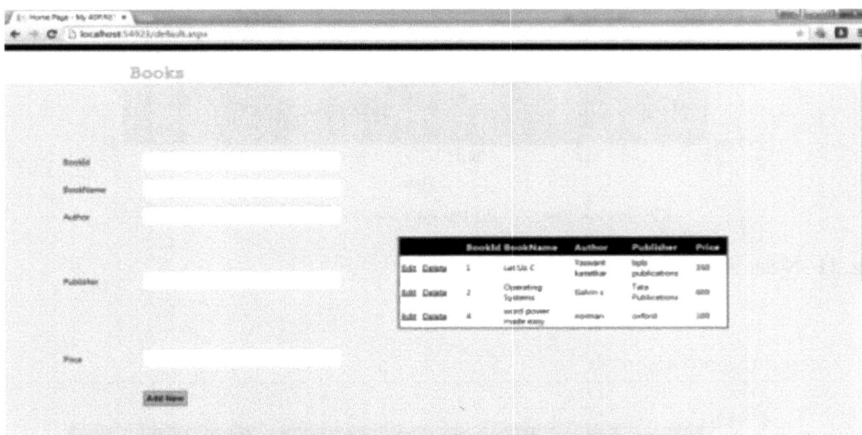

Fig. 10 DELETE operation

4 Evaluation

The performance evaluation of Web applications is done by considering various factors like Page Load Time, Request Transfer Speed, Response Transfer Speed, Server Time, etc. The evaluation is done by using LOADCOMPLETE 3 testing tool. LoadComplete is a load testing tool for Web applications. It helps in checking Web application's performance under heavy load which aids in knowing Web application's robustness and scalability.

The following graphs are generated considering few specifications and help in evaluating the Web application's performance. The graph shows two axes, *X*-axis and *Y*-axis. The *X*-axis shows the time interval and the *Y*-axis varies along the graphs. There are 20 virtual users assumed in both the scenario.

Graph of Web Application's graphs based using ASP.NET Technology.

4.1 Page Load Time

See Figs. 11 and 12.

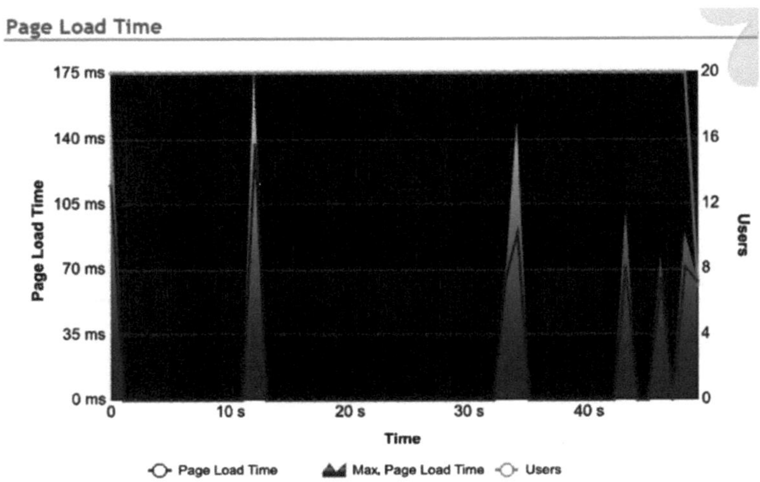

Fig. 11 N-tier architecture

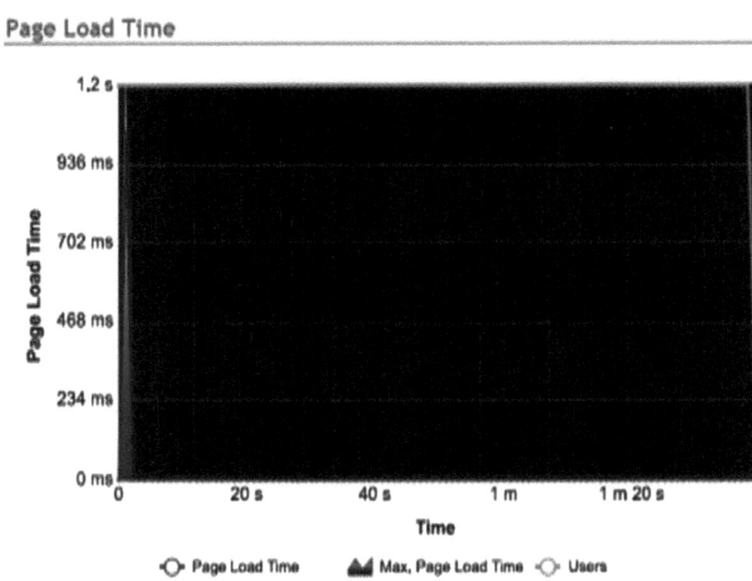

Fig. 12 Cloud architecture

4.2 Request Transfer Speed

See Figs. 13 and 14.

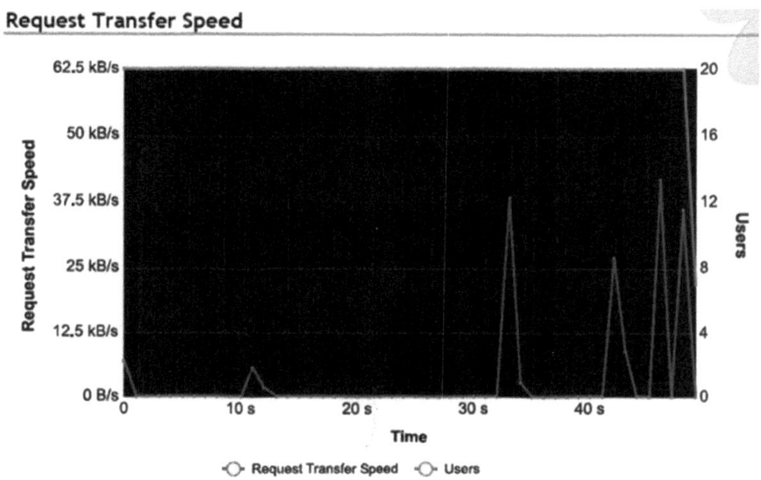

Fig. 13 N-tier architecture

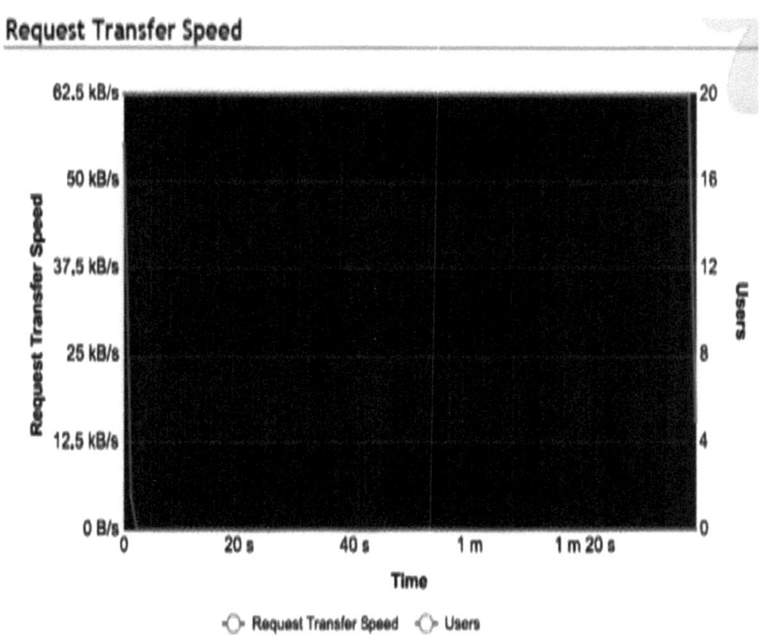

Fig. 14 Cloud architecture

4.3 Response Transfer Speed

See Figs. 15 and 16.

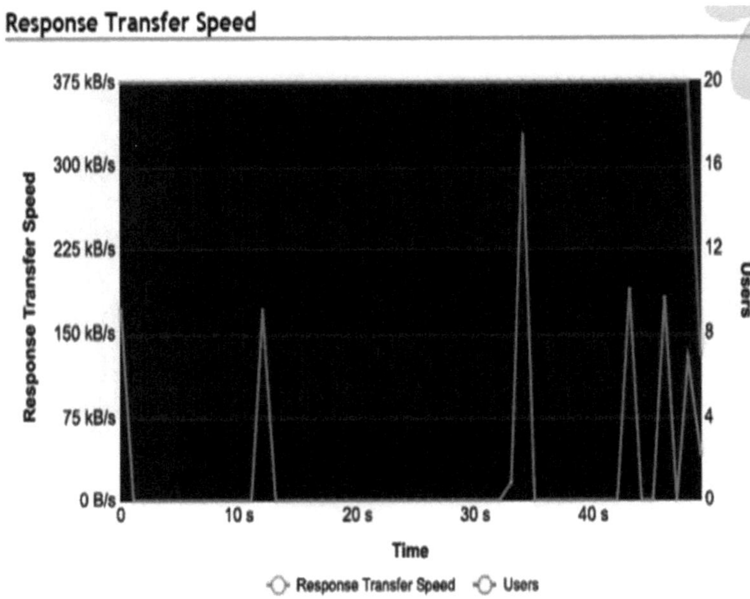

Fig. 15 N-tier architecture

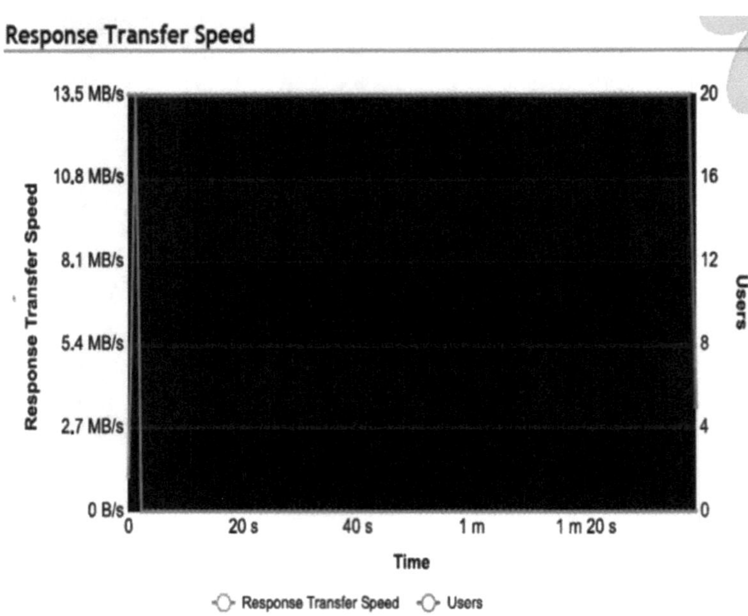

Fig. 16 Cloud architecture

Table 1 Comparison among N-tier and cloud architectures using ASP.NET

S. No.	Profile	N-tier	Cloud
1	Page load time (ms)	90	1030
2	Time to first byte (ms)	44	324
3	Time to last byte (ms)	27.23	127.25
4	Request transfer speed (kbps)	11.214	14.97
5	Response transfer speed (mbps)	0.094	3.65

The result of testing is displayed using Table 1. The results depict the differences among architectures. It shows that the Web application made using N-tier Architecture is give more promising result to page load, response transfer parameter. N-tier Architecture has greater average values in all profiles because of its variation in graph.

5 Conclusion and Future Work

The growth of Web applications in today's era is rampant. Nowadays, a Web application has opened its wings in all fields, such as education, forecasting, engineering, sports, entertainment, medical, etc. Users are so much smart that they cannot tolerate Web site which takes too much time to load because users have so many options or other Web sites which can do a same task for him. The user simply switches to other alternatives. This paper focuses mainly on how to make a Web application response quickly to retain more and more users. We compared the performance of N-tier-based Web applications and cloud-based Web application. We conclude that Cloud Architecture may utilize recourses efficiently but N-tier-based application still gives better result in page load parameter. In future, we will focus on mobile Web application with different architectures and green Web application.

References

1. Gallaugher, John, et al., "The Critical Choice of Client Server Architecture: A Comparison of Two and Three Tier Systems", Informations System Managements (New York, Auerbach Publications), 1996.
2. Subramanian, Ashok, et al., "Managing client/server implementations: today's technology, yesterday's lessons", Journal of Information Technology (1997), 12, 169-186.
3. Q. Zhang, L. Cheng, and R. Boutaba, "Cloud computing: state-of-the-art and research challenges," Journal of Internet Services and Applications, vol. 1, pp. 7–18, 2010.
4. Prasanth, Y, et al., "Oriental Journal of Computer Science & Technology, Vol. 2(2), 153–160 (2009).

5. Badurowicz, Marcin, "MVC architectural pattern in mobile web applications", Actual Problems of Economics, January 2011, 6(120):305.
6. Kumar, Ashok, et al., "Web session classes: performance metrics for business logic issues in n-tier and MVC architecture", in International Journal of Research in Computer Application & Management (IJRCM) Vol. 3, Issue 10, October 2013 ISSN: 2231-1009.
7. Badgujar j, Jailia, M., et al., "Performance metrics of Web crawler in client-server and MVC architecture", IEEE Xplore.
8. Krasner, et al., "A cookbook for using the model-view controller user interface paradigm in smalltalk-80," J. Object Oriented Program., vol. 1, no. 3, pp. 26–49, Aug. 1988. [Online].

The Design of Ultra-High Frequency (UHF) Radio Frequency Identification (RFID) Reader Antenna

Wee Fwen Hoon, Yew Been Seok, Mohamed Fareq Abdul Malek, Lee Yeng Seng, Siti Zuraidah Ibrahim and Sarah Yasmin

Abstract This paper presents the design of ultra-high frequency (UHF) radio frequency identification (RFID) reader antenna using low dielectric constant of microwave substrate. An RFID reader antenna emits electromagnetic signals to the microchip in the tag, and the microchip will be energized by modulating the wave and returns to the reader antenna. The process of wave emitting is known as backscattering due to the presence of tag been detected by the reader. High-dielectric constant substrate, for example flame-retardant-4 (FR4) which is commonly used for microstrip patch antenna, is high in dielectric constant and dielectric loss. Thus, this will lead to low gain and directivity properties of the antenna. To overcome this matter, low-dielectric constant substrate which is Taconic TLY-5 was proposed to be utilized for microstrip patch antenna design. The TLY-5 microstrip substrate thickness used is

W.F. Hoon (✉) · S.Z. Ibrahim · S. Yasmin
School of Computer and Communication Engineering,
Universiti Malaysia Perlis, Arau, Malaysia
e-mail: fhwee@unimap.edu.my

S.Z. Ibrahim
e-mail: sitizuraidah@unimap.edu.my

S. Yasmin
e-mail: saraseraa@gmail.com

Y.B. Seok
Faculty of Innovative Design and Technology,
Universiti Sultan Zainal Abidin, Kuala Terengganu, Malaysia
e-mail: bseokyew@unisza.edu.my

M.F.A. Malek
Faculty of Engineering and Information Sciences,
University of Wollongong, Dubai, United Arab Emirates
e-mail: MohamedFareqMalek@uowdubai.ac.ae

L.Y. Seng
Department of Electronic Engineering Technology,
Faculty of Engineering Technology, Universiti Malaysia Perlis,
Arau, Malaysia
e-mail: yslee@unimap.edu.my

D.K. Mishra et al. (eds.), *Information and Communication Technology*,
Advances in Intelligent Systems and Computing 625,
https://doi.org/10.1007/978-981-10-5508-9_24

1.6 mm, dielectric constant of 2.2, and loss tangent of 0.019. A high-conductivity metal which is typically a conductive copper is been used for the two layers of dielectric substrate, the top radiating patch layer and bottom ground layer where the copper thickness is 0.035 mm. Microstrip feed line is used for this UHF RFID reader antenna. The width of the feed line was tuned to obtain impedance matching of 50 Ω. The proposed antenna which is fork-shaped patch antenna was simulated using Computer Simulation Technology (CST) and Microwave Studio software at resonant frequency of 910 MHz with the outcome results of 7.985 dB gain and −11.11 dB return loss. Nevertheless, the typical value obtained for VSWR is less than 2.

Keywords Ultra-high frequency (UHF) · Radio frequency identification (RFID) · Reader antenna

1 Introduction

Among fast-developing technology nowadays includes radio-frequency identification (RFID) in wireless communication. This wireless data transmission and receiving technique had been an emerging technology for many other developing applications in different ranges such as automatic identification, electronic-related ticketing services, security surveillance, and access control [1].

In practical usage of RFID, the tags are always oriented arbitrarily. Circularly polarized (CP) reader antennas have been used in UHF RFID systems for ensuring the reliability of communications between tags and readers. Normally, the tag antennas are polarizing in linear form. However, there is a standard international UHF range of RFID for each part of the world. Various frequency ranges in Europe, Australia, Singapore, US/Canada, Korea, and Japan are 865.5–867.6, 920–926 MHz, (866–869, 923–925 MHz), 902–928, 910–914, and 952–955 MHz, respectively [2]. The implementation and cost of RFID systems can be simplified and reduced. In this paper, an UHF RFID reader antenna is designed using microstrip patch with frequency bands of UHF range, 860–960 MHz. The entire UHF frequency band (860–960 MHz) needs to be covered to reach the targeted solutions.

Meanwhile, microstrip patch antenna in telecommunications is broadly used because there are advantages that can be obtained from the design such as thin and low cost of fabrication process. The characteristics of microstrip antennas used in RFID technology are also applicable in service industries, transport systems, and distribution logistics [2, 3]. However, microstrip patch antenna has narrow bandwidth which is one of the serious limitations to the design. This will affect the sensitivity and directivity gain of the antenna. Small and thin features of antenna also will affect the gain performance, where if a high dielectric substrate is used for the microstrip antenna, the size of the patch will reduce and the gain also reduces drastically [4]. Due to this, low dielectric constant substrate material is proposed to be utilized in this project design where it is capable to obtain higher gain in antenna design [1].

2 UHF RFID Reader Antenna Design Method

2.1 Preliminary Design–Rectangular Design

Microstrip radiating patch antennas can be printed precisely onto a microwave substrate board using chemical developing, etching, and photoresist method. Microstrip antennas are becoming overwhelming in this new era technology world that requires wireless transmission of signal. Microstrip patch antennas are compact with low profile, lighter with low cost of material used, and ease to fabricate [5]. The rectangular-shaped microstrip patch antenna as shown in Fig. 1 consists of substrate width, length, patch width (*W*), patch length (*L*), and feeding impedance of 50 Ω where the feeding line width and length is introduced symmetrically with respect to the probe position.

The input impedance is controlled by microstrip patch width, *W* of the antenna. The larger widths will wider the size of bandwidth. The proposed rectangular-shaped

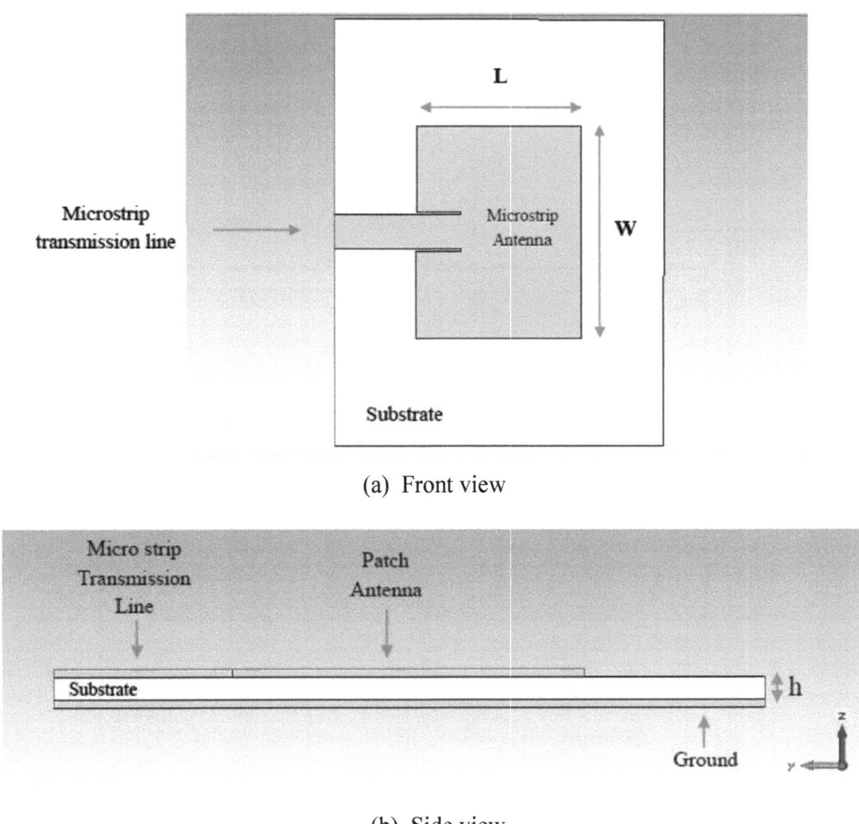

Fig. 1 Preliminary design of microstrip patch antenna

microstrip antenna uses a loss-free Taconic TLY-5 substrate as dielectric substrate materials. The substrate thickness (h) used is 1.6 mm, dielectric constant (ε_r) of 2.2, and loss tangent of 0.019. A high-conductivity metal which is typically a conductive

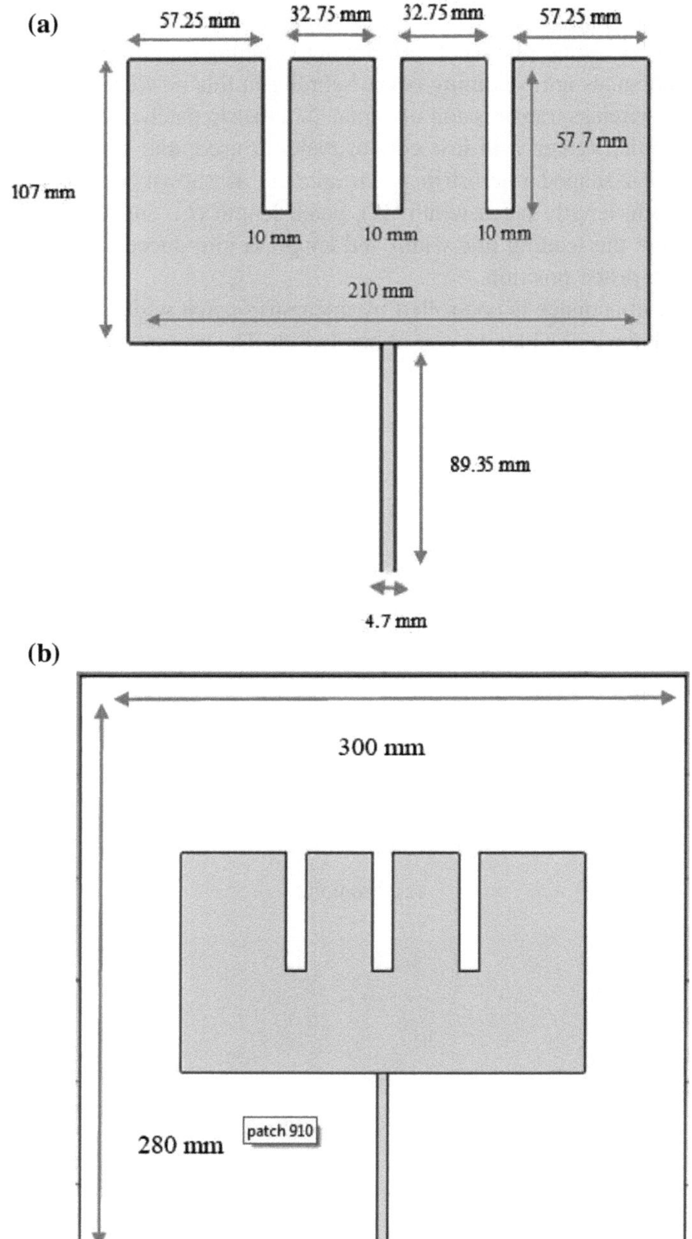

Fig. 2 Dimension of fork-shaped design **a** radiation patch, **b** substrate

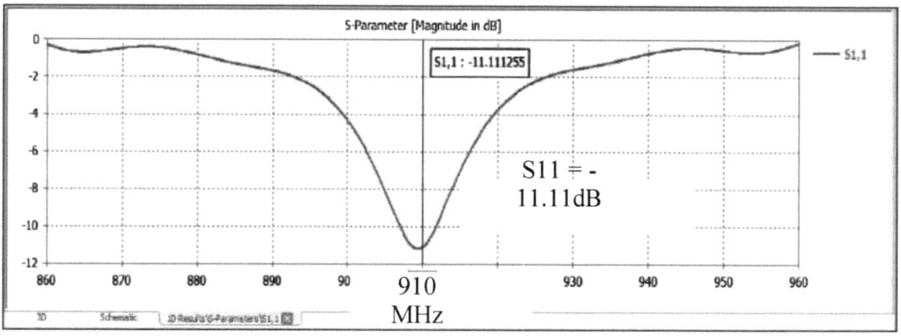

Fig. 3 S_{11} of fork-shaped microstrip patch antenna

copper is been used for the two layers of dielectric substrate, the top radiating patch layer and bottom ground layer where the copper thickness is 0.035 mm.

2.2 *Fork-Shaped Design*

The proposed shape design for this microstrip patch antenna is fork-shaped design. Fork-shaped design microstrip patch was fed by a 50-Ω impedance feed line for antenna matching purpose. The length of the radiating patch was calculated to be approximately $\lambda g/2$, where the patch was able to radiate perfectly. The proposed antenna gave good polarization when it was able to radiate along the patch width (*W*). However, the spurious radiation might become the disadvantages and the need for impedance matching needed. This is because the typical edge impedance of a microstrip antenna ranges from 150 to 300 Ω [6].

The bandwidth also can be increased by enlarging the transmission width size. Besides that, the impedance can be reduced to 50 Ω but the reduction of impedance value will affect the matching of the antenna, where UHF band can't be achieved, and it also requires a very wide patch size of antenna. This will take up a lot of valuable space. The patch width also controls the radiation pattern of antenna. As there is change of gap width between two lines of the fork shape, the polarization of the electromagnetic field will be affected by the recognition of distance change [5]. Feeding method is also applied by having microstrip transmission feed line in the design. Figures 2 and 3 show the antenna dimension of the proposed design.

Table 1 Overall results of fork-shaped antenna

Parameter	Result value
Return loss, S_{11}	−11.11 dB
VSWR	1.77
Gain, dB	7.985 dB
Directivity	8.959 dB
Line impedance, Z_0	50 Ω
Radiation pattern	Directional
Efficiency	89.1%

3 UHF RFID Reader Antenna Results and Discussion

The fork-shaped antenna parameters that consist of S_{11}, VSWR, gain, directivity, line impedance, and bandwidth have been simulated. *S*-parameters describe the input–output relationship between ports (or terminal) in an electrical system. From Fig. 3, the observed result value for return loss (S_{11}) was below −10 dB which is −11.11 dB at 910 MHz.

Table 1 result shows the obtained directivity and gain for the proposed design. The directivity and gain obtained are high which are 8.959 and 7.985 dB, respectively. The results show high sensitivity characteristic of UHF RFID antenna was obtained.

The significant transmission feed line method to the antenna will encourage spurious radiation from feed line. This happen when using thick microwave substrate. For this project, the thin and low-dielectric substrate material is used and the spurious radiation can be avoided. The overall final results of the proposed antenna, fork-shaped antenna for UHF RFID reader antenna are given in Table 1.

4 Conclusion

Based on this project, a fork-shaped microstrip antenna has been designed for ultra-high frequency, radio frequency identification UHF RFID reader application. The simulation result has been analyzed and optimized until the best performance of reader antenna for a RFID system is obtained. The return loss at UHF frequency of 910 MHz was obtained below −10 dB. This antenna utilizes low-dielectric constant substrate material to get small and compact size with high gain and directivity value with 7.985 and 8.959 dB, respectively.

References

1. Ragged, H. et. al. (2012). "Compact Half- Cylindrical Dielectric Resonator Antenna for UHF RFID Applications." IEEE 2012 International Conference on RFID. pp. 1–5.
2. Karkamar, N.C. (2010). Handbook of Smart Antennas for RFID Systems. John Wiley & Sons, Inc., Hoboken, New Jersey. pp. 144–149.
3. Syamimi, M.N., Ismarani, I & Mohd F.M.B (2012). "Designing an UHF RFID Reader Antenna" IEEE 2012 Symposium on Humanities, pp. 1–4.
4. Lee W., Oh K. & Yu J. (2011). "Design of Spiral-shaped UHF Near-Field Reader Antenna for RFID Application." Intelligent Radio for Future Personal Terminals (IMWS-IRFPT), 2011 IEEE MTT-S International Microwave Workshop Series on. pp. 1–2.
5. Rimbault N., Sharaiha A., and Collardey S. (2012). "Low Profile High Gain Helix Antenna Over a Conical Ground Plane for UHF RFID Applications."IEEE Antenna Technology and Applied Electromagnetics (ANTEM)", 15th International Symposium, Toulouse, pp. 1–3.
6. Pflaum S., Staraj R., and Kossiavas G. (2012). "Planar Inverted F Antenna (PIFA) Circularly Polarized for RFID Applications. "IEEE Antennas and Propagation Society International Symposium (APSURSI)", Chicago. pp. 1–2.

Provisioning of Healthcare Service in Cloud

Mridul Paul and Ajanta Das

Abstract Health informatics is fast expanding due to innovations in information and communication technology (ICT) field. The adoption of ICT in health care has resulted in deeper penetration of medical diagnostics and treatment among developed and developing countries. Though hospitals, diagnostic centers, and clinics are adopting smart models for delivering efficient medical services, there are areas that needs research in order to deliver medical services anywhere at anytime with the help of cloud. This paper focuses on services required to be provisioned in cloud for enabling smart medical treatment. We propose a layered service architecture for provisioning of medical services in cloud. We discuss specific application scenarios that are required from both provider and consumer side with test case implementation in Google cloud and provide graphical analysis of the performance of the implementation.

Keywords Health informatics · Smart healthcare · Cloud computing · Service provisioning · Service-level agreement

1 Introduction

Health Informatics has gained momentum in past few years due to increased adoption and application of information and communication technology (ICT). Health Informatics aims to improve life expectancy of citizens, while increasing the reach out of medical services among masses. A startling statistics from study [1]

M. Paul · A. Das (✉)
Department of Computer Science & Engineering,
Birla Institute of Technology, Mesra, Kolkata Campus,
Kolkata 700107, India
e-mail: ajantadas@bitmesra.ac.in

M. Paul
e-mail: mridulpaul2000@yahoo.com

D.K. Mishra et al. (eds.), *Information and Communication Technology*,
Advances in Intelligent Systems and Computing 625,
https://doi.org/10.1007/978-981-10-5508-9_25

reveal that even though there is increased spending by developed countries on improving and maintaining healthcare services, there has been no significant improvement in overall health (life expectancy) of its citizens. Though there are efforts made to address this problem through use of ICT, health care needs to be relooked at with ICT innovations such as cloud computing.

Cloud computing has transformed the way computing services are deployed and delivered. Prior to cloud, the infrastructure was a responsibility of service providers. Upgrade and upkeep had to be regularly done by the service providers. However, with cloud computing now in place, service providers need not to worry about infrastructure, platform, and even software and can focus on the services that need to be deployed. Today, several cloud providers such as Amazon, Google, and Microsoft have come up cost-effective cloud offerings for which consumers can choose from. Researchers and business organizations have benefited immensely from such offerings.

Cloud computing can serve as a cheaper alternative for processing information arising from different entities such as homes, hospitals/clinics, office buildings, and vehicles. Cloud provides three different service layers—infrastructure as a service (IaaS), platform as a service (PaaS), and software as a service (SaaS) [2]—that can form base for software applications for medical treatment. The applications can leverage the key cloud characteristics—on-demand access, broad network access, resource pooling, rapid elasticity, and measured service [3]. While cloud provides scalable models for service provisioning, the services that need to be provisioned for smart medical services require multiple facets to be explored. As there are several types of users accessing such services, the communication channels that connect to the services need to be accounted as a part of service delivery. The architecture proposed in this paper provides a model to allow medical services to be deployed in cloud. An important aspect pertaining to provisioning of services is service-level agreements (SLAs). SLAs bind service consumers and providers through legal document [4].

The remaining part of this paper is organized as follows: Sect. 2 presents related work. An evaluation of proposed architecture is described in Sect. 3. Section 4 details out application scenarios and expected privacy characteristics. Test case and results are discussed in Sect. 5. The paper concludes with Sect. 6.

2 Related Work

Doukas et al. [5] proposed usage of cloud computing and mobile application to create a mobile system that can manage patient's health records and medical images. Yang et al. [6] leveraged cloud for implementing medical image file accessing system that can share, store, and exchange medical images across different hospitals. The objective was to bring in synergy among hospitals and provide seamless services to the patients. Another research paper [7] attempts to connect individuals with hospitals, caregivers, and homestay through cloud solution.

The solution tries to establish interoperability among hospitals, healthcare providers, and consumers and simulates tests of basic operations such as sharing medical images among hospitals, using Google App Engine. Rashid et al. [8] proposes a ubiquitous healthcare system that uses Web services and cloud storage enabling patients to undergo specific tests without any intervention from experts. These tests pertain to blood pressure, weight and balance, body fat, and agility.

However, above research work does not deal with the service provisioning aspects of cloud. Das et al. [9] demonstrate usage of Big Data and cloud to provide medical solution through MedTravel app for mobile users. The research work proposes state-of-the-art mobile technology to create smart application for users and manages both structured and unstructured data in the cloud. For medical services, there is a need to define architecture that can lay grounds for deploying medical services. In this paper, we address this area to a major extent.

3 Provisioning of Smart Medical Service in Cloud

The basic premise for any smart medical service is the creation and management of EHR for each patient. EHRs are created based on the personal health record (PHR) which is typically documentary details of a patient. EHRs created in the system require robust data models [10] to store information that constitute of patient's medical history including disorders, ailments, allergies, historical diagnosis, medications, hospitalization, test results, clinical documents, and procedures. The medical services need to cater to different aspects of medical processes. Patients being the center of the medical processes, the other actors that are participating in the processes are hospitals, diagnostic centers, physicians, and pharmacy stores. In order to design these services in cloud, the requirements and interactions of each of these actors have to be considered.

This paper proposes a system architecture for smart medical treatment that provides medical care at the very place where it is required compared to traditional methods. The architecture takes into consideration that the flexibility of cloud computing provides in terms of software, platform, and hardware resources. The assumption of the model is that patients can connect to the system through mobile devices (voice or data), Internet, or health kiosks. The nature of treatment can range from simple queries about healthcare providers to complex diagnostics such as ECG analysis and recommendations.

The proposed architecture derives its fundamentals from healthcare architecture proposed in [11]. While the layered healthcare architecture addressed a wider audience, the architecture proposed in this paper is focused on providing a level deep into each of the service layers. The assumption is that IaaS provides hardware-level infrastructure and PaaS provides software-level platform to build different layers. Each layer has logical group of services that communicate with each other to provide required end service to the consumers. Thus, the layered architecture contains basic atomic services, orchestrates these atomic services in sequence, manages users and device communication, and finally communicates service results to the consumer. Figure 1 depicts the layered architecture in cloud. The description of each layer considered in the architecture is presented in the following.

- **Atomic Service Layer** forms the bottom-most layer that accommodates fine-grained and loosely coupled atomic services. These services perform basic functions such as storing and retrieving patient's medical data in data stores,

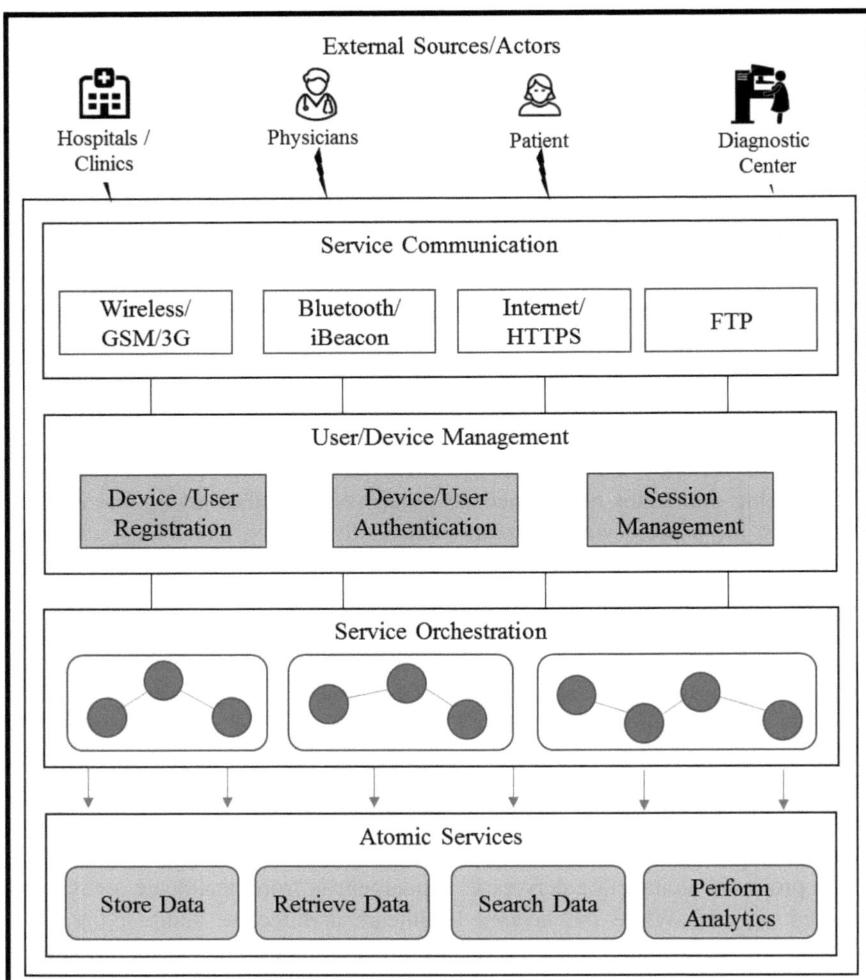

Fig. 1 Layered architecture for medical services in cloud

search for records with specific parameters, and perform analytics on historical data.
This layer forms the heart of the medical services. The services that are defined in this layer undertake tasks that are stand-alone in nature. *Store Data* service takes in values in the form of textual or visual content and location reference where to store the content. Similarly, *Retrieve Data* service fetches information using certain conditions to uniquely identify the content. Subsequently, the service data service pulls all contents that match certain query. The *Perform Analytics* service undertakes specific tasks that can churn group of records to derive patterns. For instance, this service can be called for calculating frequency of heart attacks reported in last 6 months for particular region.

- **Service Orchestration Layer** resides above atomic services. As the atomic services are discrete and are needed to be composed into forming larger functions that can be realized by the consumers. This layer manages that composition. For instance, a patient may want to view past prescriptions to view physicians visited by him and book appointment appropriately.
  The above scenario will need atomic services such as *Retrieve Data* related to prescription and then call *Search Data* for physician's details. Further for booking appointments, basic services—*Store Data*—need to be invoked after that.
- **User/Device Management** layer is placed on top of *Service Orchestration* layer. This layer is expected to manage details on the users and devices that access the services in cloud. *Device or User Registration* service takes care of the first-time registration requirements. User authentication and authorization are key for medical services. Hence, *Device or User Authentication* service manages different roles and associated access permissions. The users will be using different devices such as laptops, desktops, mobiles, and tablets to access these services. Therefore, it is equally important to maintain track of different devices used for accessing the services. Once the users are accessing different services, session management service maintains user-specific information so that user does not require to enter his information again and again.
- **Service Communication Layer** is the top-most layer that focuses on the communication protocols required for service to interact with the consumers. The layer supports wireless communication such as GSM, 3G, and 4G which manage specific request from mobile consumers. HTTPS standards [12] are used for secure internet connection which is required by the users connecting from desktops and laptops. Similarly, certain medical devices have capability to communicate using Bluetooth and iBeacon [13] protocols. Such devices can communicate via this layer with the services. At times, hospitals and diagnostic centers have requirements to upload patient files whereby file transfer protocol (FTP) can be used for communicating with the *Store Data* service.

The above architecture provides a holistic approach to deploy medical services in cloud. The key aspects of service definition, composition, and communication are addressed by different layers. The main advantage of this architecture is that the services in the layers can be changed with minimal impact to the other layers. The architecture is also flexible as the service providers can expand services in the layers as required.

4 Application Scenarios and Related Privacy

In context to smart medical services, this paper considers application scenario where patient has medical problems and books an appointment with the physician. The physician prescribes interim medicines and tests. The patient undergoes tests in a diagnostic center. The diagnostic center compiles test results so that physician can review patient condition to dispense appropriate treatment to the patient. The smart medical solution for this application scenario can leverage cloud for seamless interaction among these actors (that is patient, physician, and diagnostic center). The solution comprises of set of services presented in Fig. 2. Patients can use appointment service to book an appointment. The patient can look up list of physicians nearby to his location and select appropriate doctor or physician at suitable time. On the other hand, physician can use this service to get list of patients who will make a visit before starting their day. During the visit, the physician can use diagnostic service to diagnose patient's ailment by accessing patient's medical history (derived from EHR). Then the physician can use prescription service to recommend test and interim medicines. Once the patient undergoes recommended tests, the diagnostic center uses test-report service to upload test results. Besides, the physician can use this service to view reports to further prescribe medications.

As the services hosted on cloud deal with patient data, security and privacy of patient information is of paramount importance. Patients enter symptoms pertaining to their ailments. These symptoms assist physician during diagnosis. Hence, symptoms can only be shared between patient and a physician. Similarly, diagnosis done by the physicians can only be shared to the respective patients. In order to enable services in cloud, the access restrictions need to be maintained to ensure privacy. Table 1 provides insights to various access restrictions from different users of the service.

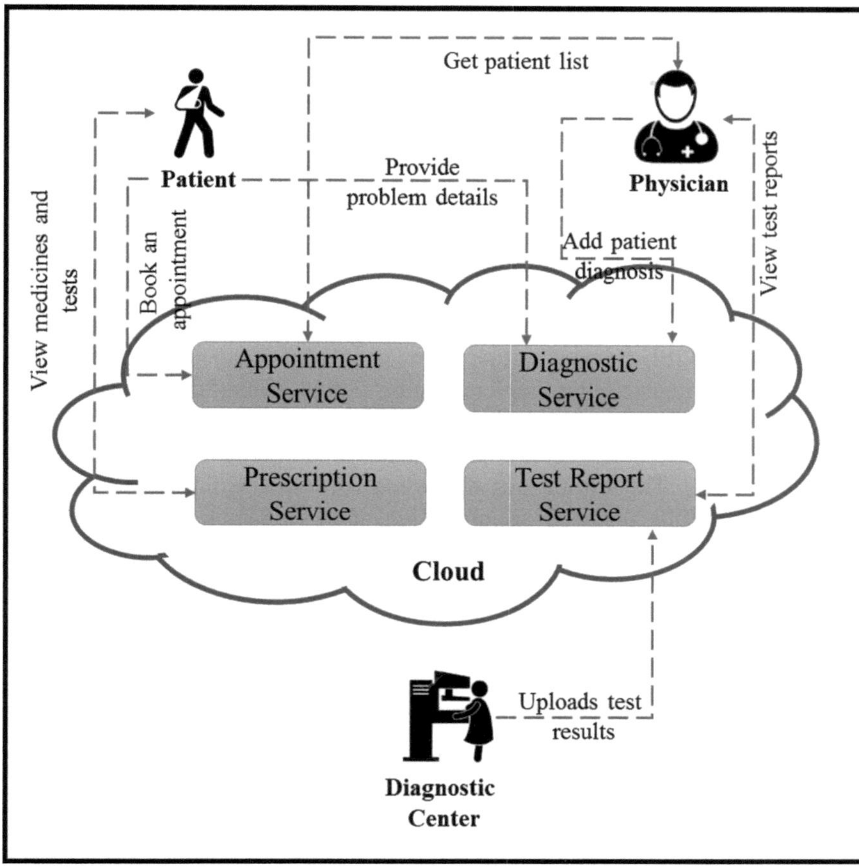

Fig. 2 Cloud services and interaction

Table 1 Access restrictions for different users

Users	Access restrictions
Patients	• The symptoms can only be added by the patients • The symptoms can only be viewed by the patients and assigned physician • The prescriptions can only be viewed by the patients and assigned physician • The test reports can only be viewed by patients and their physician
Physicians	• Physician can view list of patients that have booked appointments with him/her • Physician can write prescription for the patients that are assigned to him/her • The physician can refer test reports during diagnosis for patient assigned to him/her
Diagnostic center	• The diagnostic center can upload test results for the respective patients only • The diagnostic center cannot view any of the previous reports for that patient

5 Test Case and Discussion

This section describes test case implementation considered for smart medical treatment and discusses results achieved on Google App Engine. The implementation, which is based on the layered architecture, leverages Google Cloud storage —Data store, a NoSQL database offering from Google—that stores structured as well unstructured data. Patient information is stored in data store. This implementation considers textual details about the patient such as symptoms, history of ailments, and medicines prescribed by physicians. The first-time patients are required to register through a Web interface. This interface is a light-weight portal developed using J2EE framework that stores patient information in the data store. Once the patient registers, the implementation provides interface for various services such as booking appointments, entering symptoms, viewing diagnosis, prescriptions, and test reports. Figure 3 presents patient's interface from the service to view prescriptions. The service was monitored for response time at specific time intervals. The results obtained are promising and comparable with the average response time (0.71 s) of the top-20 global Web sites from various business domains [14]. Figure 4 presents the response time obtained through monitoring the service for at least 3 h.

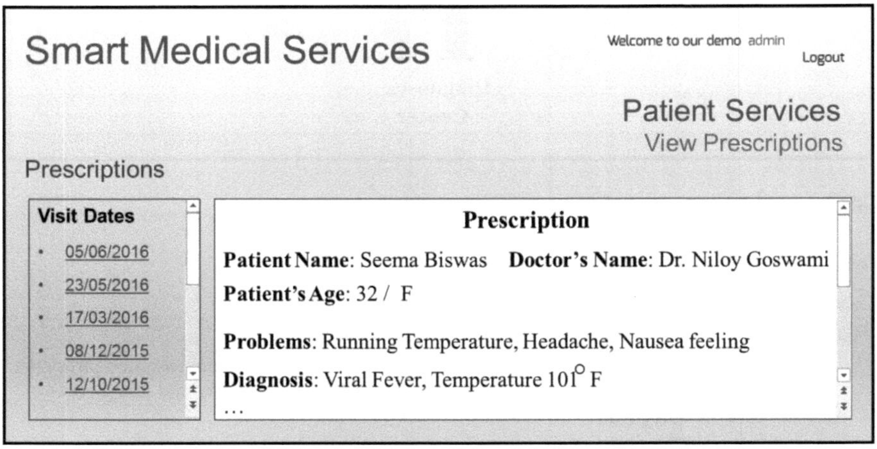

Fig. 3 Patient's interface for viewing prescriptions

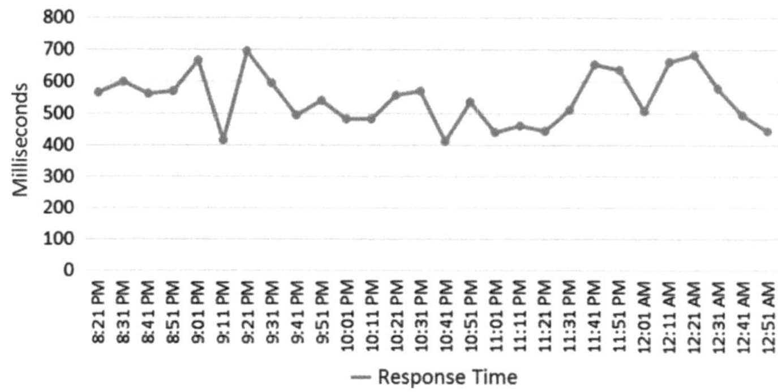

Fig. 4 Response time graph for patient's service

6 Conclusion

Medical services can become smarter with the use of cloud. Cloud computing has been a paradigm shift in the way services are deployed and delivered. It provides a cost-effective way for healthcare providers for delivering medical services. Though cloud computing provides flexibility in choosing appropriate infrastructure for deploying services, it requires critical analysis of service provisioning in cloud. This paper discusses some of the latest concepts in smart medical services and techniques where cloud can be leveraged. A comprehensive view of provisioning medical services is presented through proposed layered architecture. The paper discusses practical interactions where the architecture can be leveraged. It discusses the test case and results from specific interactions along with performance measurement of the service in cloud using graphical analysis. Further scope of this work will be aimed at elaborating on functional and non-functional requirements of use cases and defining SLA parameters for monitoring medical services in cloud.

References

1. Braunstein, M. L.: Practitioner's Guide to Health Informatics. Springer, 1–162. (2015)
2. Paul, M. & Das, A.: A Review on Provisioning of Services in Cloud Computing. International Journal of Science and Research, Volume 3 Issue 11, 2692–2698. (2014)
3. Mell, P., Grance, T.: The NIST Definition of Cloud Computing. (2009)
4. Paul, M., Das, A.: SLA Based e-Learning Service Provisioning in Cloud. Advances in Intelligent Systems and Computing. Volume 435, 49–57. (2016)
5. Doukas, C., Pliakas, T., Maglogiannis, I.: Mobile healthcare information management utilizing Cloud Computing and Android OS. Engineering in Medicine and Biology Society (EMBC). Annual International Conference of the IEEE, 1037–1040, IEEE. (2010)
6. Yang, C. T., Chen, L. T. Chou, W. L., Wang, K. C.: Implementation of a medical image file accessing system on cloud computing. Computational Science and Engineering (CSE). IEEE 13th International Conference, 321–326, IEEE. (2010)

7. Hu, Y., Lu, F., Khan, I., Bai, G.: A cloud computing solution for sharing healthcare information. 7th International Conference for Internet Technology and Secured Transactions (ICITST), IEEE. (2012)
8. Rashid, Z., Farooq, U., Jang, J. K., Park, S. H.: Cloud computing aware ubiquitous health care system. E-Health and Bioengineering Conference (EHB), 1–4, IEEE. (2011)
9. Das, A., Mitra, P., Basu, D. Provisioning of Medical Application Analysis in Cloud. 3rd International Conference on Computing for Sustainable Global Development, 1725–1729, INDIACom. (2016)
10. Rea, S., Pathak, J., Savova, G., Oniki, T. A., Westberg, L., Beebe, C. E., Tao C.: Building a robust, scalable and standards-driven infrastructure for secondary use of EHR data: the SHARPn project. Journal of biomedical informatics 45, no. 4, 763–771. (2012)
11. Paul, M., & Das, A.: Health Informatics as a Service (HIaaS) for Developing Countries. Internet of Things and Big Data Technologies for Next Generation Healthcare, Springer International Publishing, 251–279. (2017)
12. De, R., Philippe, Desmet, L., Piessens, F., Joosen, W.: Improving the security of session management in web applications. (2013)
13. Swan, M.: Sensor mania! the internet of things, wearable computing, objective metrics, and the quantified self 2.0. Journal of Sensor and Actuator Networks 1, no. 3, 217–253. (2012)
14. Website Performance Benchmarks By Industry and Geography, https://www.site24x7.com/benchmarks/

Academic Analytics Implemented for Students Performance in Terms of Canonical Correlation Analysis and Chi-Square Analysis

Aniket Muley, Parag Bhalchandra, Mahesh Joshi and Pawan Wasnik

Abstract In this research study, we were interested to test the significant association between selected variables which otherwise called as invisible and have indirect impact on the performance of the students. We have devised out our own dataset for the experimental purpose. Our study has made these variables and their relationship visible. The results enable us to determine characteristics of learning environment related to performance.

Keywords Data mining · Statistical analysis · Patterns

1 Introduction

Academic analytics is one branch of modern day's data analysis which uses statistical analysis and data mining methods to reveal and recognize hidden patterns in vast educational databases [1–6]. Such patterns enable us to throw better light on educational aspects related to student behavior, prognostication, student-centric learning, remedial aspects, and learning outcome with high accuracy. This will

A. Muley (✉)
School of Mathematical Sciences, S.R.T.M. University, Nanded 431606, Maharashtra, India
e-mail: aniket.muley@gmail.com

P. Bhalchandra · P. Wasnik
School of Computational Sciences, S.R.T.M. University, Nanded 431606, Maharashtra, India
e-mail: srtmun.parag@gmail.com

P. Wasnik
e-mail: pawan_wasnik@yahoo.com

M. Joshi
School of Educational Sciences, S.R.T.M. University, Nanded 431606, Maharashtra, India
e-mail: maheshmj25@gmail.com

D.K. Mishra et al. (eds.), *Information and Communication Technology*, Advances in Intelligent Systems and Computing 625,
https://doi.org/10.1007/978-981-10-5508-9_26

definitely increase standards of Indian higher educational system [6]. Due to digitization and effective use of computers, IT and ICT technologies, all educational organizations, institutions, and universities have generated and stored large data in their databases [7–13]. This data can be a key source for futuristic decision making processes if it is being processed through academic analytics. We took it as a challenge to see all the business intelligence, patterns, correlations, and rules embedded in this data. Our work is an interdisciplinary work undertaken by three schools of our university as performance analysis shares sphere with educational pedagogies, statistics, and computer-enabled technologies. The academic analytics was implemented using SPSS software [14, 15].

A closed questionnaire with predefined answers was used for data gathering [16] on A4 size single-sided paper sheet. Performance-related economical, social, and emotional attributes of this questionnaire were selected with the help of School of Educational Sciences and as per theory of Pritchard and Wilson [16, 17]. The questionnaire was modified number of times to reduce the complexity of understanding as well as to increase simplicity of answering. It was tested on subset of students after every revision. An Excel sheet was prepared for the answers using code such as 0, 1, 2. The confidential issues in datasets were properly addressed as dataset carried personal information of students. The error rate during preprocessing was 38% which finally reduced to 5% after proper convincing to students. The questionnaire looks like Figs. 1 and 2.

1	Course code	MSc (5) , MCA (6)
2	Your name	
3	Gender (sex)	Male (1) Female (0)
4	Marital status	Married (2) unmarried(3)
5	Age	
6	Home address	Urban(1) rural (2) foreign(3)
7	Mobile no.	
8	Personal email id	
9	Degree passer and percentage	General B.Sc. / B.Sc.(computer CS) / BCA / BCS/ Other / (1) (2) (3) (4) (5)
	Percentage	
10	Degree collage name	
11	Father's Education	Below or SSC/ HSC/ Graduate/ Post Graduate/ other (1) (2) (3) (4) (5)
12	Fathers job and annual income	Service / Business/ Agriculture/ In house/ Other/ (1) (2) (3) (4) (5)
	Income	0-1 lakh (1) , 1.1-2 lakh(2), 2.1-5 lakh(3) , 5lakh - above (4)
13	Mothers education	Below or SSC/ HSC/ Graduate/ Post Graduate/ other (1) (2) (3) (4) (5)
14	Mothers job and annual income	Service / Business/ Agriculture/ In house/ Other/ (1) (2) (3) (4) (5)
	Income	0-1 lakh (1) , 1.1-2 lakh(2), 2.1-5 lakh(3) , 5lakh - above (4)
15	Family size	
16	Family relationship	Excellent / Good/ Satisfactory/ Bad/ Very Bad (1) (2) (3) (4) (5)
17	Family support to your education	Excellent / Good/ Satisfactory/ Bad/ Very Bad (1) (2) (3) (4) (5)
18	Reason to choose this course	Career in IT/ Near to Home/ Reputation of course / Blind Decision/ Parents wish: (1) (2) (3) (4) (5)
19	Travel mode and time needed	Bus/ Railway/ City Bus/ Rickshaw/ Self Vehicle / walking (6) (1) (2) (3 but taken as 1) (4) (5)

Fig. 1 Sample questionnaire

FORM_NO	CORSNAME	YRNAME	GENDER	MARRIED	AGE	REGION	MOBNO	UG
1	6	DUGANE SEDHARTH NAGORAO	1	3	24	2	8446461553	2
2	6	SHASHANK H. BALASKAR	1	3	21	2	7709512133	3
3	6	MOTE PRADEEP KASINATH	1	3	22	2	2462229251	3
4	6	PANDIT SWAPNIL RAGHUNATH	1	3	24	2	9970403753	2
5	6	GOTRE NILESHKUMAR NAMDEV	1	3	22	1	9890402292	2
6	6	GADEWAR LEENA PRAMOD	0	3	23	2	9665360530	2
7	6	KOLHE VANITA PANDURANG	0	3	22	1	9579836606	2
8	6	BARDE NISHA P	0	3	21	1	2462229251	2
9	6	GAWANDE SANTOSH PRABHAKAR	1	3	24	2	9028064993	2
10	6	NITIN NARESH DEKATE	1	3	21	2	9011952075	2
11	6	ASHUTOSH V. DONGRE	1	3	23	1	9730834463	2
12	6	SWAMI SANTOSH SHIVLING	1	3	22	2	8793336938	3
13	6	WATHORE ANKUSH NAVNATH	1	3	21	1	2462229251	3
14	6	BAHIWAL AKSHAY BALAPRASAD	1	3	21	1	7709689015	3
15	6	EMEKAR SANDEEP SUBHASH	1	3	23	1	9881177713	2
16	6	GALIPELLI SANDHYA SHANKABABU	0	3	21	1	7709977551	2

Fig. 2 Data set in MS-Excel

2 Experimentations and Discussion

Our aim was to discover invisible attributes related to performance of students. So we had discussions with educationalist and then finally understood that the semester end marks alone cannot be taken as main indicator of student's performance. The performance is indistinct term. For proper knowledge, surveyed literatures such as Shoukat Ali et al. [4], Graetz [18], Considine and Zappala [19], and Bratti and Staffolani [20]. This analysis is helpful for identifying the personal, social and economic kinds attributes in our study.

With these preliminary investigations and understanding, we decided to identify key variables that accelerates or downgrades educational performance at large. We had thought that economical and social conditions of students can be important variables from our dataset/questionnaire as far as performance was concerned. To do so, many variables and their interrelations needed to be analyzed for proper analysis. It is always true for questionnaires as they consist of many questions, such that each question contributes for one variable [7, 21–23]. Studying all variables and their interrelation may be complicated as they may divert us from the original research focus. For such exploratory analysis, factor analysis has been invented [22]. We have used SPSS22.0v to analyze the data set. The snapshots given below show the evidence of empirical analysis. The descriptive statistics are used through MS-Excel to represent our data in the diagrammatic form. Some of the interesting facts are shown in Figs. 3, 4, 5 and 6. Further, canonical correlation analysis and chi-square testing have been done on the experimental data set.

Fig. 3 Region-wise distribution

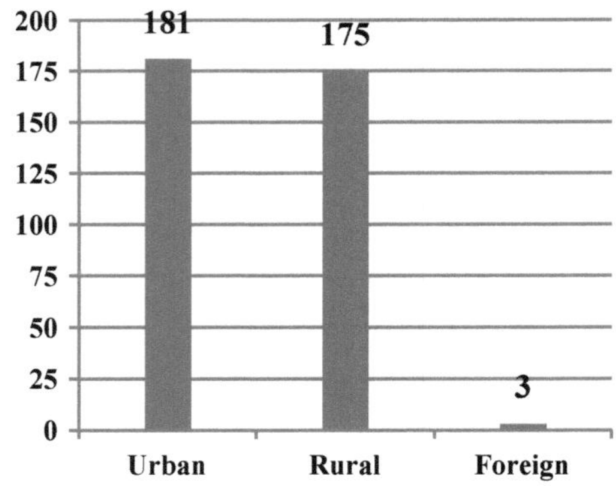

Fig. 4 Diversity in jobs among parents

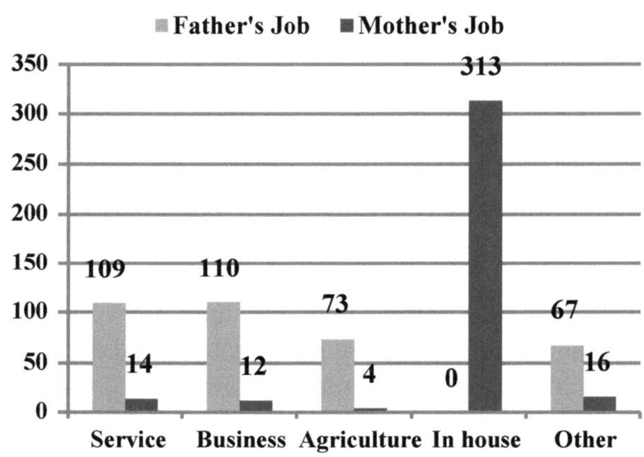

Fig. 5 Parents versus their education level

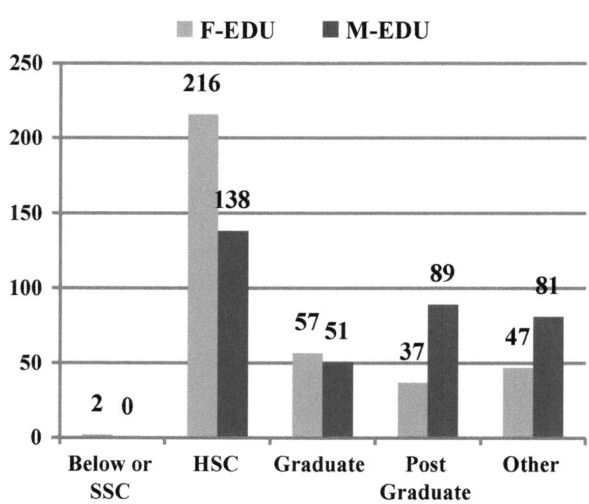

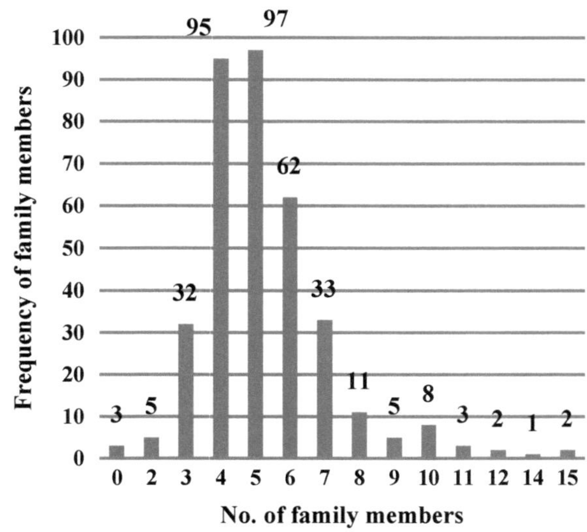

Fig. 6 Students versus their family size

2.1 Program Code

The SPSS22.0v is used to analyze the data set [16].

```
FREQUENCIES VARIABLES=GENDER MARRIED AGE REGION UG FEDU FJOB
FINCOM MEDU MJOB MINCOM FSIZE FRELATIONS FSUPPORT REASON
TMODE
TTIME STIME FAILURES TUTORIAL SCHOLERSHIP PJOB MM HARDSUB_UG
STUDY_HOME SELFLIB SELFPC PLACELVING INTERNET F_T_STUDY
F_T_FRIEND MOVIEPWEEK CAREERDREM PARALLELCOURSE OWN_NOTES
FREE_T_ACC PER_SATISF MATERIAL HLT_STATUS
/ORDER=ANALYSIS.
CROSSTABS
/TABLES=REGION GENDER BY FAILURES STIME SCHOLERSHIP PJOB
SELFLIB SELFPC PLACELVING INTERNET F_T_STUDY F_T_FRIEND
MOVIEPWEEK OWN_NOTES PER_SATISF MATERIAL
/FORMAT=AVALUE TABLES
/STATISTICS=CHISQ
/CELLS=COUNT
/COUNT ROUND CELL.
```

The use of descriptive statistics has been made using MS-Excel to represent our data in the diagrammatic form. Figures 3, 4, 5 and 6 show the distribution of the data according to region-wise classification, diversity of parents jobs, education-wise, and their family size-wise, respectively. The students came from urban and rural backgrounds are found to be approximately same of Indian students as compared to foreign students. The discrimination in the student's performance is observed according to their parent's job and educational background. Also, numbers of family members in student's family were represented through the bar plot. The interesting facts are shown in Figs. 3, 4, 5 and 6.

2.2 Canonical Correlation Analysis

The core objective is to find relationship between personal details with family background. We made two groups for proper analysis. The first group is student's details containing three parameters, viz. gender, age, and UG percentage. The second group is his/her family background and the parameters chosen are: father's education, father's job, father's income, mother's education, mother's job, mother's income, family size, and whether student does any part-time job? Here, Canonical correlation analysis is used to find the significant relationship between student's details and his family background to determine the associations among two sets of variables. Our observations gave us significant outcomes.

2.3 Chi-Square Analysis

Sample analysis using chi-square tests is mentioned here. Similar way, the results were computed and it has been represented in the form of conclusion. Below figures and tables show the use of descriptive statistics. These together show some data regarding diversity of the students according to course-wise, gender-wise, undergraduate background, father's occupation, and their family size. We have applied chi-square test to test the significance among the above objectives and assumptions that there will be significant difference among the variables under study.

Some of the parameters which show significant differences in our study are as scholarship holder students with gender-wise; difference gender-wise about their career dreams; between gender-wise percentages obtained at UG level, between region-wise percentages obtained at UG level by the students; between age group-wise obtained scholarships; between age group-wise obtained UG Percentage; students and their father's education; students and their father's job; between gender-wise and their mother's education; age-wise and their family size; age-wise and part-time job; region-wise and father's education; region-wise and family size; students place of living and self library. Further, we have made analysis using chi-square Tests with the help of SPSS 22.0 software [15] and found some significant results. These are represented in the form of tables. According to region-wise study with respect to variables like place of living, do they have their own PC? Do they use internet? How much free time they have for study? It was surprising to note that there are significant differences with respect to student's living places. These differences came because of student's awareness to use internet. Our students are from computer science field, and hence, it is expected that they must frequently use internet. From our experimental analysis, it is found to be true. While dealing with students free time for study perspective, it has been observed that there is significant enough good time is available for study. It was assumed that in due course of studentship, he/she may get sufficient time for study rather than doing any other work. This particularly holds true as the Nanded region is not a metro city or an industrial hub. When we did gender-wise study with a variable, how much scholarship they get? It is observed that there is significance difference. Male students get more scholarship than female students. Also, we found significance among gender-wise difference in their place of living. Most of the female students preferred to live at own home or in hostels due to safety issues. Tables 1, 2, 3, 4 and 5 show these results.

Table 1 Chi-square tests analysis for region versus students having self PC

	Value	df	Asymp. sig. (2-sided)
Pearson chi-square	21.366[a]	3	0.000
Likelihood ratio	22.504	3	0.000
Linear-by-linear association	15.360	1	0.000
No. of valid cases	359		

[a]2 cells (33.3%) have expected count less than 5 and the minimum expected count is 1.09

Table 2 Chi-square tests analysis for region versus place of living

	Value	df	Asymp. sig. (2-sided)
Pearson chi-square	15.703[a]	3	0.000
Likelihood ratio	17.056	3	0.000
Linear-by-linear association	13.262	1	0.000
No. of valid cases	359		

[a]2 cells (33.3%) have expected count less than 5 and the minimum expected count is 0.25

Table 3 Chi-square tests analysis for region versus free time to study

	Value	df	Asymp. sig. (2-sided)
Pearson chi-square	15.808[a]	3	0.000
Likelihood ratio	17.387	3	0.000
Linear-by-linear association	6.080	1	0.000
No. of valid cases	359		

[a]9 cells (60.0%) have expected count less than 5 and the minimum expected count is 0.01

Table 4 Chi-square tests analysis for region versus students having self PC

	Value	df	Asymp. sig. (2-sided)
Pearson chi-square	21.366[a]	3	0.000
Likelihood ratio	22.504	3	0.000
Linear-by-linear association	15.360	1	0.000
No. of valid cases	359		

[a]0 cells (0.0%) have expected count less than 5 and the minimum expected count is 78.28

Table 5 Chi-square tests analysis for gender versus place of living

	Value	df	Asymp. sig. (2-sided)
Pearson chi-square	12.996[a]	3	0.000
Likelihood ratio	13.174	3	0.000
Linear-by-linear association	0.002	1	0.000
No. of valid cases	359		

[a]0 cells (0.0%) have expected count less than 5 and the minimum expected count is 10.93

3 Conclusion

The performance of the student is fuzzy terms and it is affected by many parameters. In this study, our data reveal that it is due to the social and economical condition of students. However, no scientific evidences were there for such outcome. The study took it as challenge and it has been discovered that the student's performance mere did not depend on his/her studious nature. This paper shows effective use of academic analytics in terms of descriptive statistics. Here, we have applied canonical correlation analysis and chi-square test to test the significance among the stated objectives and assumptions. We have finally discovered new variables, which otherwise were invisible that hampers performance of students.

References

1. Dunham Margaret H.: Data Mining: Introductory and Advanced Topics, Pearson publications (2002)
2. Han, J. and Kamber, M., Data Mining: Concepts and Techniques, 2nd edition. The Morgan Kaufmann Series in Data Management Systems, Jim Gray. (2006)
3. Behrouz et.al.: Predicting Student Performance: An Application of Data Mining Methods With The Educational Web-Based System Lon-CAPA IEEE, Boulder, CO. (2003)
4. Shoukat Ali et al.: Factors Contributing to the Students Academic Performance: A Case Study of Islamia University Sub-Campus, American Journal of Educational Research, 1 (8), pp. 283–289 (2013)
5. Gordon Linoff, Michael J, et al.: Data Mining Techniques, 3Ed, Wiley Publications.
6. Eko Indrato: edited notes on Data Mining, retrieved from http://recommender-systems.readthedocs.org/en/latest/datamining.html

7. Ma, Y., Liu, B., Wong, C. K., Yu, P. S., & Lee, S. M.: Targeting the right students using data mining. Paper presented at the Sixth ACM SIGKDD International Conference Proceedings, Boston, MA, pp. 457–464 (2000)
8. Minaei-Bidgoli, B., Kashy, D. A., Kortemeyer G. and, Punch, W. F.: Predicting student performance: an application of data mining methods with the educational web-based system LON-CAPA, In Proceedings of ASEE/IEEE Frontiers in Education Conference, Boulder, CO: IEEE (2003)
9. Kotsiantis S.: Educational Data Mining: A Case Study for Predicting Dropout – Prone Students. Int. J.Knowledge Engineering and Soft Data Paradigms, 1(2), pp. 101–111 (2009)
10. Berkhin Pavel: Survey of Clustering Data Mining Techniques, Accrue Software, available at www.cc.gatech.edu/~isbell/reading/papers/berkhin02survey.pdf
11. Sasirekha K., Baby, P.: Agglomerative Hierarchical Clustering Algorithm- A Review, International Journal of Scientific and Research Publications, 3(3), pp. 83 (2013)
12. Nikhil Rajadhayx et al.: Data mining in Educational Domain, retrieved from http://arxiv.org/pdf/1207.1535.pdf
13. Murugesan Keerthiram, Zhang Jun: Hybrid Hierarchical Clustering: An Experimental Analysis, Technical Report: CMIDA-hipsccs#001–11, retrieved from www.cs.uky.edu/~jzhang/pub/techrep.html
14. Field, A.: Discovering Statistics using SPSS for Windows. London–Thousand Oaks – New Delhi: Sage publications. (2000)
15. IBM SPSS Statistics 22 Documentation on internet retrieved at www.ibm.com/support/docview.wss?uid=swg27038407
16. Cortez Paulo and Silva Alice: Using Data Mining to Predict Secondary School Student Performance, retrieved from http://www.researchgate.net/publication/ Using_data_mining _ to_ predict_secondary_ school_ student_ performance
17. Pritchard, M. E., and Wilson, G. S.: Using emotional and social factors to predict student success. Journal of College Student Development 44(1): pp. 18–28. (2003)
18. Graetz, B.: Socio-economic status in education research and policy in John Ainley et al., Socio-economic Status and School Education DEET/ACER Canberra., J Pediatr Psychol. 20 (2):205–216 (1995)
19. Considine, G. & Zappala, G.: Influence of social and economic disadvantage in the academic performance of school students in Australia. Journal of Sociology, 38, 129–148 (2002)
20. Bratti, M. and Staffolani, S.: Student Time Allocation and Educational Production Functions, University of Ancona Department of Economics Working Paper No. 170 (2002)
21. Introduction to factor analysis, web resource www.yorku.ca/ptryfos/f1400.pdf
22. Rietveld, T. & Van Hout R.: Statistical Techniques for the Study of Language and Language Behaviour. Berlin – New York: Mouton de Gruyter (1993)
23. Habing, B.: Exploratory Factor Analysis. Website: http://www.stat.sc.edu/~habing/courses/530EFA.pdf (accessed 10 May 2004) (2003)

A Pairwise Alignment Algorithm for Long Sequences of High Similarity

Chien-Tai Lee and Sheng-Lung Peng

Abstract Alignment algorithms are important in bioinformatics for comparing the similarity among sequences. The algorithm of Needleman–Wunsch is well known for globally aligning two sequences. However, this algorithm is unsuitable for sequences of long length. Many heuristic algorithms are proposed, such as BLAST and FASTA. However, they are still unsuitable for long sequences. In this paper, we study the alignment problem on highly similar sequences. By taking SARS viruses as an example, our result shows that our algorithm runs faster than Clustalx for aligning two SARS viruses. It implies that our algorithm is suitable for viruses of high similarity.

Keywords Pairwise alignment · SARS virus · Whole-genome alignment

1 Introduction

Sequence alignment is an important technique for comparing the similarity of two biological sequences in bioinformatics. In 1970, biologists Needleman and Wunsch were the first to analyze biological hereditary information by using electronic calculator. They applied the method of dynamic programming to analyze the similarity between two amino acid sequences [1]. In 1988, the US government established the National Center for Biotechnology Information (NCBI) to help in creating automatic analyzing system of biological information. This further enhances the ability for biomedical researchers to search and analyze biological information more efficiently and accurately. In 1992, the most famous nuclear acid sequence database, GenBank, began the service for the whole world. It collected more than 150 billion of base pairs counted to February 2013. Meanwhile, in Europe, the European Molecular Biology Laboratory (EMBL) was founded in

C.-T. Lee · S.-L. Peng (✉)
Department of Computer Science and Information Engineering,
National Dong Hwa University, Hualien 97401, Taiwan
e-mail: slpeng@mail.ndhu.edu.tw

D.K. Mishra et al. (eds.), *Information and Communication Technology*,
Advances in Intelligent Systems and Computing 625,
https://doi.org/10.1007/978-981-10-5508-9_27

1988. In 1990, Japan established their DNA database of Japan (DDBJ). Currently, GenBank, EMBL, and DDBJ provide best quality in the search of gene sequence globally.

DNA is the most important element in the study of life science. After knowing the sequences of DNA and RNA, the next important thing is to analyze all the information originated from the DNA sequences. There are three billion of bases to form chromosomes in human beings. It would take tremendous time to know all the sequences in human genes. It could be more difficult to analyze the underlying meaning in the DNA sequences. Recently, the rapid development in information science and technology has resolved a lot of problems. It has become more popular to combine information technology and biological science, thereby trying to have a big breakthrough in the fields of biomedicine and pharmaceutical research. With the help of computer technology, this goal may be expected in the future.

The main subjects in biological information science include prediction of a protein structure, comparison of the similarity and the homology of sequences, phylogeny construction, genome sequence analysis. In this paper, we study the DNA strains of Severe Acute Respiratory Syndrome (SARS) viruses. We propose an algorithm for aligning two highly similar sequences. In our experimental result, if these two sequences are highly similar, the expected time for alignment is almost linear.

2 Preliminaries

A virus is a simplest microorganism. All its hereditary information is located on the chains of nucleotides. Once something is changed in the nucleotide chain, it results in a variation in the descendant. Further, viral gene underwent spontaneous mutation during the process of proliferation. Although most of them died, some survived and acquired the ability to adjust themselves to the stress from the surrounding environment.

A DNA mutation has many types, e.g., insertion, deletion, substitution (replacement), duplication. Mutations can be defined in two ways:

1. Chromosomal mutations: They are modifications of a chromosome. For example, in the deletion or duplication of genes (or segments) of a chromosome, it changes the total number of chromosomes sometimes.
2. Gene mutations: These mutations occur in the nucleotide sequence and are caused by the substitution of one nucleotide for another or by insertion or deletion of one or more nucleotides in the DNA. More specifically, insertion means addition of one or many nucleotides from DNA. Deletion means removal of one or many nucleotides from DNA. Substitution (replacement) means to replace one DNA base by another DNA base.

There are two popular methods available to estimate the similarity of two sequences. One applies the principle of probability and statistics [2], and the other is to use a sequence alignment algorithm. Usually, we use indel to refer deletions or insertions of an alignment.

2.1 Global Alignment

Dynamic programming is an important technique for algorithm design. It usually solved an optimization problem by caching subproblem solutions rather than recomputing them [3–5]. Needleman–Wunsch's algorithm [1] uses dynamic programming technique to compute an optimal alignment for two sequences. It uses a two-dimensional array *D* to store the best scores for each entry. Therefore, they can guarantee to find an optimal solution, but consume resources very much, and it is not efficient for comparing with long sequences. This method has two main steps:

1. Compute similarity scores
   - A score is computed for each entry in the array according to their similarity.
   - The similarity score is usually defined as [6, 7]:

$$D(i,j) = \max \begin{pmatrix} D(i-1,j-1) + \delta(a_i, b_j) \\ D(i-1,j) + \delta(a_i, -) \\ D(i,j-1) + \delta(-, b_j) \end{pmatrix}.$$

2. Construct an alignment
   - Backtrack the matrix to obtain an optimal alignment.

In the optimization of the alignment algorithm, we may need to add some gaps to the sequence. The frequency of "insertion" or "deletion" events causes gap insertions. Thus, we can use different penalty parameters to define a gap insertion.

In Needleman–Wunsch's algorithm, we need to store $(n + 1)$ $(m + 1)$ numbers for aligning a sequence of length n with another sequence of length m. By the above-mentioned formula, each number takes a constant number of computations. Thus, the algorithm runs in $O(nm)$ time and requires $O(nm)$ memory. For whole-genome alignment, this running time is still too slow. In other words, it is not feasible for real biological applications.

2.2 Heuristic Algorithms

Generally speaking, aligning two sequences with length at most n, the time complexity is $O(n^2)$. This is by no means the best time bound. However, when aligning three sequences, it consumes longer execution time, e.g., $O(n^3)$. How to reduce the time complexity is the most important work while performing the whole-genome alignment. Many heuristic methods are proposed. The kind of these methods is an algorithm that usually but not always gives an optimal answer.

For example, to use the most famous tools such as FASTA or BLAST (Basic Local Alignment Search Tool) programs to speed up the performance, the problem is to compromise its sensitivity. Therefore, they do not guarantee to find an optimal solution, but can improve the efficiency.

- FASTA series [8]: The method of these series is quite precise, but the speed is slower. It can only compare a nucleic acid sequence to the nucleic acid information bank, or compare a protein sequence to the protein information bank [8–11].
- BLAST series [12]: The method of these series has a very quick search speed but has the fault when the sequences have low similarity. It contains a group of programs and can automatically execute according to the information bank type of input sequences [12, 13].

2.3 Genome-Scale Alignment: MUMmer

MUMmer is a tool for aligning entire genomes [14]. It has three main steps:

1. Computing MUMs: A MUM (maximal unique match) for two sequences x and y is a pair of subsequences (x', y'), that is, an exact match, and there is no other matching subsequence pair containing x' and y' simultaneously. During the computation, MUMmer first constructs a suffix tree for sequence x. Then, the suffixes of y are inserted to the constructed tree. It is called a generalized suffix tree. Finally, all the MUMs can be computed by traversing this generalized suffix tree.
2. Finding the backbone of the alignment: All the MUMs in x are arranged in an increasing order. Then, we find a longest increasing subsequence of MUMs for y with respect to the MUMs of x. These MUMs define the backbone of the alignment.
3. Closing gaps: To obtain a final result, the gaps between consecutive MUMs of the backbone are aligned by using the Smith–Waterman algorithm [15].

MUMmer is also suitable for two sequences with high similarity. However, it has two flaws: One is that it needs a very large space to save the suffix tree, and the other is that MUMs cannot be found when the same substring appeared more than once. Recently, by using the computational power of GPUs, an algorithm for very long sequence is proposed [16].

3 The Proposed Algorithm

In bioinformatics, pairwise sequence alignment is the most important tool. Many tools use it as a core for approximating the multiple sequence alignment or comparing the similarity of two genes. The Needleman–Wunsch alignment is the best pairwise sequence alignment algorithm for finding an optimal solution. It uses a dynamic programming approach and runs in $O(mn)$ time where m and n are the lengths of the two input sequences, respectively. However, while considering the whole genome as an input, this algorithm is impractical. Many heuristic algorithms occur for aligning two whole-genome sequences. For example, MUMmer uses the maximal unique matching sequences as a base for aligning two entire genomes. MUMmer uses the data structure of suffix tree to find all the maximal unique matching sequences. Although genomes have high similarity, it still needs a very large space. Thus, we propose a new method for saving space and time to align two entire highly similar sequences.

SARS virus, one kind of coronavirus, is a highly transmissible and virulent virus that caused a disastrous outbreak in the Southeast Asia in 2003. Many scientists and physicians wanted to know more about different strains of SRAS viruses and tried their best to prevent the spread of this disease. Its DNA sequence is approximately 29,700 base pairs. Here, we propose an algorithm to do the alignments for SARS viruses. Our algorithm can be divided into three phases.

1. The first phase is Dot Matrix Phase: We use dot matrix to find the longest common substring (LCS) in a sliding window. The segment in sliding window is broken into three parts by the LCS: the LCS itself, prefix string that before the LCS, and the suffix string that after the LCS.
2. The second phase is Needleman–Wunsch Phase: Let the prefix strings be aligned by the Needleman–Wunsch's algorithm, the follow-up for getting optimal alignment with appending the LCS.
3. The third phase is the Liner Comparison Phase: If there is only one remaining suffix string, then we compare it with the other uncompared string. If the two sequences are highly similar, then we obtain a linear alignment.

We define three variables SW, SR, and MES as follows:

- Sliding window (SW): determination of a length for Dot Matrix Phase.
- Similarity rate threshold (SR): a threshold of similarity ratio for the two subsequences in sliding window.
- Max continuous equal substring length threshold (MES): a threshold for the length of LCS in sliding window.

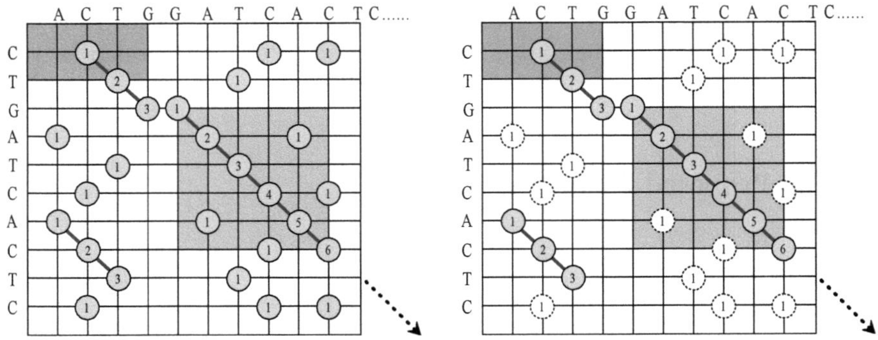

Fig. 1 Dot matrix

First, let $S1$ and $S2$ be two sequences for doing an alignment. From the beginning of the two sequences, determine a sliding window size SW to perform dot matrix. An example is shown in Fig. 1. The plots in dot matrix provide an easy and efficient way to find similarity between two sequences.

The alignment should be a diagonal of continuous dots. Sometimes, it is broken at a point of mutation and shifting to another diagonal because of indels. In this method, we find the LCS and determine whether the similarity is greater than SR and the length of LCS is greater than MES. If the answer is no, then we double the window size. We then repeat the procedures until one of the previous two conditions is satisfiable. Then, we do an alignment using Needleman–Wunsch's algorithm on the prefix strings.

After finishing a partial alignment, we then extend the alignment from LCS by comparing it with the follow-up sequences until they are different. Finally, we repeat the above procedure for the remaining subsequences. Figure 2 shows the concept of a phase of our algorithm.

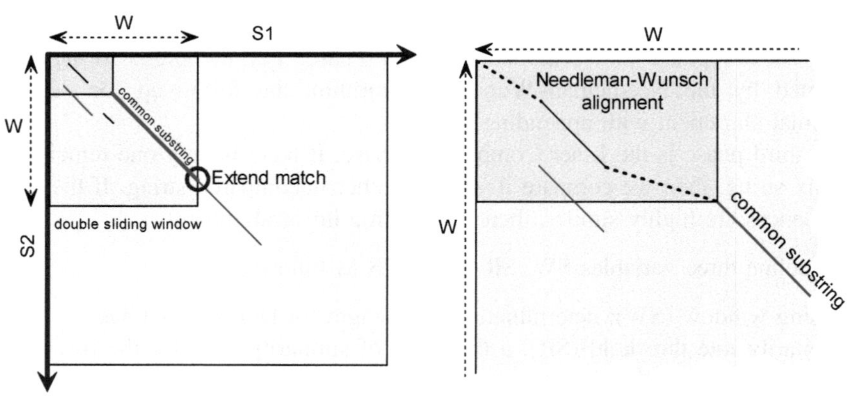

Fig. 2 Complete alignment in sliding window

For any two sequences with high similarity, the alignment is very efficient because the liner comparison does a favor. However, this method is only suitable for highly similar sequences.

4 Experimental Results

For obtaining a quantitative comparison, we use an old computer system to run the proposed algorithm. Our experimental environment is as follows:

- Machine: Compaq
- CPU: IntelR Xeon™ CPU 3.06 GHz (2CPUs)
- Memory: 1024 MB
- Operating system: Microsoft Windows 2000

We use 102 SARS sequences of FASTA format as an input (which are downloaded from NCBI). The FASTA format begins with a single-line description and is followed by lines of sequence data. The description line starts with ">" symbol for distinguishing from the other sequence data. It is recommended that each line of text should be shorter than 80 characters. An example of FASTA format is given as follows:

```
>gi|31416306|gb|AY279354.2|SARS coronavirus BJ04
TACCCAGGAAAAGCCAACCAACCTCGATCTCTTGTAGATCTGTTCTCTAAACGAACTTTAAAATCTGTGT
AGCTGTCGCTCGGCTGCATGCCTAGTGCACCTACGCAGTATAAACAATAATAAATTTTACTGTCGTTGAC
......
CGGCCACAAGGTCGTTGAGCTGGTTGCAGAAATGGACGGCATTCAGTACGGTCGTAGCGGTATAACACTG
GGAGTACTCGTGCCACATGTGGGCGAAACCCCAATTGCATACCGCAATGTTCTTCTTCGTAAGAACGGTA
ATAAGGGAGCCGGTGGTCATAGCTATGGCATCGATCTAAAGTCTTATGACTTAGGTGACGAGCTTGGCAC
```

A SARS of whole genome has about 29,700 base pairs. We use our algorithm to align and analyze similar degree according to the matching rules of IUPAC (International Union of Pure and Applied Chemistry).

Clustalx is famous software in bioinformatics. It performs alignments and calculates distances for all pairs of sequences. We randomly take out 10 sequences from 102 SARS sequences to do a comparison with SARS NC004718.3. Table 1 shows the results.

Table 1 A comparison with Clustalx

Time in mm:ss	SinP5	Sin852	TW4	LC3	HZS2-E	Shang QXC2	PUMC 03	TWC3	TWK g RNA	BJ02
Clustalx 1.83	9:49	9:49	9:50	9:42	9:50	9:36	9:50	9:50	9:50	9:90
Our method	0:02	0:18	0:01	4:36	0:01	5:19	0:02	0:01	0:01	0:02
Seq. length	29,713	29,670	29,729	29,350	29,736	29,013	29,745	29,727	29,727	29,745

5 Conclusions

Bioinformatics has become the key to answer the mystery of life in the post-genomic era. This paper proposes a sequence alignment algorithm for those DNAs of high similarity. Although the Needleman–Wunsch alignment is the best pairwise sequence alignment algorithm for finding an optimal solution, it is impractical for whole genomes. MUMmer's algorithm needs a very large extra space for saving the suffix tree, and it is also time-consuming for searching MUMs in the suffix tree. We provide another kind of alignment algorithms. It cannot avoid the limitation of $O(n^2)$, but it has a good performance for highly similar sequences.

Acknowledgements This work was partially supported by the Ministry of Science and Technology of Taiwan, under Contract No. MOST 104-2221-E-259-006.

References

1. Needleman, S.B., Wunsch, C.D.: A general method applicable to the search for similarities in the amino acid sequences of two proteins. J. Mol. Biol. 147, 195–197 (1970)
2. Smith, T.F., Waterman, M.S., Burks, C.: The statistical distribution of nucleic acid similarities. Nucleic Acids Res. 13, 645–656 (1985)
3. Caho, K.M., Pearson, W.R., Miller, W.: Aligning two sequences within a specified diagonal band. Comput. Appl. Biosci. 8, 481–487 (1992)
4. Chao, K.M.: Dynamic-programming strategies for analyzing biomolecular se- quences. http://www.ims.nus.edu.sg/Programs/genome/files/sg2002.pdf.
5. Pearson, W.R., Miller, W.: Dynamic programming algorithms for biological sequence comparison. Methods Enzymol. 210, 575–601 (1992)
6. Fitch, W.M., Smith, T.F.: Optimal sequence alignments. Proc. Natl. Acad. Sci. USA 80, 1382–1386 (1983)
7. Gotoh, O.: An improved algorithm for matching biological sequences. J. Mol. Biol. 162, 705–708 (1982)
8. Pearson, W.R., Lipman, D.J.: Improved tools for biological sequence comparison. Proc. Natl. Acad. Sci. USA 85, 2444–2448 (1988)
9. Fickett, J.W.: Fast Optimal alignment. Nucleic Acids Res. 12, 175–180 (1984)
10. Pearson, W.R.: Rapid and sensitive comparison with FASTP and FASTA. Methods Enzymol. 183, 63–98 (1990)
11. Pearson, W.R.: Searching protein sequence libraries: Comparison of the sensitivity and selectivity of the Smith-Waterman and FASTA algorithms. Genomics 11, 635–650 (1991)
12. Altschul, S.F., Gish, W., Miller, W., Myers, E.W., Lipman, D.J.: Basic Local Alignment search Tool. J. Mol. Biol. 215, 403–410 (1990)
13. Altschul, S.F., Madden, T.L., Schaffer, A.A., Zhang, J., Zhang, Z., Miller, W., Lip-man, D.J.: Gapped BLAST and PSI-BLAST: a new generation of protein database search programs. Nucleic Acids Res. 25, 3389–402 (1997)
14. Delcher, A.L., Kasif, S., Fleishmann, R.D., Peterson, J., White, O., Salzberg, S.L.: Alignment of whole genomes. Nuc. Acids. Res. 27, 2369–2376 (1999)
15. Smith, T.F., Waterman, M.S.: Identification of common molecular subsequences. J. Mol. Biol. 147, 195–197 (1981)
16. Li, J., Ranka, S., Sahni, S.: Pairwise sequence alignment for very long sequences on GPUs. Int. J. Bioinform. Res. Appl. 10(4–5), 345–368 (2014)

Information Extraction Approaches: A Survey

Monia Mannai, Wahiba Ben Abdessalem Karâa and Henda Hajjami Ben Ghezala

Abstract In the recent years, the amount of available information in the Web is growing. Thereby, the search of pertinent information through those large documents has become a difficult task. That's why, we need to develop information extraction systems in order to facilitate the treatment and the representation of data according to the user's need. These systems should adopt an extraction approach for its implementation. In this paper, we provide an overview of the basic information extraction (IE) approaches used in the developed systems. We survey a specific class of IE approaches based on semantics, due to the importance of semantic processing of the data.

Keywords Information extraction · Ontology-based knowledge · Vector space model

1 Introduction

In the recent years, the amount of available information in the Web is growing. Thereby, the search of pertinent information through those large documents has become a difficult task. In order to reduce this "information overload" problem, we bring into play the information extraction process.

M. Mannai (✉) · W.B.A. Karâa
Computer Science Department, High Institute of Management,
University of Tunis, Bouchoucha, Bardo-Tunis, Tunisie
e-mail: moniamannai@gmail.com

W.B.A. Karâa
e-mail: wahiba.bak@gmail.com

H.H.B. Ghezala
Computer Science Department, National School of Computer Sciences,
University of Manouba, Manouba, Tunisie
e-mail: henda.benghezala@ensi.rnu.tn

D.K. Mishra et al. (eds.), *Information and Communication Technology*,
Advances in Intelligent Systems and Computing 625,
https://doi.org/10.1007/978-981-10-5508-9_28

In the literature, information extraction was used as an automatic training approach [1]. The information extraction (IE) as defined by [2] is a kind of natural language system, a system designed for understanding ambiguous natural text. Some occurrences of entities or relations between them in natural language text could be identified, extracted, and transformed into a structured format (e.g., a database) [1, 3]. This new extracted information will be easier to handle and to interpret by the users [4]. Information extraction is a discovery of pertinent information without any learning [5].

The information extraction task is founded on retrieving documents from a collection of Web documents, then, tagging particular terms in the text that seems relevant to the user's need according to his query. Thus, by scanning large bodies of text written in natural language and filtering semantic information from it, the information extraction could be seen as the activity of natural language processing (NLP).

As defined above, the information extraction applications include the simplification of the text scanned in order to have a structured representation of the information drawn in the text.

To achieve this purpose, IE relies on many methods and approaches which can be divided into two fundamental classes. We will depict this classification in the second section. In Sect. 3, we present two chosen (IE) techniques based on semantics. In Sect. 4, we will summarize the work with a light comparison between these techniques. Finally, we finish with a conclusion in the fifth section.

2 Information Extraction Basic Approaches

Any information extraction system must carry on a specific method of extracting relevant information to satisfy the user's need. In the literature, we have many information extraction methods that can be classified into two ultimate divisions which are the knowledge engineering approach and the automatic training approach (Fig. 1).

2.1 Knowledge Engineering Approach

Some complex problems appearing in the computer systems require an intervention from experts to remove these complications. This field of research is named by [6] knowledge engineering (KE). Fox [7] describe the KE as a discipline that argue the knowledge integration into computer systems in order to solve issues normally requiring a high level of human expertise.

This knowledge-based method occurs in many computer science domains such as expert systems, geographic information systems, and data mining in the form of production rules. That's why the KE is also called as a rule-based approach.

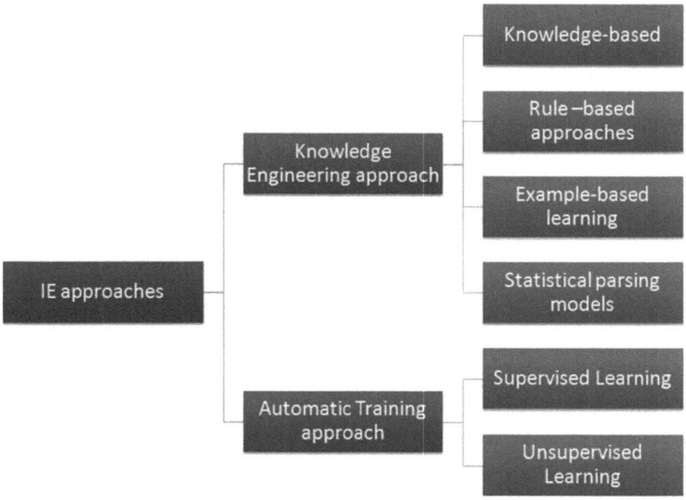

Fig. 1 Information extraction basic approaches

However, it is a simple, fast, and language-independent and is easy to retarget. This approach is highly time- and effort-consuming owing to its iterative process. First, a domain expert called also a knowledge engineer must define the set of rules that will be used for extracting relevant information from a text by applying his knowledge and intuitions. The design of the system is widely dependent on human expertise. Then, these rules will be rewritten by the designer in particular system language according to such pattern. Moreover, in the KE approach, the task of collection and maintenance's lists seems to be tough, and this method cannot handle the ambiguity problem and has difficulties in dealing with fact variants [8, 9].

The knowledge engineering approach can be divided into sub-models as shown in Fig. 1:

Knowledge based [10]: Hand-written Patterns, Gazetteers.
Rule-based approaches: FASTUS [11], Proteus [12].
Example-based learning: AutoSlog (UMASS 1993), CRYSTAL (UMASS 1996).
Statistical parsing models [13]

One of the most known IE systems which are designed using KE is Finite-State Automaton Text Understanding System (FASTUS) which is a rule-based approach. It is an extracting information system from natural language text that uses nondeterministic finite-state mechanisms. Since it was developed back in 1992, it has undergone huge change, that is, coding structure; however, the logic remains identical: A set of cascaded automata is applied to row data, each pass will serve as input to the next one, and the end results are combined. Several assessments to this system showed the reliability and efficiency of FASTUS in information extraction tasks. This ability allows FASTUS to be more oriented toward information extraction applications, especially information retrieval [11].

2.2 Automatic Training Approach

The second category of information extraction approaches is the automatic training approach which is also called as machine learning approach as a result of the use of learning techniques and the implementation of machine learning algorithms that generate the rules for the information extraction system. Hence, there is no need of a knowledge engineer to manually extract these rules. All that we need in this approach is someone familiar with the domain and the system functionalities [9]. For automatic extraction process, the implemented algorithms need an annotated corpus and a set of training annotated texts as an input of the system. By running these texts, the algorithms learn and give rise to the extraction rules [14].

Many machine learning algorithms can be used in this approach such as hidden Markov model [15], maximum entropy models [16], conditional random fields, naive Bayes networks, decision trees. These algorithms can be used for any domain if we have its relevant corpus. Thus, the information extraction systems attain less domain independency.

The machine learning approach can be divided into two classes: supervised learning and unsupervised learning [17]. The supervised learning is defined as the activity of building a predictor model from training dataset which is a set of training examples including a set of input data and desired output response. This developed predictor model generates the correct output for any new data, and that's why it is termed a classifier. Thus, the algorithms learn to extract the information from the input documents.

One of the supervised extraction systems is CRYSTAL that focuses on texts handled by a syntactic parser. This system employs a thesaurus and labeled training documents producing by an expert in order to create the extraction rules. Also, it has recourse to inductive learning to obtain the restrictive constraints covering the most similar pair or rules merged together before [18].

However, the unsupervised learning is the task of seeking hidden structure from unlabeled input data and learning to generate input patterns as the known responses or outputs are unnoticed. The unsupervised learning uses many data mining techniques to preprocess the unlabeled data [19].

Many information extraction systems apply the unsupervised learning, and we present in this paper AutoSlog-TS which is an extension of AutoSlog. This system called for a training corpus to provide extraction patterns for the input data using heuristics. Then, it states the reliable patterns by means of statistics, evaluates it, and then ranks it according to its relevant statistics [20].

3 Semantic-Based IE Methods

Among the IE approaches, we interested, in this paper, in semantic-based methods thanks to the added value that the IE systems outperform with the semantic analysis solving many problems related to the meaning like the ambiguity.

3.1 Formulas

Ontology plays an important role in information extraction. In the literature, the starting point of ontology is Greek. It is a branch of philosophy [21]. Ontology is a combined word: ontos for being and logos for word [22]. It was used many years in the laboratories of artificial intelligence (AI), and it has just left to be used on workstations' experts in the field. Ontology is used for information extraction; it is a part of the comprehensive process of text for the extraction of pertinent information. It determines the interpretation of the text and explores the different relationships between the components. In the literature, many systems of information extraction based on the ontology are developed in several domains such as biochemical domain such as Genome Information Extraction (GenIE) [23]; medical [24]; business intelligence such as ontology-based information extraction (OBIE) [25]; biology, commerce, and marketing.

Ontology represents a new tool for extracting information that allows to access quickly information based on its initial modeling domain. Several studies use ontology as a tool for textual interpretation and understanding because it adds knowledge to comprehension process of information extraction [26]. The heart of an IE system is the semantic and the understanding section, and ontology is conceders as a part of this understanding process. So we can say that ontology is necessary to process IE. In other studies, ontology was used as a guide for the syntactic and semantic analysis in the IE process. The use of ontology in IE process has undergone evolutions to Ontology-based IE (OBIE). The major difference between these two approaches is the use of a formal ontology rather than the use of a dictionary [27, 28]. The OBIE is a representation of a specific domain and collects the experts' knowledge [25]. The OBIE is a formal presentation. The main idea [25] is to explain how OBIE does not just extract entity but identifies it by binding its semantic description of the ontology; this approach allows us to extract more relevant textual data better than the conventional process of IE.

In several works of information extraction, ontologies are developed with the help of domain experts to build a database knowledge that will be used after that in order to enrich the process of information extraction. Also, we can nominate ontological annotation which is driven by ontology [29]. It is considered among the most important tasks for semantic analysis in information extraction process, precisely in the identification of the entities. In this context, there are many systems that have been put in place.

3.2 Vector Space Model (VSM)

The vector space model (VSM) was used, for the first time, in SMART Information Retrieval System [30]. The interesting point in this algebraic model is the ability to represent documents as vectors of terms in information retrieval [31]. The VSM is applied also in information filtering and relevancy ranking.

By using the VSM, we can handle a collection of documents as a matrix D with columns each one representing a single document vector d_i. We can count the specific word occurrences appearing in a specific document. The number of occurrences is called the weight D_{ij} of the word i in document j. In the information retrieval, the query q is represented in the same manner as the documents. We calculate the similarity between the query and the document using the inner product of their vectors dTq which is a simple weighted match between the coinciding terms of the query and the document [30]. This method is adopted by the known search engines.

Some of information retrieval models could be used in the information extraction process like the VSM technique that not only helps in extracting information from unstructured data but also can filter out the results and provide the needed occurrences with the desired meaning. So, the information extraction systems become more and more semantic based.

We will present in this paper two different vector space models that help disambiguate the word sense during the extraction process. Tsatsaronis and Panagiotopoulou [32] presented the Generalized Vector Space Model (GVSM) which is a standard Vector Space Model (VSM) extension incorporating added semantic information from WordNet, the known

English word thesaurus improves the previous contradicting results by employing the semantic relatedness measure. The authors treat the polysemy and synonymy problems with the GVSM by calculating the inner product of the terms indexing the documents in the collection. In other words, the semantic links presented in the WordNet graph are used for measuring the semantic relatedness by calculating the maximum relatedness between any two connected nodes (query and document terms). The experiments showed that the use of the semantic information with the GVSM enhanced the retrieval performance.

The vector space model mentioned above is a traditional word sense representation that uses a single prototype vector for representing the word meaning. This standard method may encounter complications during the word meaning programming because of the lexical ambiguity such as homonymy and polysemy problems. Thus, the single prototype vector space model is context independent when the extraction of word meaning widely depends on the context [33]. Hence, a set of prototype vectors are bring to play. They are determined using a word sense discovery (WSD) which clusters the context of the appropriate word. After having a group of similar context vectors, the authors assign for each cluster centroid a prototype vector according to separate computations. So, for each word, a set of vectors are generated. These prototype vectors measure the semantic similarity of words whatever isolated or in context by computing the minimum distance between one of the words *N*'s vectors and one of the words *M*'s vectors. This point lets this approach outperform other vector space models such as prototype and exemplar based.

The vector space model presents many advantages compared to Standard Boolean Model: First from structural angle, it is simpler as it is based on linear algebra and hence the term weights are not binary. As a technique, it allows computing a continuous degree of similarity between queries and documents; also, it has the ability to rank documents according to their possible relevance as well as partial matching for the queries. On the other hand, some limitations need to be noticed here; long documents are poorly represented because they have poor similarity values (a small scalar product and a large dimensionality). VSM is semantically oriented, and hence sensitive to that, documents having similar context but using different term vocabulary will not be associated, then resulting in a "false-negative match". Besides, words sub-strings might lead to a "false-positive match" if the search keywords are not matching precisely with document terms. Other considerable points here can be summarized as follows: VSM assumes that terms are statistically independent; the weighting is intuitive but not very formal; and finally that order of appearance of terms in the document will be lost in the vector space representation. Many of these difficulties can, however, be addressed by the integration of various tools including mathematical techniques (i.e., singular value decomposition) or use of lexical databases such as WorldNet. Ontology is a promising area in IE, especially when associated with semantic-based techniques; primarily, it helps permit searching within different contexts and represents a background knowledge exchange and reuse. It helps improve the accuracy of representations and overcome the limitations of access to information; as well it can be used to support gathering disassembled information and reaffirming systems interoperability.

4 Conclusion

The information extraction is an activity of the natural language processing consisting of retrieving a collection of documents and then tagging particular terms that seem to be relevant and satisfy the user's request.

As it is a wide field of research, the IE approaches can be classified according to the need of a domain expert intervention into knowledge engineering-based approaches and machine learning approaches. Thus, each information extraction system could choose the appropriate approach according to its requirements.

Out of the commonly used techniques for IE, we opted in this paper for semantic-based approaches due to the added value of providing reliable results and accurate information. Among these approaches in information extraction process, we presented the VSM and ontology-based approaches and their related work and compared them by numerating their advantages and limits.

References

1. R Gishman, “Information extraction: Techniques and challenges.” In: Pazienza, M.T. (ed.) Information Extraction: A Multidisciplinary Approach to an Emerging Information Technology. Berlin, Heidelberg: Springer-Verlag, pp. 10–27, 1997.
2. E. Douglas, Appelt, R. Jerry, Hobbs, J. Bear, D. Israel, and M. Tyson “FASTUS: A Finite-state Processor for Information Extraction from Real-world Text.,” In Proc. 13th Int’l Joint Conf. Artificial Intelligence (LJCAI-93), pages 1172–1178. 1993.
3. L. Eikvil,. “Information Extraction from World Wide Web A Survey”. Norwegian Computing Center, ISBN: 82-539-0429-0. 1999.
4. J. Cowie and W. Lehnert. “Information extraction In Special natural language processing” issue of the communications of the ACM (Vol. 39, pp. 80{91}). New York, NY, USA. 1996.
5. W. Gatterbauer, P. Bohunsky, M. Herzog, B. Krupl, and B. Pollak. “Towards Domain independent Information Extraction from Web Tables”., Proceedings of the 16th International Conference on World Wide Web, Canada, pp. 71–80, ISBN: 978-1-59593- 654-7.2007.
6. E. Feigenbaum and P. McCorduck. The Fifth Generation. artificial intelligence and japan’s computer challenge to the world, Addison-Wesley.1983.
7. J. Fox. “Formalizing knowledge and expertise: where have we been and where are we going?”. In The Knowledge Engineering Review, 26 (1), pp. 5–10 Cambridge university press. 2011.
8. H. Ji,. “Information Extraction: Techniques, Advances and Challenges.” Invited Lecture at the North American Chapter of the Association for Computational Linguistics (NAACL) Summer School. 2012.
9. K. Kaiser and S Miksch,. “Information Extraction a Survey”. Vienna University of Technology, Institute of Software Technology and Interactive Systems, Vienna, Technical Report, Asgaard-TR 2005.
10. G. La Rocca, “Knowledge based engineering: Between AI and CAD”. Review of a language based technology to support engineering design. Advanced Engineering Informatics Volume 26, Issue 2, Pages 159–179 Knowledge based engineering to support complex product design 2012.
11. Hobbs, Appelt, E; Douglas., R. Jerry, B. John, D. Israel, M. kameyama and M. Tyson. “FASTUS: A Finite-State Processor for Information Extraction from Real-World Text”, Proceedings. IJCAI-93, Chambery, France, August 1993.
12. J. Miller, D. Job and V. Vassilev, “Principles in the evolutionary design of digital circuits” - part I. Genetic Programming and Evolvable Machines, 1: 7–36, April 2000.
13. D, Collins. “Three Generative, Lexicalized Models for Statistical Parsing. In Proceedings of the 35th Annual Meeting of the Association for Computational Linguistics”, pages 16–23, Madrid, Spain 1997.
14. M. Ipalakova. “Information Extraction,” Initial background report, University of Manchester school of computer science. 2010.
15. P. Blunsom. “Maximum Entropy Markov models for semantic role labelling.” Proceedings of the Australasian Language Technology Workshop 2004. Pages 109–116. Sydney, Australia 2004.
16. A. Ratnaparkhi, “A Simple Introduction to maximum Entropy Models for Natural Language” Processing. Institute for Research in Cognitive Science IRCS Technical Reports Series. 2007.
17. M; Moens. “Information Extraction: Algorithms and Prospects in a Retrieval Context”. New York: Springer Netherlands. 2006.
18. S. Soderland, D. Fisher, J. Aseltine and W. Lehnert. 1995. CRYSTAL: inducing a conceptual dictionary. In Proceedings of the Fourteenth International Joint Conference on Artificial Intelligence (IJCAI’95), pages 1314–1319, 1995.
19. Duda, O. Richard, Hart, E. Peter, Stork and G. David, “Unsupervised Learning and Clustering”, Chapter 10 in Pattern classification (2nd edition), p. 571, New York, NY: Wiley,

ISBN 0-471-05669-3.2001G. O. Young, "Synthetic structure of industrial plastics," in Plastics, 2nd ed. vol. 3, J. Peters, Ed. New York: McGraw-Hill, 1964, pp. 15–64.
20. E. Riloff,1996. "An Empirical Study of Automated Dictionary Construction for Information Extraction in Three Domains", AI Journal, Vol. 85. W. D. Doyle, "Magnetization reversal in films with biaxial anisotropy," in Proc. 1987 INTERMAG Conf., 1987, pp. 2.2-1–2.2-6.
21. G. Morente, Historia universal Contiene tablas cronológicas p. 609–659 • Indíce alfabético p. 661–736;1964.
22. FJohn. Sowa:,. "Ontology, Metadata, and Semiotics". On Conceptual Structures: Logical Linguistic, and Computational Issues ICCS, 2000, page 55–81.2000.
23. Jonquet, C,. Mark A. Musen, Nigam H. Shah (2009). Help will be provided for this task: Ontology-Based Annotator Web Service. 2009.
24. Cimianoa, P,. Reyleb, U., Šarićc, j,. 2005. Ontology-driven discourse analysis for information extraction.
25. H. Saggion, A. Funk, D. Maynard,. and Bontcheva K 2007. Ontology-based Information Extraction for Business Intelligence; ISWC/ASWC 2007: 843–856.
26. Labsky, M,. 2008. Information Extraction from Websites using Extraction Ontologies. A dissertation submitted in partial satisfaction of the requirements for the degree Doctor of Philosophy in Computer Science. University of Economics Prague.
27. Maynard, D., Yankova, M,. Kourakis, A., Kokossis, A., Ontology-based information extraction for market monitoring and technology watch 2013.
28. Amit V. Deokar, Sagnika Sen, 2010 Ontology-Based Information Extraction for Analyzing IT Services. In proceeding of: Proceedings of the International Conference on Information Systems, ICIS 2010, Saint Louis, Missouri, USA.
29. I. Oren, EO Mann, O Paulsen, Hajos N Synaptic currents in anatomically identified CA3 neurons during hippocampal gamma oscillations in vitro. J Neurosci 26:9923–9934.2006.
30. Salton, G. (1971). The SMART retrieval system: Experiments in automatic document processing.
31. H. Prentice, Upper Saddle R, NJ. R. Baeza, Y and Ribeiro-Neto,B.1999. Modern Information Retrieval. Addison Wesley.
32. G. Tsatsaronis, and V. Panagiotopoulou, A Generalized Vector Space Model for Text Retrieval Based on Semantic Relatedness 2009.
33. J. Reisinger,. J. Raymond Mooney,. "Multi-Prototype Vector-Space Models of Word Meaning". search engines. 2010.

The Role of IoT-Based Devices for the Better World

Ajay Chaudhary and Sateesh K. Peddoju

Abstract In last 15 years, wireless sensor network (WSN)-based and internet of things (IoT)-based system effect human on every aspect of our life. WSN grows in rapid pace as it emerges as one of the most important technological developments. Since its emergence, the wireless sensor network (WSN) constitutes one of the most important technological developments in the last decade. It has the potential to affect our lifestyle deeply. However, its success relies greatly on a well-defined architecture that will provide scalable, dynamic, and secure basement to its deployment. The IoT-based architectures are intelligent applications that make energy, logistics, industrial control, retail, agriculture, and many other domains "smarter." With emergence of wireless sensor network as Internet of Things is a new revolution of the Internet, that is, rapidly gathering momentum driven by the advancements in sensor networks, mobile devices, wireless communications, and networking and cloud technologies. With rapidly increasing wireless sensor network (WSN)- and internet of thing (IoT)-based services; a lot of data is generated. It is becoming very difficult to manage power constrained small sensors and other data-generating devices. WSN or IoTs enables anything can become part of the Internet and generate data. Moreover, data generation needs to be managed according to its requirements, to create more valuable services. For this purpose, integration of WSN or IoTs with cloud computing is becoming very important. The small IoT sensors deployed in agricultural fields measuring the vast amount of information using sensors like air pressure, environmental temperature, relative humidity, solar radiation, soil moisture, soil temperature, wind speed, leaf wetness, CO concentration and N, P, and K concentration. These sensors continue to monitor and generate huge data which need to process sensibly to extract key factors.

Keywords Component · Internet of things · Wireless sensor network · Cloud computing · IoT · WSN · Agriculture · Environmental monitoring

---

A. Chaudhary (✉) · S.K. Peddoju
Indian Institute of Technology, Roorkee, Roorkee 247667, India
e-mail: ajaychaudhary@acm.org

S.K. Peddoju
e-mail: sateesh@ieee.org

D.K. Mishra et al. (eds.), *Information and Communication Technology*,
Advances in Intelligent Systems and Computing 625,
https://doi.org/10.1007/978-981-10-5508-9_29

1 Internet of Things (IoT) Devices

In the past few years, WSN devices have been gaining increasing attention because of their potential of enabling of the novel and attractive solutions in areas such as industrial automation, environmental monitoring, transportation business, healthcare, disaster, natural hazards. They are deployed everywhere from the battlefield to forest fire detection, from structural health monitoring to human health monitoring, from smart cities to landslide monitoring, from air quality monitoring to water quality monitoring. Even it's widely used in the critical process of chemical reaction monitoring to control reactions in nuclear power plants. The wireless sensors also widely used in habitat monitoring, biodiversity monitoring, active volcano monitoring, underground mine monitoring like coal mines and precision agriculture. The sensor networks or its variants like body sensor network are deployed almost everywhere we can think off. The WSN-based sensor nodes are deployed widely almost in all fields, due to its capacity to sense pressure, motion (accelerometers), temperature sensors, humidity sensors, chemical sensors, biosensors, luminosity, gyroscope, gasses (Co_2, O_2), acoustic, GPS, etc.

The IoT devices like wireless sensor networks consist of the infrastructure-less wireless system. They deployed ranges from tens to thousands without any physically wired link and layout. The WSN devices had greatly deployed and served in adverse conditions in last decade. The device nodes capabilities and smartness are enhanced with the technologies like micro-electro-mechanical systems (MEMS) help in speeding up the processing capacity of many folds. These devices are used in variety of fields due to their capabilities to sense surroundings and provide required data to the user or gateway. The WSN devices are low-power devices equipped with one or more sensors, a processor, memory, a power supply, a radio which enable decision and sensing capabilities. A variety of mechanical, thermal, biological, chemical, optical, and magnetic sensors may be attached to the WSN devices to measure properties of the environment. These are low cost and inexpensive as compared to traditional sensors devices with limited processing and computing capacity. Due to low cost, these devices can be deployed in the large area at the mass level in applications which requires sense and transmitting data toward the server or gateways. Most of the time these devices are deployed in a difficult-to-access and extreme locations like mine tunnels, battlefield, forest, a war zone, earthquake, and climatic natural sites. They deployed in an unstructured, or structured ad hoc fashion was they are supposed to stay awake and provide services for the long duration of time. These devices require a small power source which is a small battery or harvested energy like solar power to continue its services. The devices must use smart scheduling and transmission algorithms to optimize the use of available power source [1]. Other than the issue with battery or power, the devices also had issues like short range communication, limited processing capabilities, near nil or very less storage capacity, and also low bandwidth.

The traditional communication protocols not fit well within WSN environments due to its limited processing capabilities. Still these devices need to provide a minimum set of quality of service (QoS) paradigm like congestion control, minimum Packet drop ratio, active buffer monitoring, and packet loss recovery. The WSN devices might not guarantee to provide QoS as desired for video and audio transmission, but it should be able to provide minimum. The amount of QoS like low delay, enough bandwidth availability and most important reliable data transmission. But for a sensor network, it is a tough task to provide reliable data transmission as any QoS assurances are only as good as the weakest chain between sender and receiver as its all depends on the communicating network and as long as network stands the QoS is stands. Hence, any solution to support real-time traffic should take into consideration the overall QoS architecture that spans the entire network. The required bit rate, delay, jitter, packet dropping probability, and bit error rate may guarantee; Sensor network needs to able to deliver these QoS despite its shortcoming in communication link and lack of available bandwidth.

The innovations in pervasive computing-based technologies for healthcare system made it easy for doctors to monitor the patient health from anywhere in the world and the feedback systems help to check more physiological statistics as and when required. Small devices known as body sensors help to monitor and measure the physiological condition of the patients along with other general purpose sensors that are used to measure the surrounding environmental condition helps the doctor to treat the patients as if the patient admitted to the hospital. Nowadays, there are a lot of commercially available kits for body sensor networks. There are several types of sensors available for agriculture purpose, list of such sensors are given in Table 1 [2].

Table 1 Typical list of IoT sensors

S. No.	Sensor type	Sensor sub type	Operations and functionality
1	Pressure sensors	Piezo resistive pressure	In this, piezo resistor integrated into a membrane. Pressure or force applied directly to member causing it to deform hence pressure or force is measured. It measures atmospheric pressure or any force directly applied to it
2		Capacitive pressure	In it if pressure/force applied to the sensor surface causing a membrane to deflect resulting capacitance to change and pressure or force is measured. It can measure pressure with great sensitivity but had high production cost
3	Temperature sensors	Electromechanical temperature sensors	It measures the temperature based on expanding and contracting properties of materials using a bi-metal thermostat
4		Resistive temperature sensors	It measures the temperature based on a property of resistant changes with temperature

(continued)

Table 1 (continued)

S. No.	Sensor type	Sensor sub type	Operations and functionality
5	Humidity	Capacitive RH sensors	In a capacitive RH sensor, humidity measures in term of change in dielectric constant which is directly proportional to relative humidity
6		Resistive humidity sensors	It measures humidity in term of resistance changes in the environment as resistance is inversely proportional to the environmental humidity. Resistive humidity sensors are small size, low cost, and are usable from remote locations
7	Image	active pixel sensor (CMOS)	It detects and conveys information in order to constitute an image by converting waves in signals
8	Light	Ambient light sensor	It approximates the human eye response under a variety of light conditions
9	GPS	GPS	Devices have a base station which measures the position based on geostationary GPS satellites, at least, three satellites are required to measure exact position, work with great precision in outdoor environments
10	Acoustic, sound	Microphone	Measure sound wave by converting it into signals

2 Wireless Sensor Devices (WSN) Deployment Models

With the increase in population, there is immense pressure on production increase for agriculture with limited land availability, Other than land other resources are also like freshwater is also scarce to overcome such problems the IoT-based devices like WSN (wireless sensor network) are widely used to provide support and help to the farmers. The researchers are doing extensive research to address the issue of the use of IoT-based devices or endpoints in agriculture field to maximize crop yield with minimum wastage of resource and minimum use of fertilizers. IoT-based devices having a group of sensor nodes deployed in farm to capture key parameters are related to crops. Other than agriculture due to its pervasive computing capabilities, WSN is widely used in health monitoring, structural monitoring, military, landslide monitoring, volcano monitoring, habitat monitoring, and smart cities. The IoT-based devices are an effective device for data acquisition or real-time data acquisition. The data acquisition can be stated as collecting, processing, and transmitting data in predetermined latency boundaries. It mainly includes sampling, MAC layer operations, network layer routing, data aggregation, and some additional processes. It is a process in which the data from the sensor node is collected, preprocessed, represented, and finally transmitted with predefined latency boundaries, i.e., with an upper time limit, which includes a sampling of signal/data, preprocessing, i.e., removing of noise, etc., MAC layer operations, routing at the network layer, data aggregation, etc. [3]. But there are many constraints at sensor node like environmental limitations, energy limitation, limited memory, and storage space, etc. the data acquisition also suffers from an issue with a communication channel, routing protocols, channel error rate, unreliable communication, etc. [4].

Routing plays a key role in IoT-based devices like WSN for efficient data acquisition. According to K. Akkaya and M. Younis, based on their implementation and goal, the routing algorithm can be divided into following schemes [5].

(i) Data-centric protocols: In this routing in general, data is transmitted to every node within a deployment region, i.e., data is flooding in the region. In this, data is transmitted with significant redundancy. When data is required sink node sends queries to specific region and waits for data from sensors in that region, then all nodes in that region flooded the data with great redundancy and it provides high reliability. SPIN (sensor protocols for information via negotiation) [6], directed diffusion [7] are the two most classical data-centric routing protocols.

(ii) Hierarchical protocols: As all sensors node sends data toward the gateway, with an increase in sensor nodes number of nodes tries to access and send data to the gateway is a big issue in such cases the gateway is overload with an increase in sender's density. A highly overloaded gateway leads to high latency in communication, a drop of the active communication link, etc., which is creating a problem with real-time data acquisitions. Also, sensors are deployed in the large area in such cases single gateway might not serve the purpose as sensors had very short communication range. To overcome these issues, a hierarchical routing algorithm is used. The hierarchical routing is an efficient routing scheme helps in energy saving provides multi-hop communications and adaptive routing. LEACH (low-energy adaptive clustering hierarchy) [8], threshold energy efficient sensor network protocol (TEEN) [9], adaptive threshold sensitive energy efficient sensor network protocol (APTEEN) [10] are some of the classical hierarchical scheme-based routing protocols.

(iii) Location-based protocols: Sometimes location of the sensor is required to derive decision logics. Location of sensors also required to do efficient routing also to provide energy-efficient routing. The location of nodes is also used to do region-based sensing. Geographic adaptive fidelity (GAF) [11], Geographic and energy-aware routing (GEAR) [12] are well notable location-based routing scheme protocols.

(iv) QoS-aware protocols: The quality of service (QoS) is always a key challenge for real-time data acquisition; a QoS-aware routing algorithm may do the needful. It is not feasible to meet all QoS parameter but to provide real-time services at least end-to-end delay if calculated in advance before data transmission starts then it reduces overall delay within the network and provides reliable QoS at least. SPEED [13] is one of few initial works to provide QoS for WSN.

In order to deliver consistent data over the long period from a remote location and to provide continued monitoring of crops, the data acquisition and routing algorithms must be energy and power aware and must support real-time recovery mechanism. Over the time, there is lots of advancement in the data acquisition schemes and routing protocols.

ZigBee-based technology can be used for implementing WSN in precision irrigations. ZigBee is one widespread wireless communication technology which is a smart, self-configuring, cost-effective, and energy-efficient mesh networking proprietary standard. The low cost allows the technology to widely deployed in wireless control and monitoring applications. ZigBee based on the IEEE 802.15.4 standard for wireless personal area networks. Koubaa et al. [14] suggested a hierarchical algorithm named superframe duration scheduling (SDS) algorithm based on cluster tree network, in this algorithm Zigbee coordinator may allow other special nodes, called Zigbee routers (ZR) or coordinators, to send periodic beacon frames to synchronize the nodes in their vicinity. This algorithm supports mobility and provides the highly connected network.

LEACH-EP [15] is hierarchical multi-hop energy-aware adaptive clustering algorithm, in which all cluster heads send its location and residual energy to the sink, then sink selects the channel whose residual energy is higher than that of all other nodes average residual energy. It is a centralized control algorithm in which the decision of sink is propagated toward source nodes.

LEACH-LPR [16] is hybrid algorithm based on LEACH protocol, and they use genetic algorithms to prolong network uptime, in this algorithm, the selection criteria of cluster head depend on three parameters, i.e., neighborhood density, the distance of the node from the sink, and residual energy of the node. But this algorithm is supported static nodes only.

Application-specific low power routing (ASLPR) [17] is a hybrid distributed optimization algorithm based on genetic algorithm and to achieve the best performance the cluster head is selected by application-specific parameter tuning genetic algorithm that helps to do optimization, but this algorithm is suitable for static nodes only.

Gao et al. [18] suggested a multicast distributed rechargeable energy source-based scheduling algorithm in which data acquired by source node and packet send by the source node to at least or at most or exact k number of nodes toward sink in Anycast group are identified by Anycast addresses. The nodes can be mobile, and it supports multi-hop transmission toward the sink.

Bagaa et al. [19] suggested a scheduling distributed algorithm for integrated tree construction and data aggregation (DICA) based on tree formation and node scheduling to reduce the time latency. DICA form an aggregation tree, and it tries to maximize the available choices for parent selection at every node on the basis of hop count toward the base station. Each node in the tree waits for data from all its predecessors before transmitting the data toward base station.

Liu et al. [20] suggested a hierarchical, reactive, centralized controlled algorithm in which to increase the battery life of nodes, they adopted concept of sleeping and active nodes and its multilayer node recovery algorithm in which if any node fails is replaced by its neighbor having least distance from gateway. It is a location-aware algorithm and failed node is replaced dynamically in order to provide optimal QoS.

Researcher implemented it in different geographical aspects and variety of crops. They also suggested architecture and routing algorithms for its implementation in pervasive agriculture. Kelly et al. have proposed an effective implementation of IoT in environmental condition monitoring. They design a system to provide a greater control over routing of packets in order to provide a QoS; they determined QoS in term of throughput (1.55 bytes/s) and reliability (97%) [21]. Ochiai et al. suggested a DTN (delay-distruption tolenrant networking)-based agricultural monitoring system; they deployed 39 DTN nodes which generate and send packets in every 30 s and able to achieve 99.8% success rate for data gathering [22]. Chen et al. [23] use temperature and humidity sensor in order to determine moisture content of tea, with their system they achieved accuracy of moisture determination in range of 0.5% wet basis at RH < 70%. In their work, Aquino-Santos et al. [24] suggested platform for implementation of the sensor for precision agriculture which improves the performance of its motes. The wireless sensor network based on stationary nodes and mobile base station which helps in simplified routing and localization is also used for precision farming [25]. Emmi et al. [26] in their paper proposed an integration of sensor and actuators in agricultural vehicles; they provide a framework for selection, arrangement, integration, and synchronization of sensors to form an autonomous agricultural vehicle for agricultural applications. They successfully tested their framework in a guidance and weed control task in a maize field to prove its utility. Shah et al. [27] deployed WSN network in control area as well as the open area in vineyards of Nasik, India, and showed substantial improvement in moisture content and crop water requirement after deployment of WSN over the range of period. Garcia-Sanchez et al. [28] in their paper try to integrate WSN with RFID and found that it could be a great idea to merge two or more technology to improve the performance of the overall system for pervasive irrigation as RFID helps in identification while WSN help in sensing. Kim et al. [29] deployed a site-specific irrigation control for linear move irrigation using WSN. They low-cost Bluetooth wireless radio communication for sensor network and irrigation controller to the base station. They design a sprinkler system to pinpoint target required based on a communication received from WSN. López Riquelme et al. designed a WSN-based system which measures soil parameters like temperature, volumetric percentage, electrical conductivity, and salinity [30]. Valente et al. [31] in their research paper suggested a solution to overcome the problem of a large region of interest in the wide area where the single network is not possible. They suggested UAV as airborne communication nodes which communicated with WSN nodes based on the ad hoc communication protocol. Lee et al. [32] in their research paper do the survey of various technologies used for precision irrigation. They state that more reliable, accurate, rugged, and less expensive sensing systems are required for better and efficient site-specific management of crops.

El-kader and El-Basioni [33] in their paper deployed MICA-based sensors to monitor potato crop; they deployed around 85 sensor nodes to monitor overall potato fields divided in tubs and in deployment they used APTEEN routing protocol.

Díaz et al. [34] deployed sensor-based architecture to monitor grape/vineyard in northern California vineyard using ZigBee MicaZ sensors for significant productivity gains and facility of maintenance.

3 Role of WSN in Precision Agriculture/Farming

The traditional farming system lags behind if it continues with traditional knowledge and without any involvement of technologies. There is a need for optimal use of pesticides and management of water resources. In developing countries like India, agriculture is also suffering from proper farm management, crop surveillance, and maintenance. The illiteracy in farmers and large geographical area of cultivated land are the two key factors that affect the crop management badly. Due to illiteracy, the farmers are not able to monitor the farm properly by timely and proper use of pesticide and fertilizers, these chemicals are either underused or overused which is either affect the crop yield or human health in either case. Second, due to the large geographical area it is not feasible for agriculture scientist to visit each and every farm and check the health of the crop. Most agriculture universities, agriculture research centers, testing labs, etc., are set up in cities. As a result, it restricts the active reach of scientist or agriculture officer to the farm. Due to lack of awareness, the farmers are not able to visit these facilities in time, as the plant life is very short and very vulnerable toward environmental conditions and diseases, a delay of a day or two may completely destroy the plan or affects its yield drastically. For a layman like farmers without any latest equipment or sensors, the early detection of issues with the crops is nearly impossible. If detection is delayed by day or two, the crop's condition affects drastically. To meet global food demands and overcome these issues faced by most of the developing countries, the Food and Agriculture Organization of USA in their report on building a common vision for sustainable food and agriculture—Principles and approaches—suggested the key principles of sustainability in food and agriculture. These principles are keys to fulfill future demands and to make efficient use of water and pesticide management in crops. The principles are conservation agriculture, judicious use of organic and inorganic fertilizers, improved soil moisture management, improved water productivity, precision irrigation, integrated pest management (IPM) [8]. These principles highlight the future need of precision farming or agriculture with the help of latest state-of-the-art technologies. The perfect combination of technologies may serve the purpose very well and deliver the required goals. The state-of-art technologies like IoT devices and cloud computing resources may play a vital role to improve agriculture with little effort. Precision farming also plays a vital role in the restricted use of pesticides. It also helps in maintenance of proper soil moisture required for a crop with minimum water wastage.

The smart high-tech farming using wireless sensor network and cloud computing is in need of time, and it can act a support system of traditional farming. With zero or minimal human involvement, the automated systems using cloud computing or wireless sensor network helps to fill the gaps.

4 Conclusion

In this paper, we have discussed various issues regarding the role of WSN applications in monitoring. The researchers have suggested numerous approaches, but most of them are for the purpose of farm management, marketing, disease detection, labor management, etc. In most of the applications of cloud computing in precision farming, the input for the framework is either provided by the farmer himself or through some equipment which requires active and sophisticated handling. The wireless sensor network devices are also being used in agriculture to implement either cattle management at farm houses or to monitor limited crops, etc. The WSN-based system had great capabilities to capture and acquire data related to crops, but the real-time data acquisition is suffered due to hardware and software issues. The real-time data acquisition consists of data capturing, data preprocessing, and routing algorithms. To improve the overall performance of the device and prolong its battery life, there is a requirement of real-time data acquisition algorithm. The WSN-based sensor devices have great capabilities to sense and capture a huge amount of data from the remotest areas, but they have very limited data processing capabilities to develop decision support system.

References

1. J. Yick, et al., "Wireless sensor network survey," Computer Networks, vol. 52, pp. 2292–2330, 2008.
2. G. Rakocevic, "Overview of sensors for wireless sensor networks," Transactions on Internet Research, vol. 5, pp. 13–8, 2009.
3. M. Soyturk, H. Cicibas, and O. Unal, "Real-Time Data Acquisition in Wireless Sensor Networks," in Data Acquisition by Michele Vadursi, ed: www.intechopen.com, November 28, 2010, pp. 63–84.
4. I. F. Akyildiz, W. Su, Y. Sankarasubramaniam, and E. Cayirci, "Wireless sensor networks: a survey," Computer Networks, vol. 38, pp. 393–422, 2002.
5. K. Akkaya and M. Younis, "A survey on routing protocols for wireless sensor networks," Ad hoc networks, vol. 3, pp. 325–349, 2005.
6. W. R. Heinzelman, J. Kulik, and H. Balakrishnan, "Adaptive protocols for information dissemination in wireless sensor networks," in Proceedings of the 5th annual ACM/IEEE international conference on Mobile computing and networking, 1999, pp. 174–185.
7. C. Intanagonwiwat, R. Govindan, and D. Estrin, "Directed diffusion: a scalable and robust communication paradigm for sensor networks," in Proceedings of the 6th annual international conference on Mobile computing and networking, 2000, pp. 56–67.

8. W. R. Heinzelman, A. Chandrakasan, and H. Balakrishnan, "Energy-efficient communication protocol for wireless microsensor networks," in Proceedings of the 33rd annual Hawaii international conference on System sciences, 2000, p. 10 pp. vol. 2.
9. A. Manjeshwar and D. P. Agrawal, "TEEN: a routing protocol for enhanced efficiency in wireless sensor networks," in Proceedings 15th International Parallel and Distributed Processing Symposium. San Francisco, CA, USA 23–27 April 2000 pp. 2009–2015.
10. A. Manjeshwar and D. P. Agrawal, "APTEEN: a hybrid protocol for efficient routing and comprehensive information retrieval in wireless," in Proceedings of IEEE International Parallel and Distributed Processing Symposium (IPDPS 2002), Ft. Lauderdale, FL 15–19 April 2001 pp. 1–8.
11. Y. Xu, J. Heidemann, and D. Estrin, "Geography-informed energy conservation for ad hoc routing," in Proceedings of the 7th annual ACM international conference on Mobile computing and networking (ACM SIGMOBILE, 2001), Rome, Italy, 2001, pp. 70–84.
12. Y. Yu, R. Govindan, and D. Estrin, "Geographical and energy aware routing: A recursive data dissemination protocol for wireless sensor networks," UCLA Computer Science Department. Technical report ucla/csd-tr-01-0023, May, 2001.
13. T. He, J. A. Stankovic, C. Lu, and T. Abdelzaher, "SPEED: A stateless protocol for real-time communication in sensor networks," in Proceedings of the 23rd International Conference on Distributed Computing Systems, 2003, 2003, pp. 46–55.
14. A. Koubaa, A. Cunha, and M. Alves, "A time division beacon scheduling mechanism for IEEE 802.15. 4/ZigBee cluster-tree wireless sensor networks," in the Proceedings of 19th Euromicro Conference on Real-Time Systems, 2007(ECRTS'07), 2007, pp. 125–135.
15. J.-G. Jia, Z.-W. He, J.-M. Kuang, and Y.-H. Mu, "An energy consumption balanced clustering algorithm for wireless sensor network," in the proceedings of 6th International Conference On Wireless Communications Networking and Mobile Computing (WiCOM, 2010), 2010, pp. 1–4.
16. M. Shokouhifar and A. Hassanzadeh, "An energy efficient routing protocol in wireless sensor networks using genetic algorithm," Advances in Environmental Biology, vol. 8, pp. 86–93, 2014.
17. M. Shokouhifar and A. Jalali, "A new evolutionary based application specific routing protocol for clustered wireless sensor networks," AEU-International Journal of Electronics and Communications, vol. 69, pp. 432–441, 2015.
18. D. Gao, H. Lin, and X. Liu, "Routing protocol for k-anycast communication in rechargeable wireless sensor networks," Computer Standards & Interfaces, vol. 43, pp. 12–20, 2016.
19. M. Bagaa, M. Younis, D. Djenouri, A. Derhab, and N. Badache, "Distributed low-latency data aggregation scheduling in wireless sensor networks," ACM Transactions on Sensor Networks (TOSN), vol. 11, p. 49, 2015.
20. H.-Y. Liu, Y.-N. Guo, M.-R. Chen, and Y.-S. Zhu, "A Hierarchical Scheduling Scheme in WSNs Based on Node-Failure Pretreatment," International Journal of Distributed Sensor Networks, vol. 2015, pp. 1–12, 2015.
21. S. D. T. Kelly, N. K. Suryadevara, and S. C. Mukhopadhyay, "Towards the implementation of IoT for environmental condition monitoring in homes," IEEE Sensors Journal, vol. 13, pp. 3846–3853, 2013.
22. H. Ochiai, H. Ishizuka, Y. Kawakami, and H. Esaki, "A DTN-based sensor data gathering for agricultural applications," IEEE Sensors Journal, vol. 11, pp. 2861–2868, 2011.
23. A. Chen, H.-Y. Chen, and C. Chen, "Use of temperature and humidity sensors to determine moisture content of oolong tea," sensors, vol. 14, pp. 15593–15609, 2014.
24. R. Aquino-Santos, A. Gonzalez-Potes, A. Edwards-Block, and R. A. Virgen-Ortiz, "Developing a new wireless sensor network platform and its application in precision agriculture," sensors, vol. 11, pp. 1192–1211, 2011.
25. S. Thaskani and G. Rammurthy, "Application of topology under control wireless sensor networks in precision agriculture," Centre for Communications, International Institute of Information Technology, Hyderabad, 2010.

26. L. Emmi, M. Gonzalez-de-Soto, G. Pajares, and P. Gonzalez-de-Santos, "Integrating sensory/actuation systems in agricultural vehicles," sensors, vol. 14, pp. 4014–4049, 2014.
27. N. Shah, U. Desai, I. Das, S. Merchant, and S. Yadav, "In-Field Wireless Sensor Network (Wsn) for Estimating Evapotranspiration and Leaf Wetness," International Agricultural Engineering Journal, vol. 18, pp. 43–51, 2009.
28. A. -J. Garcia-Sanchez, F. Garcia-Sanchez, and J. Garcia-Haro, "Wireless sensor network deployment for integrating video-surveillance and data-monitoring in precision agriculture over distributed crops," Computers and electronics in agriculture, vol. 75, pp. 288–303, 2011.
29. Y. Kim, R. G. Evans, and W. M. Iversen, "Evaluation of closed-loop site-specific irrigation with wireless sensor network," Journal of Irrigation and Drainage Engineering, vol. 135, pp. 25–31, 2009.
30. J. A. López Riquelme, F. Soto, J. Suardíaz, P. Sánchez, A. Iborra, and J. A. Vera, "Wireless Sensor Networks for precision horticulture in Southern Spain," Computers and electronics in agriculture, vol. 68, pp. 25–35, 2009.
31. J. Valente, D. Sanz, A. Barrientos, J. D. Cerro, Á. Ribeiro, and C. Rossi, "An Air-Ground Wireless Sensor Network for Crop Monitoring," sensors, vol. 11, p. 6088, 2011.
32. W. S. Lee, V. Alchanatis, C. Yang, M. Hirafuji, D. Moshou, and C. Li, "Sensing technologies for precision specialty crop production," Computers and electronics in agriculture, vol. 74, pp. 2–33, 2010.
33. S. M. A. El-kader and B. M. M. El-Basioni, "Precision farming solution in Egypt using the wireless sensor network technology," Egyptian Informatics Journal, vol. 14, pp. 221–233, 2013.
34. S. E. Díaz, J. C. Pérez, A. C. Mateos, M.-C. Marinescu, and B. B. Guerra, "A novel methodology for the monitoring of the agricultural production process based on wireless sensor networks," Computers and electronics in agriculture, vol. 76, pp. 252–265, 2011.
35. FAO. (2014), "Building a common vision for sustainable food and agriculture Principles And Approaches," Rome, 2014.
36. A. Chaudhary, "A cluster based wireless sensor network deployment for precision agriculture in dried and arid states of India," presented at the Proceedings of the 2014 International Conference on Information and Communication Technology for Competitive Strategies, Udaipur, Rajasthan, India, 2014.
37. A. Chaudhary, et al., "Cloud Based Wireless Infrastructure for Health Monitoring," Cloud Computing Systems and Applications in Healthcare, p. 19, 2016.

Feature Extraction of Protein Contact Maps from Protein 3D-Coordinates

K. Suvarna Vani and K. Praveen Kumar

Abstract This work mainly proposes an alternate way of solving challenging problems of computational biology like protein secondary structure assignment, protein fold identification/recognition, protein fold signatures, and contact map overlap problem by exploiting the idea that proteins belonging to the same protein fold have similar contact maps. Pattern mining of contact maps is conducted to extract features that pertain to fold information. Using the work in the literature that predicts contact maps from the primary amino acid sequence, we propose that using pattern features from predicted contact maps would lead to an Ab-Initio method. Hence, instead of extracting features from the primary amino acid sequence, we propose to extract pattern features from the protein contact maps. Protein secondary structure assignment is achieved with an accuracy of 76% on RS126 data set, on par with best of algorithms up to 10% of noise, and then the performance falls to 66% by 15% noise.

Keywords Protein structure · Contact maps · Features · Secondary structure elements

1 Introduction

Protein contact map is a two-dimensional representation of the protein tertiary structure. Prediction of protein contact map from the primary sequence of a protein is a highly challenging problem that has been addressed in the literature, and

---

K. Suvarna Vani · K. Praveen Kumar (✉)
Department of Computer Science and Engineering,
VR Siddhartha Engineering College, Kanuru, Andhra Pradesh, India
e-mail: praveen@vrsiddhartha.ac.in

K. Suvarna Vani
e-mail: suvarnavanik@vrsiddhartha.ac.in

D.K. Mishra et al. (eds.), *Information and Communication Technology*,
Advances in Intelligent Systems and Computing 625,
https://doi.org/10.1007/978-981-10-5508-9_30

software is available now to predict contact maps [1]. Hence, instead of extracting features from the primary amino acid sequence, we propose to extract pattern features from the protein contact maps. We demonstrate the fact that analysis of contact maps can yield important insights for protein structure identification. It is well known that the secondary structure elements of a protein are transparently laid out in the contact map, though no one framed rules to extract them from the contact map.

The protein secondary structure prediction problem is a well-known problem in bio-informatics community for last few decades. Several machine learning methods have been used for protein secondary structure prediction including neural networks (NN), support vector machines (SVM), hidden Markov models (HMM), and cascading models. A common underlying approach for all the secondary structure prediction methods is to extract statistical properties of amino acid distribution, physico-chemical properties of the residues of the protein sequences and then build models for classification. Initially Chou-Fasman and others [2–4] utilized neural networks and by using single amino acid sequence features achieved an accuracy of 50–63%. Jones et al. [5] consider multiple sequence features from the evolutionary information in the form of position-specific scoring matrices (PSSM) generated by PSI-BLAST using PSIPRED algorithm to achieve an accuracy of 70%. Rost and Sander [6] consider the Profile network from Heidelberg (PHD) and by using a two-layer neural network with evolutionary information increased the accuracy by 1%. Karypis [7] used cascaded support vector machine (SVM)-based predictor using PSI-BLAST profiles and proposed YASSPP algorithm. To summarize, average accuracy of secondary structure prediction has been in the range 71–80% so far protein secondary structure assignment is a problem which assumes 3D-coordinate information. This problem too is a challenging problem with earlier works assigning secondary structures based on backbone dihedral angles [8–10].

Other popular method is P-SEA which is based on distance as well as dihedral angles. Dictionary of Protein Secondary Structure (DPSS) [11] is considered to have set a standard for this problem that computes hydrogen bond energy between backbone atoms and structural identification method (STRIDE) modifies the energy calculation and also uses backbone dihedral angles [12]. More recently, information theoretic concepts like minimum message length inference have been used for secondary structure assignment [13] and indirect methods that compute approximation of the protein backbone using piecewise lines have been proposed [14, 15]. In this paper, we propose to extract features from both the diagonal region as well as the off-diagonal region of the contact map. Instead of using the standard clustering algorithms from the literature, as contact maps are very sparse matrices, we propose to use heuristics on the diagonal and a naive algorithm to extract rectangular/non-rectangular regions of connected pieces of contacts from the off-diagonal region.

2 Feature Extraction Along the Diagonal Region

A contact map can be divided into two regions: Diagonal region and the region obtained by masking the diagonal region referred to as the off-diagonal region.

2.1 Secondary Structure Assignment

Many researchers including Hu et al. [16] emphasize that thick bands along the diagonal denote helices, and those that are away from the diagonal correspond to beta sheets. But actual extraction of these SSE from the contact maps has not been reported in the literature.

2.2 Extract SSP Algorithm

The idea underlying prediction of specific secondary structures of helix, beta, and coils/turns is to extract bands of width W and length l along the diagonal in the upper/lower triangular matrix of the contact map with parameters tuned to different secondary structure elements.

Fixing parameters typically since one turn of the helical structure is made up of 3.6 residues, and the minimum predicted length for an α-helix should be three or four residues. Consecutive $C\alpha$ atoms are farthest apart since, in a β-strand.

```
Algorithm  Extract SSP(A)

Input: Contact Map A[r × c]
Output:Secondary structure positions of A
Input parameters:_ row width parameters: a, b, 0 < b < a
Variables: Helix Length, Beta Length, Coil  Length,
 # of helices: NH, #of betas:NB, # ofturns coils:NC all
initialized to zero.
for i ← 0 to r - 1 do
    row width = 0;
for j ← 0 to c - 1 do
if (A[i][j] == 1) then
    row width++;
    A[i][j] ← 0;
end if
end for
if (row width ≥ a) then
    Helix_Length++, Beta Length++, Coil Length++;
else if (Helix_Length ≥ 3) //(row width < a) then
    Print Helix Found;
    NH++;
Reset Helix Length=0,Beta Length=0,_Coil   Length=0;
if (b ≤ row width < a) then
    Beta Length++, Coil  Length++;
else if (Beta Length ≥ 3)// (row width < b) then
    Print Beta Found;
    NB++;
Reset Beta Length=0, Coil Length=0; 21: end if
if (0 < row width < b) then
    Coil  Length++;
    Reset Helix_Length=0, Beta Length=0; 25:
else if (Coil Length > 0) then
    Print Coil Found;
    NC++;
Reset Coil Length=0, Helix_Length=0,Beta Length=0;
end if
end if
end for
```

Relatively few residues cross the protein core with a strand. Therefore, the number of residues in a β-strand is usually limited to two or three amino acids [15, 17].

We conducted an initial scan of the data set for fixing the parameters on 10% of the data set, by fixing minimum helix length as 3 and varying helix width between the values 3, 4, and 5, potential helix regions have been extracted from the contact maps. For $W = 3$, we obtained many false positives with residues annotated as H, and for $W = 5$ in the overall protein, lesser number of helices were obtained.

Table 1 1SW8 protein has 4 helices, with original locations given in columns 2 and 3. Predicted helices for widths 3, 4, 5 show that only two helices are predicted for $W = 3$

Protein Id	Original		Predicted helices					
			$W = 3$		$W = 4$		$W = 5$	
1SW8	4	20	1	60	4	18	4	9
	28	39	64	77	28	37	10	18
	44	56			44	53	28	35
	65	79			64	76	48	52
							64	75

Table 2 Secondary structure prediction data set [6]

Class	Number of proteins
All-Alpha	21
All-Beta	44
Alpha + Beta	15
Alpha/Beta	28
Small proteins	13

A sample result can be seen in Table 1. Hence, we decided to fix the helix-width parameter as 4. Similarly, width of beta is set to 3 and minimum beta length as 3 using inputs from the literature [15, 17].

We propose *Extract SSP* algorithm, in which we set the parameters of row width a as 4 and minimum helix length as 3. The beta strand prediction is also carried out by setting row width b to be 3 and minimum beta length as 3. All the remaining contacts are labeled as belonging to coil/turn. In order to validate the algorithm, we run *Extract SSP(A)* on bench mark data set used in secondary structure assignment literature and compare the results with those obtained by some of the latest algorithms in the literature.

3 Experimentation on Bench Mark Data Sets

We consider the gold standard data set RS126 of Rost and Sander [6], given in Table 2, which has been designed for the secondary structure prediction. This protein data set contains proteins that maintain pair-wise sequence similarity of less than 25%; we use the standard evaluation measure for performance evaluation, the details of which are given below.

3.1 Evaluation Measure

To calculate performance measure have been frequently used in protein secondary structure prediction $Q3$ or accuracy (3 for the three types of secondary structures) based measure [6]. $Q3$ is a residue-based measure which calculates the overall percentage of correctly classified residues for all the three structures and is computed as follows:

$$Q3 = (H_{\text{Pre}} + E_{\text{Pre}} + C_{\text{Pre}})/N_t \quad (1)$$

where $\boldsymbol{N_t}$ is the total number of predicted residues, $\boldsymbol{H_{\text{pre}}}$ is the number of correctly predicted residues for helix, $\boldsymbol{E_{\text{pre}}}$ for sheet, and $\boldsymbol{C_{\text{pre}}}$ for coil.

3.2 *Results on Bench Mark Data Set*

Since secondary structure prediction tools consider protein sequence information and not coordinate information, we need to test the proposed algorithm on contact maps that are predicted or contact maps that are highly noisy. Hence, we present three different experimental results in this section, on the actual bench mark data RS126, as well as on the contact maps that are predicted for proteins in RS126; and additionally by introducing different levels of noise into RS126.

Results on RS126 *Extract SSP* are run on the data set RS126. We compare the results obtained by our algorithm to the existing algorithms in which the results have been reported for the data set RS126. We can see from Fig. 1 that our algorithm is performing on par with the other algorithms with prediction for helix being much higher than the results reported and $Q3$ being on par with the results. The algorithm is seen to perform very poorly for turn/coil prediction. In protein structures, one finds very long coils which are referred to as loops. A loop connects two secondary structural elements. We observe that many turns/coils have been misclassified as beta in our test.

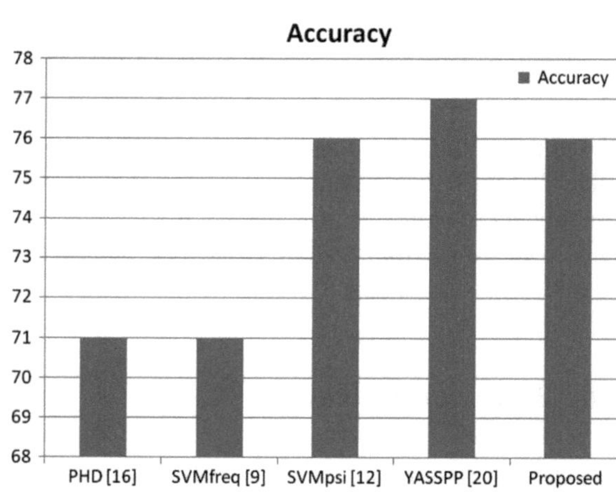

Fig. 1 Comparison between proposed and other existing methods

Table 3 Performance of secondary structure prediction algorithm on noisy contact maps

Class	SS	Noise level %						
		0	5	10	15	20	25	30
All-Alpha	H	88	88	88	88	88	88	88
	E	–	–	–	–	–	–	–
	C	77	77	77	72	70	58	55
All-Beta	H	74	74	74	72	72	68	65
	E	79	79	79	78	76	73	73
	C	47	47	47	34	33	32	32
Alpha + Beta	H	91	91	91	89	83	80	80
	E	86	86	86	72	71	70	69
	C	83	83	83	32	32	30	28
Alpha/Beta	H	88	88	88	88	84	79	75
	E	66	66	66	56	56	50	50
	C	48	48	48	33	33	28	25
Small proteins	H	83	83	83	82	81	79	75
	E	73	73	73	70	60	65	64
	C	46	46	46	43	40	23	22

3.3 Results on Noisy Contact Maps

Results on noisy contact maps contact map prediction are a highly challenging problem, and the current status of contact prediction is relatively poor. Hence it is important to validate the algorithm on highly noisy contact map, which has certain loss of information. Hence, we conduct the following experiment. The contact maps of proteins from RS126 are taken. Randomly $k\%$ of contacts (1's) are eliminated (made into 0's), and the algorithm is run on these noisy contact maps. A random noise of $k = 5$, 10% up to 30% is introduced into contact maps, and the results are provided in Table 3.

4 Experimentation for Protein Secondary Structure Assignment

Protein structure prediction problem does not have any coordinate information available and considers features from the primary amino acid sequence. Hence, even if we experiment with noisy contact maps, it is also important to test the algorithm with respect to the algorithms in the literature of protein structure assignment. We test our algorithm with the two popular standard algorithms of DPSS and STRIDE [11, 12]. Since the data set considered for these algorithms are performance of secondary structure prediction algorithm on noisy contact maps

Table 4 Performance of secondary structure assignment algorithm w.r.t. DPSS and STRIDE on SCOP data set

Class	DPSS			STRIDE			Proposed		
	H	E	C	H	E	C	H	E	C
All-Alpha	83	79	61	84	80	88	82	70	52
All-Beta	81	78	61	83	70	87	68	79	54
Alpha + Beta	81	80	61	82	71	87	82	80	51
Alpha/Beta	82	79	61	82	71	88	73	58	55
Average	82	79	61	83	73	88	76	72	53

Table 5 Performance of secondary structure assignment algorithm on mean and standard deviation

Measures	DPSS			STRIDE			Proposed		
SSE	H	E	C	H	E	C	H	E	C
Mean	18.6	14.1	12.2	17.5	12.4	10.4	18.8	20.3	28.4
Standard deviation	8.6	6.1	4.2	7.4	5.3	3.9	8.6	9.1	11.3

structure data, we consider the entire data set of ASTRAL SCOP containing 40 domains with roughly 11,000 proteins [18]. The results for the secondary structure assignment for *H*, *E*, and *C* are provided in Table 4. It can be seen that, as before, comparable results with DPSS and STRIDE are obtained for the classes of All-Alpha and Alpha + Beta and due to the low performance on coils; the overall average shows a lower value.

DPSS and STRIDE utilize the values like hydrogen bond energies between the atoms of the backbone as well as dihedral angular information. The proposed algorithm using only the contact pattern information from the contact matrix is able to achieve comparable performance for helix and sheet assignment. The proposed algorithm is very poor in coil information which needs further investigation. The proposed algorithm achieves a comparable accuracy of 79% for betas for All-Beta class and 82% for helices for Alpha + Beta class in comparison to the existing methods. All-Beta class and 82% for helices for Alpha + Beta class in comparison to the existing methods.

The results for the secondary structure assignment with low performance of mean and standard deviation for *H*, *E*, and *C* are provided in Table 5. It can be seen that, as before, comparable results with DPSS and STRIDE are obtained for the secondary structure assignment low performance on coils; the overall average shows a lower value.

5 Conclusions and Future Work

This work validates the hypothesis that contact maps contain useful information that can be utilized to understand the problem of protein fold prediction. Secondary structure elements of helices and beta strands have been successfully extracted using the pattern information in the contact map. On the other hand, coils could not be extracted well. Hence, we do not use the statistics of coil/turn in our work. This issue needs to be looked into further. Several useful features relating to both secondary structures viz, number of helices, minimum helix length, maximum helix length, number of beta sheets, minimum beta sheet, and maximum beta sheet, as well as pattern features like number of patterns, minimum and maximum density as well directional features have been extracted. These are going to be used as features for the fold identification problem in the future.

References

1. http://distill.ucd.ie/distill.
2. P. Y. Chou and Gerald D Fasman. Empirical predictions of protein conformation. *Annual review of biochemistry*, 47(1):251–276, (1978).
3. J. Garnier, D. J. Osguthorpe, and B Robson. Analysis of the accuracy and implica- tions of simple methods for predicting the secondary structure of globular proteins. *J. of molecular biology*, 120(1):97–120, (1978).
4. B. Rost, C. Sander, and Reinhard Schneider. Redefining the goals of protein secondary structure prediction. *J. of molecular biology*, 235(1):13–26, (1994).
5. D. T. Jones, Daniel WA Buchan, Domenico Cozzetto, and Massimiliano Pontil. Psicov: precise structural contact prediction using sparse inverse covariance es- timation on large multiple sequence alignments. *Bioinformatics*, 28(2):184–190, (2012).
6. B. Rost and C. Sander. Prediction of protein secondary structure at better than 70% accuracy. *J. Mol. Biol.*, pages 584–599, (1993).
7. G. Karypis. Better kernels and coding schemes lead to improvements in svm based secondary structure prediction. Technical report, DTIC Document (2005).
8. C. A. Andersen and Burkhard Rost. Secondary structure assignment. *Methods of biochemical analysis*, 44:341 (2003).
9. S.M. King and W.C. Johnson. Assigning second- ary structure from protein coordinate data. *Proteins*, 35:313320, (1999).
10. Rajgopal Srinivasan and George D Rose. A physical basis for protein secondary structure. *Proceedings of the National Academy of Sciences*, 96(25):14258–14263, (1999).
11. http://swift.cmbi.ru.nl/gv/dssp/.
12. http://webclu.bio.wzw.tum.de/stride/.
13. A. M. Konagurthu, A. S. Lesk and Allison L. Minimum message length inference of secondary structure from protein coordinate data. *Bioinformatics*, 28:i97–i105, (2012).
14. J.G Lees, J. Miles, Andrew, Frank Wien, and B.A. Wallace. A reference database for circular dichroism spectroscopy covering fold and secondary structure space. *Bioinformatics*, 22(16): 1955–1962, (2006).
15. K. J. Won, Thomas Hamelryck, Adam Pru¨gel-Bennett, and Anders Krogh. An evolutionary method for learning hmm structure: prediction of protein secondary structure. *BMC bioinformatics*, 8(1):357, (2007).

16. J. Hu, X. Shen, Y. Shao, C. Bystroff, and M. J. Zaki. Mining protein contact maps. In *2nd BIOKDD workshop on data mining in bioinformatics*, pages 196–204, (2002).
17. V. A. Simossis and Jaap Heringa. Local structure prediction of proteins. In *Com- putational Methods for Protein Structure Prediction and Modeling*, pages 207–254, (2007).
18. http://astral.berkeley.edu/.

An Assessment Report on: Statistics-Based and Signature-Based Intrusion Detection Techniques

Latika Mehrotra and Prashant Sahai Saxena

Abstract With the growing size of data, its security has become a great challenge, and security of data is a major issue in most of the research areas. A detailed study of existing IDS is presented in the current paper so as to detect threats or intrusions on the data residing on system/network. It is a bit difficult to stop security threats and breaches entirely using present security technologies. Detecting the presence of intruder is very crucial for maintaining the network security. It is found that intrusion detection systems (IDSs) that are signature-based are restricted in their areas of detecting intrusions, because of the fact that the signature-based intrusion detection system is based on matching a signature with the network details. The system using signatures or patterns can detect only known attacks and threats, but they mostly fail when it comes to novel attacks. Thus preventing/detecting the new or special types of attracts whose signature is not specified. Although signature-based IDS does not give false alarms at genuine cases, but still is inept for unknown attacks or masked attacks. Later in the paper, another category of IDS is discussed which is statistical-based intrusion detection system (SBIDS). The statistical-based intrusion detection systems have an upper hand when it is compared with the signature-based intrusion detection system. During the study, it has been found that many researchers have solved this problem by data mining classification algorithms.

Keywords IDS · NIDS · HIDS · SBIDS · Data mining

L. Mehrotra (✉) · P.S. Saxena
Department of Computer Science & Engineering,
Jaipur National University, Jaipur, India
e-mail: latika19mehrotra@gmail.com

P.S. Saxena
e-mail: sahai.prashant@gmail.com

D.K. Mishra et al. (eds.), *Information and Communication Technology*,
Advances in Intelligent Systems and Computing 625,
https://doi.org/10.1007/978-981-10-5508-9_31

1 Introduction

Over the years, the intrusion detection has become one of the most popular fields of research. The main reason is that most of the organizations have become automated and they also make use of the Internet and the network technology for the transmission of the data. So it is clear that the security of the data sent and received has become trivial. To avoid the intruders to fetch all important data, there is a need of some kind of mechanism which can prevent this unauthorized access. The data mining techniques play a vital role in intrusion detection systems. These techniques have the capability to deal with the voluminous data. Data mining techniques have been implemented on the network and host audit data for establishing an IDS. An IDS is a process in which data is analysed carefully with the help of data mining techniques in order to automatically detect user's normal or intrusive behaviours. A few common data mining techniques normally involve regression, classification, clustering and association rule learning [1].

1.1 Intrusion Detection System (IDS)

IDS helps in analysing the network or system activity and prepares systems for all possible attacks be it an incoming request or an outgoing request. They help in detecting suspicious activities and gather information from a variety of sources for possible threats and attacks (Fig. 1).

There are two main components to the intrusion detection system.

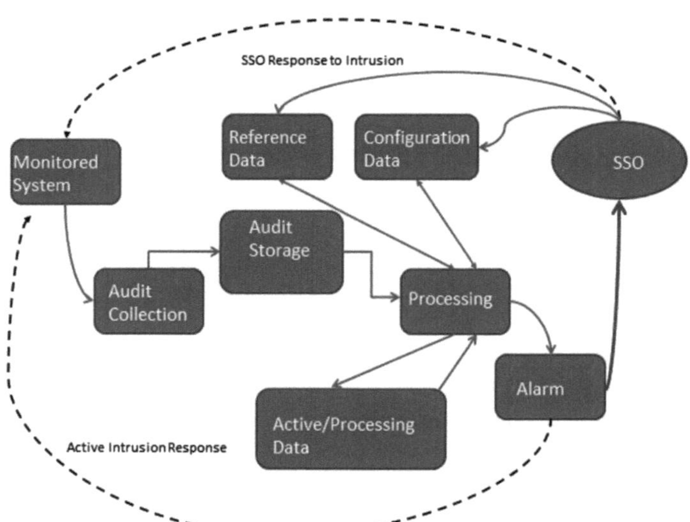

Fig. 1 Organization of a general IDS

1.1.1 Network-Based IDS

Network-based IDS (NIDS) helps in scrutinizing network traffic, but they are less beneficial as they do not assist when running on the host only [2]. On dedicated machines that observe network flow along with firewalls, NIDS is best suited option. That is why the security threats of the clients do not affect the monitors. One of the major disadvantages of NIDS is that it is restricted to local root attacks. If an authorized or legitimate user tries to attain extra privileges by establishing an encrypted channel while using machine remotely, the NIDS would not alarm.

1.1.2 Host-Based IDS (HIDS)

An HIDS tracks attacks on the machine it is implemented [2, 3]. HIDS mostly detected and logged any action taking place on machine it is monitoring. That is HIDS can only be trustworthy in some specific conditions. As network attacks have increased in number severity over the past few years, intrusion detection system (IDS) is becoming a critical component to secure the network. Due to the large volumes of security audit data, complicated and dynamic properties of intrusion behaviours, the optimization of the performance of IDS becomes a crucial problem which is receiving high attention from the research community. Uncertainty to explore if certain algorithms perform better for a few attack classes establishes the motivation for the reported herein. Intrusion detection systems (IDSs) constructed on the concepts on data mining have shown high accuracy and are also good simplification to unusual types of intrusion and vigorous behaviour in a changing environment in recent years. One of the major hurdles of IDSs is the intensive computation necessary in the model generation phase.

2 Literature Survey

In [4], Ryan et al. have worked on the concept of the neural network. It performs learning on the basis of the training data, and then it performs predictions. It clarifies the behaviour of the node as normal or abnormal. Denning [5, 6] have presented a sequential rule-based model for the prediction of abnormal behaviour. Sequential rules are constructed on the sequential data base. The training data is stored in the sequence in which they occur in the sequence data set, and then the sequence rule mining algorithm is applied on the training data set to identify the patterns of the normal behaviour and the abnormal behaviour. The system developed [7] has more accuracy in recognizing if the record is ordinary or breached. Dewan et al. [8] proposed an improved version of the self adaptive Bayesian algorithm (ISABA). It is based on the concept of the Bayesian network but accuracy rate is below expectation. Sathyabama et al. [9] presented a method based on the clustering. In this method, the similarity-based records are stored in the clusters. Also the dissimilar records are called the outliers. For the outliers, the alarm is raised and the record is checked for the abnormal behaviour. Azimi et al. [10] proposed a self-organizing map-based method for the intrusion detection. The map has been used successfully to classify the data records. In this method, the false positive alerts have been reduced up to a good extent. Bivens et al. [11] have proposed a self-organizing map-based classification technique. The false negative has been reduced in this model. The term "false negative" is a result which comes as negative when it must not. The authors of [12] proposed the ensemble approach. The mentioned approach is a combination of many present algorithms. The experimental results have demonstrated that it has outpaced many in effect techniques. It has even outpaced support vector machine. This paper [13] presents a method which is based on the concept of the dimension reduction. The dimension reduction is achieved by the feature extraction technique of the data mining. Ei-Semary et al. [14] have proposed a fusion of the a priori and the kuok algorithm. This proposed model also uses the concept of the fuzzy set, i.e. partial membership function. Shon and Moon [15] proposed a novel framework. This novel framework contains the updated support vector machine. It also performs the packet profiling. The concept of clustering is also applied in brief. Overall, it is a very complex process. In this paper [16], the author has proposed a genetic algorithm-based intrusion detection system. The accuracy is better but the solution is not guaranteed at all. Norouzian and Merati [3] also proposed genetic algorithm-based IDS. The GA is used to update the membership function. It also contains a detailed survey of the IDS. Jin-Ling Zhao et al. [17] defined multi-layer perceptron (MLP)-based technique for the intrusion detection and classification. The back propagation algorithm has been used for the error rate deduction. The work done in [18] also proposes a GA and clustering-based method for the IDS. Lunt [19], a survey of intrusion detection expert system is proposed. The proposed method is fusion of two methods to create a strong intrusion detection system. Todd et al. [20] proposed an intrusion detection system called network system monitor. This system is based on the concept of analysing network instead of the system log entry. Teng et al. [21] proposed time-based inductive machine to capture or store user behaviour. Inductive generalization is also a part of the process. Anderson and Mohan [22] proposed a network intrusion detection expert system. This system learns from the training data and predicts the test data. Lane and Brodley [23] applied the concept of the instant-based learning. Lee et al. [24] propose a novel data-mining-based framework for intrusion detection. This model is based on the concept of utilizing the contents of the audited programs. Debar et al. [25] propose taxonomy of the intrusion detection systems. This classification is done according to the property of the intrusion detection system. Axelsson [26], also proposed taxonomy of the IDS. They also presented the background details of some IDS. Yu et al. [27, 28] proposed a novel signal processing-based intrusion detection system.

Table 1 Comparative studies of a few intrusion detection systems

Authors name	Publication year	Detection principle	Granularity	Audit source	Type of response	Data processing	Data collection	Security
Jake Ryan	1987	Neural network	Batches (training data set)	Host	Passive	Centralized	Centralized	Low
Denning D.E.	1990	Sequential rule mining	Continuous	Host	Passive	Distributed	Centralized	Low
Norouzian	2005	MLP	Batches	Host	Passive	Centralized	Distributed	Higher
Dewan M.	2010	Bayesian algorithm	Continuous	Network	Passive	Centralized	Centralized	Below expectation
S. Sathyabama	2011	Clustering	Batch (clusters)	Network	Passive	Centralized	Distributed	High

3 Conclusion

Overall, any kind of intrusion is destructive for the information and data available on the Internet, which also reduces the authenticity and security. In order to make sure that all kinds of information or data are safe from intruders and an unauthorized access, the IDS is must. It identifies the system that is trying to gain access and perform malicious activity. In this paper, the comprehensive survey over the most effective and most popular intrusion detection system has been performed. Each method working together with the related merits and demerits has been discussed. This literature review will help other researchers in their further research or study.

Based on the above literature survey, a comparative table is illustrating various aspects of existing IDS (Table 1).

References

1. Rajni Tewatia, Asha Mishra, "Introduction to Intrusion Detection System: Review", International journal of scientific & technology research volume 4, issue 05, May 2015, ISSN 2277-8616, pp. 21–223.
2. Litty Lionel, "Hypervisor-based Intrusion Detection", Master of Science Graduate department of computer Science University of Toronto, 2005.
3. Norouzian. M. R., Merati. S., "Classifying Attacks in a Network Intrusion Detection System Based on Artificial Neural Networks", in the Proceedings of 13th International Conference on Advanced Communication Technology (ICACT), 2011, ISBN: 978-1-4244-8830-8, pp. 868–873.
4. Jake Ryan, Meng-Jang Lin, Risto Miikkulainen, "Intrusion Detection With Neural Networks", Advances in Neural Information Processing System 10, Cambridge, MA: MIT Press, 1998, DOI:10.1.1.31.3570.
5. Denning. D. E., "An Intrusion Detection Model", Transactions on Software Engineering, IEEE Communication Magazine, 1987, SE-13, PP. 222–232, DOI:10.1109/TSE.1987.232894.
6. Teng. H. S., Chen. K. and Lu. S. C., "Adaptive Real-Time Anomaly Detection using Inductively Generated Sequential Patterns, in the Proceedings of Symposium on research in Computer Security and Privacy, IEEE Communication Magazine, 1990, pp. 278–284.
7. Sekeh. M. A., Bin Maarof. M. A., "Fuzzy Intrusion Detection System Via Data Mining with Sequence of System Calls", in the Proceedings of International Conference on Information Assurance and security (IAS) 2009, IEEE Communication Magazine, pp. 154–158, ISBN:978-0-7695-3744-3, DOI:10.1109/IAS.2009.32.
8. Dewan. Md., Farid, Mohammed Zahidur Rahman, "Anomaly Network Intrusion Detection Based on Improved Self Adaptive Bayesian Algorithm", Journal of Computers, Vol. 5, pp. 23–31, Jan 2010, DOI:10.4.304/jcp5.1.
9. Sathyabama. S., Irfan Ahmed. M. S., Saravanan. A., "Network Intrusion Detection Using Clustering: A Data Mining Approach", International Journal of Computer Application (0975-8887), Sep-2011, Vol. 30, No. 4, ISBN: 978-93-80864-87-5, DOI:10.5120/3670-5071.
10. Amir Azimi, Alasti, Ahrabi, Ahmad Habibizad Navin, Hadi Bahrbegi, "A New System for Clustering and Classification of Intrusion Detection System Alerts Using SOM", International Journal of Computer Science and Security, Vol: 4, Issue: 6, pp. 589–597, 2011.
11. Alan Bivens, Chandrika Palagiri, Rasheda Smith, Boleslaw Szymanski, "Network-Based Intrusion Detection Using Neural Networks", in Proceedings of the Intelligent Engineering

Systems Through Artificial Neural Networks, St. Louis, ANNIE-2002, and Vol: 12, pp. 579–584, ASME Press, New York.
12. Srinivas Mukkamala, Andrew H. Sung, Ajith Abraham, "Intrusion Detection Using an Ensemble of Intelligent Paradigms", Journal of Network and Computer Applications, pp. 1–15, 2004.
13. Shilendra Kumar, Shrivastava, Preeti Jain, "Effective Anomaly Based Intrusion Detection Using Rough Set Theory and Support Vector Machine" (0975-8887), Vol. 18, No. 3, March 2011, DOI:10.5120/2261-2906.
14. Aly Ei-Semary, Janica Edmonds, Jesus Gonzalez-Pino, Mauricio Papa, "Applying Data Mining of Fuzzy Association Rules to Network Intrusion Detection", in the Proceedings of Workshop on Information Assurance United States Military Academy 2006, IEEE Communication Magazine, West Point, NY, DOI:10.1109/IAW.2006/652083.
15. Taeshik Shon, Jong Sub Moon, "A Hybrid Machine Learning Approach to Network Anomaly Detection", Information Sciences 2007, Vol. 177, Issue. 18, Publisher: USENIX Association, pp. 3799–3821, ISSN:00200255, DOI:10.1016/j.ins-2007.03.025.
16. Sachin Patil, Deepak Kapgate, P. S. Prasad, "Effective concept for detection of web based attacks using ID3 algorithm", IJCSMC, Vol. 3, Issue. 5, May 2014, pp. 321–330.
17. Jin-Ling Zhao, Jiu-fen Zhao, Jian-Jun Li, "Intrusion Detection Based on Clustering Genetic Algorithm", in Proceedings of International Conference on Machine Learning and Cybernetics (ICML), 2005, IEEE Communication Magazine, ISBN: 0-7803-9091-1, DOI:10.1109/ICML.2005.1527621.
18. Anderson. J. P., "Computer Security Threat Monitoring and Surveilance", Technical Report, James P Anderson Co., Fort Washington, Pennsylvania, 1980.
19. Lunt, T. "Detecting intruders in computer systems," in Proceedings of Conference on Auditing and Computer Technology, pp. 1–17, 1993.
20. Todd, H. L., Gihan V. D., Karl N. L., Biswanath, M., Jeff, W. and David, W. "A network security monitor," in Proceedings of Symposium on Research in Security and Privacy, Oakland, CA, pp. 296–304, 1990.
21. Teng, H., Chen, K. and Lu, S. "Adaptive real time anomaly detection using inductively generated sequential patterns', IEEE Computer Society Symposium on Research in Security and Privacy, California, IEEE Computer Society, pp. 278–84, 1990.
22. Anderson, J. B. and Mohan, S. "Sequential coding algorithms: A survey and cost analysis", IEEE Transactions on Communication, Vol. 32, pp. 169–176, 1984.
23. Lane, T. and Brodley, C. E. "Temporal sequence learning and data reduction for anomaly detection", ACM Transactions on Information and System Security, Vol. 2, No. 3, 1999.
24. Lee, W., Stolfo, S. and Mok, K. "Adaptive intrusion detection: A data mining approach", Artificial Intelligence Review, Kluwer Academic Publishers, Vol. 14, No. 6, pp. 533–567, 2000.
25. Debar, H., Becker, M. and Siboni, D. "A neural network component for an intrusion detection system," in IEEE Symposium on Research in Computer Security and Privacy, pp. 240–250, 1992.
26. Axelsson, S. Research in intrusion-detection systems: A survey, Technical Report, Department of Computer Engineering, Chalmers University of Technology, Goteborg, Sweden, 1998.
27. Yu, Z., Chen, J. and Zhua, T. "A novel adaptive intrusion detection system based on data mining," in Proceedings of the Fourth International Conference on Machine Learning and Cybernetics, Guangzhou, pp. 2390–2395, 2005.
28. Sadiq Ali Khan, "Rule-Based Network Intrusion Detection Using Genetic Algorithm", International Journal of Computer Applications, No: 8, Article: 6, 2011, DOI:10.5120/2303-2914.

OTCA Approach Towards Blurred Image Feature Estimation and Enrichment

Sandeep Kumar Sharma, C.S. Lamba and V.S. Rathore

Abstract This paper investigates a new and effective method for image edge feature estimation and enrichment based on cellular automata framework. Objective of this paper is find an area where arbitrary changes in intensity of an image and enrich the features of that area. The inner and outer neighbors of a blurred cell are used to accumulate the pixels or cells of an image whose features are better than the targeted cell. Therefore, the accumulated cells or pixels are passes to OTCA method for obtaining target image. OTCA produce threshold pixels using passes pixels for enrichment of blurred image features. During this process, one or more attributes of the image are customized and afford a massive amount of choices for humanizing the visual feature of images. The experimental results express efficiency of proposed method.

Keywords OTCA · Image features · BIFEE · Blurred · Cell

1 Introduction

Image processing has wide variety of operations like gaining, solidity, storeroom, broadcast, and imitation that may degrade the illustration features of an image. This paper anticipated a new method for blurred image feature estimation and enrichment (BIFEE). The objective of feature enrichment is to get better visual facade of an image or to switch the image to a form better suited for analysis by a human or

S.K. Sharma (✉) · C.S. Lamba · V.S. Rathore
RTU, Kota, India
e-mail: mcasandy2006@gmail.com

C.S. Lamba
e-mail: lamba5@rediffmail.com

V.S. Rathore
e-mail: vijaydiamond@gmail.com

D.K. Mishra et al. (eds.), *Information and Communication Technology*,
Advances in Intelligent Systems and Computing 625,
https://doi.org/10.1007/978-981-10-5508-9_32

machine. The principal objective of BIFEE is to discover and amend attributes of an image to make it more apposite for a given assignment and a precise observer.

Usually, various methods were produced for image feature enrichment such as graph-based methods, model-based methods, gradient-based methods, and wavelet-based methods. But the experimental result shows the efficiency of proposed method.

Cellular automata provide a framework for image processing applications in which image pixels can be represented in the form of matrix cell. In this paper, we proposed a method which used cellular automata framework to achieve the prescribed objective.

This paper focuses on that particular area of an image where blurred pixels are existing or where the intensity of an image changed. The proposed method is capable of improving features of all types of images such as color image, gray image, and black and white images. Basically, there are two approaches used to feature enrichment; one is inner enrichment which uses the nearest neighbor, and other is outer enrichment which uses neighbors of nearest neighbor. Both approaches find improved threshold value for each pixels and use the threshold to improve image features.

1.1 Related Work

Conservatively, the graph-based methods [1, 2] are demonstrated to enhance the image to amend the inclusive luster and distinction of digital image. The various existing operators like Soble, Robert and Cany operator [3–5] are capable for features estimation and gradient based methods [6] and wavelet based methods [7, 8] are also estimate the features of an image. The experimental results show the comparison between existing methods with proposed OTCA method using image quality parameters such as mean square error (MSE), peak signal noise ratio (PSNR), average difference (AD), structural content (SC), mean absolute error (MAE), and number of objects (NoBJ) [9, 10].

1.2 Notations

In this paper, R and B denote reference image and blurred image, T is used for threshold value of pixels, $D_{i.j}$ denote binary matrix of blurred image, whereas ON, IN, $\Psi_{i.j,}$ and $\Phi_{i.j}$ denote the methods to accumulate inner and outer neighbors.

The residual section of this paper is organized as follows: Sect. 2 represents the framework which is used in the proposed method, Sect. 3 represents solution of problem while formulating the proposed method, and Sect. 4 describes how the proposed method is better than other methods.

2 Framework

Cellular automata (CA) have wide framework for different types of applications. CA also play an important role in image processing. CA can define the number of rules based on neighbor pixels [11–13]. Each pixel of an image has atleast 8 neighbors, so CA can apply $2^8 = 256$ number of rules for image processing not only that particular pixel but also all the other pixels of image. CA provide more rules as neighbors are increase of a pixel. CA represent all the pixels with 3-tuple $\langle S, N, F \rangle$, where S is set of finite states, N is set of neighbors, and F represents a transition function which consigns a new state to a pixel [14, 15]. The structure of the pixels' neighbor including Von Neumann, Moore neighborhood, and outer totalistic is shown in Fig. 1.

Von Neumann include four neighbors: left, right, above, and below with radius of 1[16, 17] Moore neighborhood include all the four neighbors of Von Neumann [18–20], four corner neighbors and itself with radius 1, whereas outer totalistic include all the outer neighbors of Moore neighbors with radius 2. The total neighbors of Von Neumann and Moore neighborhood are given in Eqs. 1 and 2.

$$N_{i,j} = R_{i-1,j-1} + R_{i,j-1} + R_{i,j+1} + R_{i+1,j+1} \tag{1}$$

$$N_{i,j} = \sum_{k=i-1}^{i+1} \sum_{l=j-1}^{j+1} R_{k,l} \tag{2}$$

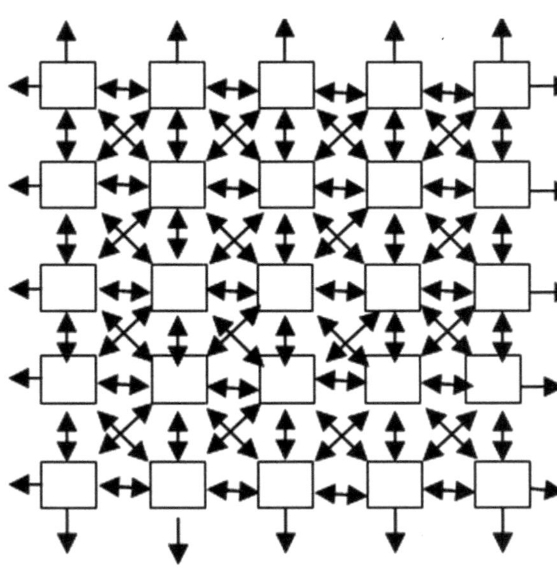

Fig. 1 CA neighbor structure of an image

The total neighbors of outer totalistic with radius 2 are given in Eq. 3.

$$N_{i,j} = \sum_{k=i-2}^{i+2} \sum_{l=j-2}^{j+2} R_{k,l} - \sum_{k=i-1}^{i+1} \sum_{l=j-1}^{j+1} R_{k,l} \quad (3)$$

3 Proposed Methods

Some work has been done on image feature estimation and enrichment. All the existing methods demonstrate input-dependent behavior [21–23]. The aim of the proposed method is to work with all type of images and produce effective and robust image features with less time-consuming. To overcome the problems of existing methods, a new method has been proposed. The proposed method has five stages for improving the features of image. The block diagram of image feature estimation and enrichment is shown in Fig. 2.

As discussed above, the proposed method can work on any type of images and CA framework is used in proposed method to achieve the objective. While working with CA, we need to convert an input image into binary matrix. Binary matrix has cell value 0 and 1 only, so to convert image pixels into 0's and 1's, we set a threshold value. Suppose B is an input image, generally each types of image contain two regions; one is object pixels and other is background. The probability distribution is used to accumulate the threshold value T.

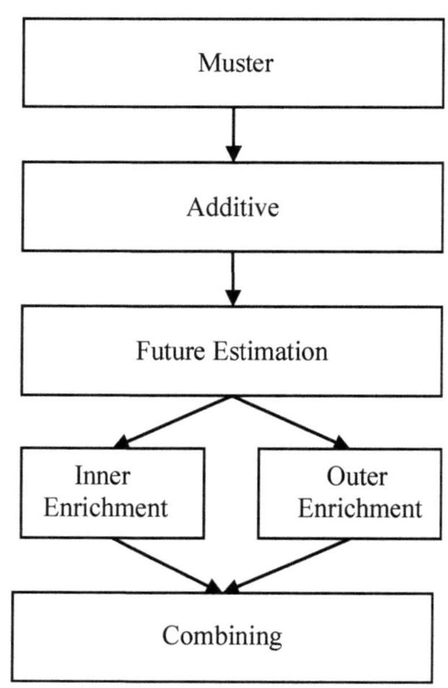

Fig. 2 Block diagram of proposed method

The probability distribution of an image is given by the following equation:

$$P(B) = P_b(B) + P_o(B)$$

$P_b(B)$ is probability of background
$P_o(B)$ is probability of object pixel

Let T_b is image background threshold, and T_o is object pixels' threshold.

$$T_b = \int_{-\infty}^{T} P_b(B)$$

$$T_o = \int_{T}^{\infty} P_o(B)$$

The threshold value of image B can be retrieved using minimization of the background T_b and object pixel T_o threshold.

$$T = \text{minimize}\,(T_b + T_o)$$

Using T, we can generate binary matrix according to following equation:

$$B_{i,j} = \begin{cases} 1 & \text{if pixel value greater than or equal to } T \\ 0 & \text{if pixel value less than to } T \end{cases}$$

To make image feature estimation and enrichment effective, we need to adjust or transform intensity, contrast, and brightness of an image using the tool box imadjust() of MATLAB. To estimate the features of an image, we need to find inner path and outer path of pixels which used to improve the quality of pixels. Let $P_{i,j}$ is a pixel of an image B size M × N, and BM is a binary matrix of input image B. Φ and Ψ used to locate horizontal and vertical locations of an image.

Case 1: $i \in \Phi(1, M)$

$$\text{ON}(P,Q,R)\begin{cases} (1,1,1) & \text{if } j \in \Psi(1,N) \\ (1,2,1) & \text{if } j \in \Psi(2,N-1) \\ (2,2,1) & \text{if } j \in \Psi(3,N-2) \end{cases}$$

Case 2: $i \in \Phi(2, M-1)$

$$\text{ON}(P,Q,R)\begin{cases} (1,1,2) & \text{if } j \in \Psi(1,N) \\ (1,2,2) & \text{if } j \in \Psi(2,N-1) \end{cases}$$

Case 3: $i \in \Phi(3, M - 2)$

$$\mathrm{ON}(P, Q, R) = \{\,(2, 1, 2) \quad \text{if}\, j \in \Psi(1, N)$$

Otherwise

$$\mathrm{ON}(P, Q, R) = (4, 2, 2)$$

Outer and inner paths are accumulated using the following equation:
The total outer neighbors (T_{ON}) of a pixel p are given as:

$$T_{\mathrm{ON}} = \sum_{k=1}^{P} 3p + \sum_{l=1}^{Q} p + \sum_{m=1}^{R} p$$

There are total outer neighbors of any pixel are maximum 16 and total inner neighbors of any pixel are maximum 8. Hence, we use 16 bits to map OTCA rule. The 16 bits of OTCA are as follows:

0 0 0 0 0 0 1 1 1 0 0 0 0 0 0 0

The above 16 bits contain 1's at 7th, 8th, and 9th position, and remaining bits are 0's. So we use minimum threshold 7 and maximum threshold 9 to calculate global threshold T which can be used for image feature enrichment.

Let

$$T_{\min} \rightarrow \text{Minimum Threshold}$$
$$T_{\max} \rightarrow \text{Maximum Threshold}$$

$$T = \frac{(T_{\min} + T_{\max})}{2}$$

Now using T, we can improve features of an image. The threshold function of binary image $G(x, y)$ is defined as:

$$G(x, y) = \begin{cases} a, & \text{if } f(x, y) > T \\ b, & \text{if } f(x, y) \leq T \end{cases}$$

Pixels labeled 'a' correspond to object pixels, and pixels labeled 'b' correspond to background pixels.

4 Experimental Results

Figures 3 and 4 show the experimental results on existing methods [24–27]and proposed method.

Figure 3 shown the experimental results on canny image and Fig. 4 and Table 1 depict comparison between existing method and OTCA methods for blurred image edge feature estimation and enrichment including image quality enhancement parameters.

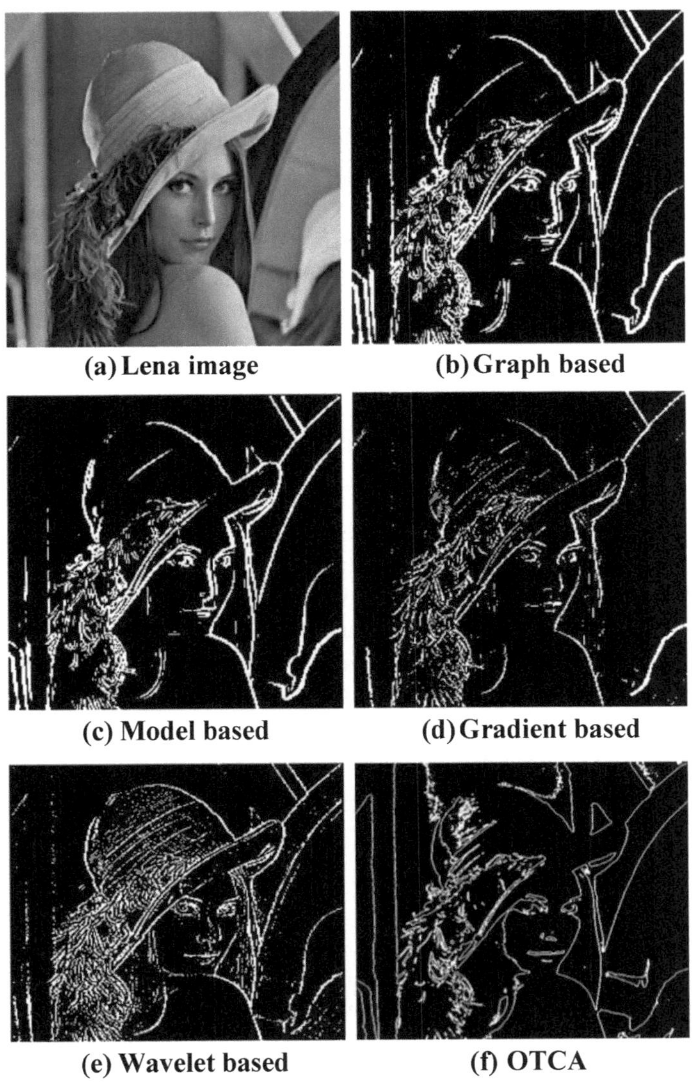

Fig. 3 Experimental result

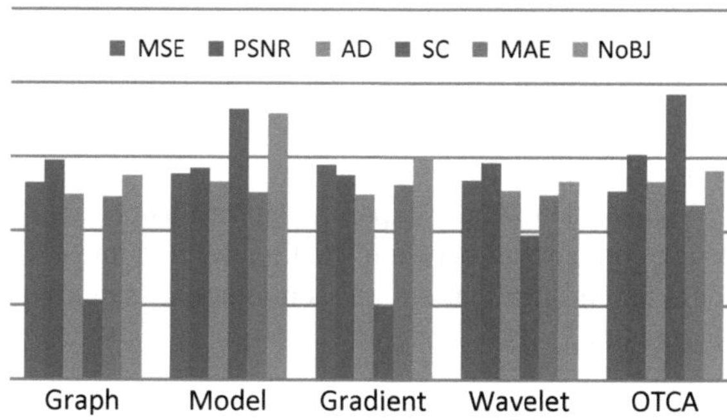

Fig. 4 Comparison of OTCA with existing methods using enrichment parameters

Table 1 Image quality enrichment parameters for different image feature enrichment methods

Enrichment methods	Feature enrichment parameters						
	MSE	PSNR	NCC	AD	SC	MAE	NoBJ
OTCA	5.079	11.07	0.06	53.3	18.7	0.97	564
Graph	5.288	10.89	0.11	49.8	5.12	0.99	549
Model	5.363	10.83	0.08	50.8	6.87	0.99	534
Gradient	5.536	10.69	0.02	53.2	18.2	1.05	717
Wavelet	5.782	10.50	0.07	49.8	4.97	1.02	600

5 Conclusion

In this paper, we present optimistic method for image features estimation and enhancement. Different feature assessment and enhancement methods have been experimented and compared with proposed method using the help of image quality assessment parameters. The proposed method has been tested on different types of images and produced very effectiveness and robustness results. Table 1 shows MSE, PSNR, SC, AD, MAE, and NoBJ values of existing methods and OTCA method. According to their values, OTCA produced better results. The inner neighbors and outer neighbors provide better choice to select threshold in less computation time.

References

1. Ming Xu, Jun Wang and Zeyun Yu "Image Edge Enhancement and Segmentation via Randomized Shortest Paths", 5th International Conference on Biomedical Engineering and Informatics (BMEI 2012), 978-1-4673-1184-7/12/$31.00 ©2012 IEEE.
2. Qingsong Zhu, Jiaming Mai, and Ling Shao "A Fast Single Image Haze Removal Algorithm Using Color Attenuation Prior", IEEE Transactions on Image Processing, Vol. 24, No. 11, November 2015.

3. Zhou Wang, Alan Conrad Bovik, Hamid Rahim Sheikh and Eero P. Simoncelli "Image Quality Assessment: From Error Visibility to Structural Similarity", IEEE Transactions on Image Processing, Vol. 13, No. 4, APRIL 2004.
4. Jianchao Yang, John Wright, Thomas S and Yi Ma "Image Super-Resolution Via Sparse Representation", IEEE Transactions On Image Processing, Vol. 19, No. 11, November 2010.
5. K Ilya Pollak, Alan S. Willsky and Hamid Krim "Image Segmentation and Edge Enhancement with Stabilized Inverse Diffusion Equations", IEEE Transactions on Image Processing, Vol. 9, No. 2, February 2000.
6. Gang Cao, Yao Zhao, Rongrong Ni, and Xuelong Li "Contrast Enhancement-Based Forensics in Digital Images". IEEE Transactions on Information Forensics and Security, Vol. 9, No. 3, March 2014.
7. Mayank Agrawal, Ratnakar Dash, "Image Resolution Enhancement using Lifting Wavelet and Stationary Wavelet Transform", 2014 International Conference on Electronic Systems, Signal Processing and Computing Technologies, 978-1-4799-2102-7/14 $31.00 © 2014 IEEE DOI 10.1109/ICESC.2014.61.
8. Benjamin Hell, Marc Kassubeck, Pablo Bauszat, Martin Eisemann, and Marcus Magnor "An Approach Toward Fast Gradient-Based Image Segmentation", IEEE Transactions on Image Processing, Vol. 24, No. 9, September 2015.
9. Javier Galbally, Sébastien Marcel and Julian Fierrez "Image Quality Assessment for Fake Biometric Detection: Application to Iris, Fingerprint and Face Recognition", IEEE Transactions on Image Processing, Vol. 23, No. 2, February 2014.
10. Jun-Jie Huang, Wan-Chi Siu and Tian-Rui Liu "Fast Image Interpolation via Random Forests", IEEE Transactions on Image Processing, Vol. 24, No. 10, October 2015.
11. Diplaxmi R. Waghule and Dr. Rohini S. Ochawar "Overview on Edge Detection Methods", 2014 International Conference on Electronic Systems, Signal Processing and Computing Technologies, 978-1-4799-2102-7/14 $31.00 © 2014 IEEE DOI 10.1109/ICESC.2014.31.
12. Senthilkumaran N and Thimmiaraja J "Histogram Equalization for Image Enhancement Using MRI brain images", 2014 World Congress on Computing and Communication Technologies, 978-1-4799-2876-7/13 $31.00 © 2013 IEEE DOI 10.1109/WCCCT.2014.45.
13. Mitra Basu "Gaussian-Based Edge-Detection Methods—A Survey", IEEE Transactions on Systems, Man, and Cybernetics—Part C: Applications And Reviews, Vol. 32, NO. 3, August 2002.
14. Sos S. Agaian, Karen Panetta and Artyom M. Grigoryan "Transform-Based Image Enhancement Algorithms with Performance Measure", IEEE Transactions on Image Processing, Vol. 10, No. 3, March 2001.
15. Miguel A. Duval-Poo, Francesca Odone, and Ernesto De Vito "Edges and Corners With Shearlets", IEEE Transactions on Image Processing, Vol. 24, No. 11, November 2015.
16. Huanjing Yue, Xiaoyan Sun, Jingyu Yang and Feng Wu "Image Denoising by Exploring External and Internal Correlations", IEEE Transactions on Image Processing, Vol. 24, No. 6, June 2015.
17. Miagquan, Zhou, Guohua Geng "Detail Enhancement and Noise Reduction with True Color Image Edge Detection based on Wavelet Multi-scale" IEEE, pp. 1061–1064, 2011.
18. Aroop Mukherjee and Soumen Kanrar "Image Enhancement with Statistical Estimation", the International Journal of Multimedia & Its Applications (IJMA) Vol. 4, No. 2, April 2012.
19. Kailash Sinha, G. R. Sinha "Efficient Segmentation Methods for Tumor Detection in MRI Images", 2014 IEEE Conference on Electrical, Electronics and Computer Science.
20. Andac Hamamci, Nadir Kucuk, Kutlay Karaman, Kayihan Engin, and Gozde Unal "Tumor-Cut: Segmentation of Brain Tumors on Contrast Enhanced MR Images for Radio surgery Applications", IEEE Transactions on Medical Imaging, Vol. 31, No. 3, March 2012.
21. Xianfeng Yang, Jing Zhang, Bo Peng, Shutao You "An Adaptive Edge Enhancement Method Based on Histogram Matching for Ultrasound Images", 2010 International Conference on Computational and Information Sciences, 978-0-7695-4270-6/10 $26.00 © 2010 IEEE DOI 10.1109/ICCIS.2010.333.

22. Soo-Chang Pei and Li-Heng Chen "Image Quality Assessment Using Human Visual DOG Model Fused With Random Forest", IEEE Transactions on Image Processing, Vol. 24, No. 11, November 2015.
23. Roshni Ravi and Josemartin M. J. "A Novel Image Processing Filter Designed Using Discrete Fourier Invariant Signal", 2014 International Conference on Electronic Systems, Signal Processing and Computing Technologies, 978-1-4799-2102-7/14 $31.00 © 2014 IEEE DOI 10.1109/ICESC.2014.88.
24. Tao Wang, Irene Cheng and Anup Basu "Fluid Vector Flow and Applications in Brain Tumor Segmentation", IEEE Transactions on Biomedical Engineering, Vol. 56, No. 3, March 2009.
25. Josef Strom Bartunek, Mikael Nilsson, Benny Sällberg, Ingvar Claesson "Adaptive Fingerprint Image Enhancement with Emphasis on Pre-processing of Data", IEEE Transactions on Image Processing, Vol. 22, Issue 2, 2013.
26. G. T. Shrivakshan and Dr. C. Chandrasekar "A Comparison of various Edge Detection Techniques used in Image Processing", IJCSI International Journal of Computer Science Issues, Vol. 9, Issue 5, No 1, September 2012.
27. A. Djerouni, H. Hamada and N. Berrached "MR imaging contrast enhancement and segmentation using fuzzy clustering", IJCSI International Journal of Computer Science Issues, Vol. 8, Issue 4, No 2, July 2011.

Author Index

D.K. Mishra et al. (eds.), *Information and Communication Technology*,
Advances in Intelligent Systems and Computing 625,
https://doi.org/10.1007/978-981-10-5508-9

MIX
Papier aus verantwortungsvollen Quellen
Paper from responsible sources
FSC® C105338

If you have any concerns about our products,
you can contact us on
ProductSafety@springernature.com

In case Publisher is established outside the EU,
the EU authorized representative is:
Springer Nature Customer Service Center GmbH
Europaplatz 3, 69115 Heidelberg, Germany

Printed by Libri Plureos GmbH
in Hamburg, Germany